W9-CMN-011

Lectures in Applied Mathematics

Proceedings of the Summer Seminar, Boulder, Colorado, 1960

Proceedings of the Summer Seminar, Ithaca, New York, 1963

Proceedings of the Summer Seminar, Ithaca, New York, 1965

Proceedings of the Summer Seminar, Stanford, California, 1967

LARGE-SCALE COMPUTATIONS IN FLUID MECHANICS

PART 2

Volume 22-Part 2
Lectures in Applied Mathematics

LARGE-SCALE COMPUTATIONS
IN FLUID MECHANICS

Edited by
Bjorn E. Engquist
Stanley Osher
Richard C. J. Somerville

1985
American Mathematical Society, Providence, Rhode Island

The proceedings of the Summer Seminar were prepared by the American Mathematical Society with partial support from National Science Foundation Grant MCS 82-17817, National Aeronautics and Space Administration Grant NASW-3830 and Army Research Office Grant DAAG29-83-M-0194. The views, opinions, and/or findings contained in this report are those of the authors and should not be construed as an official Department of the Army position, policy, or decision, unless so designated by other documentation.

1980 *Mathematics Subject Classification.* Primary 35-XX, 65-XX, 76-XX, 86-XX; Secondary 39-XX, 49B36, 80A25, 85A35, 92A09.

Library of Congress Cataloging in Publication Data
Main entry under title:
Large-scale computations in fluid mechanics.
    (Lectures in applied mathematics; v. 22)
    Papers presented at the Fifteenth AMS–SIAM Summer Seminar on Applied Mathematics, held at Scripps Institution of Oceanography, June 27–July 8, 1983.
    1. Fluid mechanics–Congresses. 2. Numerical analysis–Congresses. I. Engquist, Björn, 1945–   . II. Osher, S. (Stanley) III. Somerville, Richard. IV. Summer Seminar on Applied Mathematics (15th: 1983: Scripps Institution of Oceanography) V. Series: Lectures in applied mathematics (American Mathematical Society); v. 22.
QA901.L37   1985   532'.01'51   84-24534
ISBN 0-8218-1122-3

MathSci
cat. as
Sep.

Contents

Part 1

CONTENTS

Part 2

Lectures in Applied Mathematics
Volume **22**, 1985

The Use of Spectral Techniques
in Numerical Weather Prediction

M. Jarraud and A. P. M. Baede

1. Introduction. Numerical Weather Prediction (NWP) is a relatively young branch of meteorology. Its purpose is the prediction of the future state of the atmosphere by solving numerically the equations governing its physical behaviour. From the point of view of physics, NWP is part of classical fluid dynamics and, as such, its physical principles are known since long ago. Nevertheless it was only in 1904 that the famous meteorologist V. Bjerkness proposed that the laws of fluid dynamics

1980 *Mathematics Subject Classification*. Primary 86-02; Secondary 35-04, 76-04, 76U05.
Key words and phrases. Numerical weather prediction, expansion in spherical harmonics, spectral method, intercomparison between spectral and grid point method.

could be used in principle to predict the weather. It was L. F. Richardson who had the courage to actually try it. In 1922 he published his famous book *Weather prediction by numerical process*. His experiment was a complete failure, as we know now for both physical and numerical reasons, but his work still stands as a landmark in the history of NWP.

It was not until the early fifties that a happy coincidence of circumstances made it possible for NWP to regain momentum and to grow out to a mature branch of the meteorological sciences. First, of course, there was the advent of the electronic computer. Furthermore, essential developments had taken place since Richardson's brave attempt in the field of numerical mathematics and the theory of atmospheric dynamics. Moreover, during and after the second world war the observational network had greatly improved. This made it possible to obtain a better definition of the initial state of the atmosphere, an absolutely necessary prerequisite for successful weather prediction by any method.

But probably most important of all was the fact that the brilliant and very versatile mathematician John von Neumann recognised NWP as an ideal project for his newly developed electronic computer. Together with a team of excellent meteorologists he produced in 1950 the first successful numerical weather prediction. A very readable account of this project can be found in a paper by Platzman [1979]. After this first success, NWP has been developed rapidly into an instrument without which modern operational meteorology is inconceivable.

The numerical methods used by Richardson and later by Von Neumann and his colleagues were finite difference methods. These have kept their dominant position until very recently for various reasons which will be discussed in this paper. Only recently other methods such as finite elements or spectral techniques have been accepted as viable alternatives, and in fact, in recent years atmospheric models based on spectral techniques have taken over from their finite difference predecessors in many operational and research institutes.

In this paper it is our purpose to present the spectral technique with emphasis on its application to meteorological problems. We shall first present in §2 the general problem of NWP, with emphasis on the design and the constraints of an operational forecasting system. Two important monographs are already available on spectral methods (Machenhauer [1979], Gottlieb and Orszag [1977]). We have leaned heavily, in particular, on Machenhauer's monograph when discussing the theory of the spectral method in §3.

What distinguishes this paper from Machenhauer's monograph is its emphasis on an operational intercomparison of spectral and finite difference atmospheric models. The two models used for this intercomparison were developed at the European Centre for Medium Range Weather

Forecasts (ECMWF) for medium range forecasting purposes. Such models are highly complicated and the numerical technique used is just one of the many factors determining the results. A careful design of an intercomparison is therefore necessary. In §4 we present this intercomparison and show that the spectral model is superior at least for short and medium range forecasting.

2. The general problem of numerical weather prediction. Until the second world war, the methods used by meteorologists to forecast synoptic scale motions (several hundred kilometers) were mainly empirical: they consisted of detecting the perturbations and extrapolating their trajectories in more or less a linear way. This proved reasonably satisfactory up to 24 hours or even beyond when there was no dramatic change in the weather pattern. But forecasting the deepening or filling of depressions, which is of a more nonlinear nature was a serious problem. The other major obstacle was to forecast the changes in the weather patterns over periods of 4 to 7 days. The empirical method offered no solution to this problem. As an alternative, meteorologists also tried statistical methods: it consisted in looking into past records to find analogous weather types and hoping that nature would more or less reproduce its own evolution. This method was rather popular but also admittedly rather unreliable. It was due to the fact that the analogies were almost always very local in both space and time and the records notoriously insufficient. Another possibility was to go back to the source of the problem and regard the atmosphere as what it is: a continuous fluid, the motions of which are governed by the well-known laws of fluid mechanics and thermodynamics. However, despite the pioneering attempt by Richardson in the twenties, this was to remain fancy as long as adequate computational help was not available. The situation changed dramatically when the first numerical computers became available after the second world war.

The hypothesis made is that the atmosphere as a continuous medium must obey the following laws:

Newton's law, which states in our case that the acceleration of an air parcel in an inertial system, multiplied by its mass, equals the sum of all the forces acting on it (mainly friction, pressure and gravitational forces). This law governs basically the modification of the wind field.

The first law of thermodynamics, which establishes some equivalence between work and heat. It governs basically the modifications of the temperature field.

The equation of state of an ideal gas (Avogadro's law), when applied to dry air, states: pressure = constant · density/temperature.

The continuity equations, which state the conservation of the mass of the atmosphere.

The evolution of humidity is also treated through a conservation law.

Hence we are apparently faced with a classical evolution problem with an initial condition corresponding to the state of the atmosphere at a given instant and boundary conditions corresponding to constraints at the upper and lower limits of the atmosphere. Unfortunately, this system of equations deals with all scales of motions (except the molecular scale) from the smallest eddies and the sound waves to the atmospheric tides. It

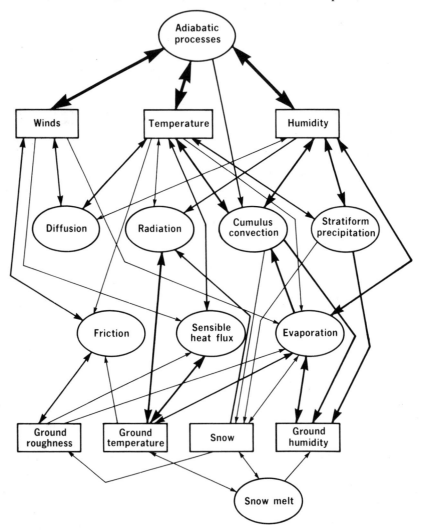

FIGURE 1. Schematic representation of the physical quantities and processes included in the ECMWF models. Interactions are indicated by arrows, the thickness of which is representative for the importance of the process.

is thus not surprising that it has no analytical solution (although this has not been formally proven).

Approximations are therefore necessary which both simplify the mathematical problem and filter unwanted phenomena like sound waves. Most simplifications are based on a scale analysis, and a whole hierarchy of mathematical models has been designed to deal with the various degrees of sophistication required. The ECMWF model, like most other numerical models of the atmosphere, is based on the so-called "primitive" equations, which, despite their name, still include a large degree of filtering (they are written in Annex A in the form used by the operational ECMWF spectral model).

All models can be broadly divided in two parts: the "adiabatic" part and the "diabatic" or "physical" part, to use the meteorologists jargon. The adiabatic part is the core of the model and corresponds to the model equations without any sources or sinks. It describes the changes in the atmosphere due to mainly horizontal and vertical advection. All the other

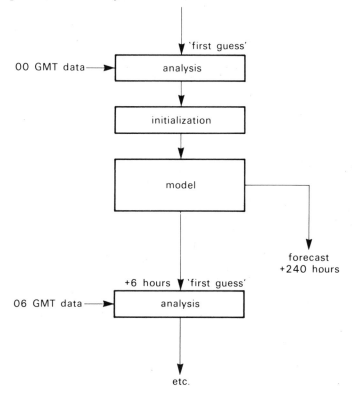

FIGURE 2. Diagram of an atmospheric model in an operational environment.

many "physical" processes such as radiation, precipitation, evaporation, friction, etc. are represented in the source and sink terms. A schematic representation of all this is shown in Figure 1. The interaction between all the different processes is indicated by arrows, the thickness of which is representative for the importance of the process.

But the model is only part of the whole operational process of numerical weather forecasting. A typical layout of such a system is shown in Figure 2:

The first step is to produce a so-called analysis, i.e. a representation of the state of the atmosphere at a given moment in some digital format. This is produced by combining an estimate ("first guess") of the state of the atmosphere with real data collected from all kinds of sources: radio soundings measuring pressure, humidity and temperature at many atmospheric levels; satellites probing the atmospheric temperatures from above; aeroplanes measuring winds along their track; etc. These data are interpolated to the levels and grid points used by the model or expanded in terms of some basis functions (e.g. spherical harmonics) to correct the "first guess". The analysis so produced must be "initialized", i.e. corrected such that the model accepts the analysis as a proper specification of the initial state. Otherwise a large fraction of the atmospheric motion might incorrectly be recognized by the model as gravity waves and be dissipated in the early stages of the integration. This initialized analysis is then used as the initial state for the model to produce forecasts by numerical integrations of the equations. So the model serves, in fact, two purposes: it produces a short forecast (3 or 6 hours) serving as "first guess" for the next analysis and it produces the main forecast out to 10 days (for ECMWF).

3. The spectral method.

3.1. *Introduction and history.* From 1950, when the first very simple models were developed, until 1975, the numerical models for simulating the large scale behaviour of the atmosphere were based almost exclusively on grid point (finite differences) techniques. The other possible techniques were regarded more as mathematical recreations than as realistic potential alternatives to grid point methods. However, in the last twelve years there has been a renewed interest and a rapid development of two other techniques: the finite elements method (which has also been widely used in many other fields of fluid dynamics), and spectral methods. In this presentation, we shall concentrate on the latter.

The use of spectral methods in numerical atmospheric models can be traced back to 1943 in the USSR (where Blinova made a proposal for long-range forecasting, using a linearised model). In 1952 Haurwitz and Craig also proposed to represent atmospheric flow patterns using spherical harmonics expansions. In 1954 Silberman solved the nondivergent barotropic vorticity equation (i.e. a highly simplified model) in spherical geometry. It was followed by several theoretical studies and applications. They demonstrated a number of attractive properties but also major drawbacks: the amount of computation and the size of storage required was rapidly becoming prohibitive when resolution was increased. This was a consequence of the method (interaction coefficients) used to compute the nonlinear terms. Another major obstacle was the inclusion of the so-called "physical processes" like convection, precipitation, etc. which did not seem feasible. The general feeling was thus that spectral models could be attractive and accurate tools for some categories of theoretical problems, mainly the ones dealing with low order (very coarse resolution) systems, but that they could not compete with operational routine numerical weather prediction models.

The real breakthrough was the adaptation of transform methods to numerical spectral models worked out independently by Orszag [1970] and Eliasen et al. [1970]: the idea is to evaluate all main quantities on an associated grid where nonlinear terms are computed, thus making possible the inclusion of "physical" processes in a way similar to grid point models. It also considerably reduced the number of computations required as well as the storage needed. It then became possible to envisage spectral models with substantially higher resolutions and having an efficiency at least comparable to that of the most efficient grid point models for a similar accuracy. After some preliminary attempts (Eliasen et al. [1970], Bourke [1972], Machenhauer and Rasmussen [1972]), several groups developed more complex multilevel hemispheric or global spectral models: Machenhauer and Daley [1972], Bourke [1974], Hoskins and Simmons [1975], and Daley et al. [1978].

This led to the implementation of spectral models for routine forecasts in Australia and Canada in 1976. Now spectral models are used operationally by most national weather centres: in USA (NMC) since 1980, in France since 1982, and in Japan and ECMWF since 1983. Several research groups also make use of spectral models for general circulation studies (numerical simulation over several months or years with coarser resolution models).

So, less than a decade after the introduction of the first spectral models for routine weather forecasting they have become the most widely used numerical tool for treating the horizontal part of the equations. However, this does not mean that they represent the ultimate step in numerical techniques for numerical weather prediction: first, they are not suitable for applications over limited areas. Second, other techniques are being developed which could turn out to be more efficient methods for a comparable accuracy even for global problems.

3.2. *General presentation.* This section is largely inspired by the presentation by B. Machenhauer in the GARP series [1979].

The equations used in numerical weather prediction can be written (see Annex A) in the general form

$$\partial F_i/\partial t = A_i(F_j), \qquad j = 1,\ldots,J, \tag{1}$$

where

$$F_i = F_i(x, t), \tag{2}$$

x representing the 3-dimensional space coordinates, t representing the time coordinate. For all meteorological applications the F_i are supposed to be as smooth (meaning differentiable) as needed, at least away from the boundaries. The operators A_i are usually nonlinear and involve partial space derivatives as well as possibly space integrals.

When finite differences ("grid point") techniques are used, an ensemble of grid points (x_p, t_q) is chosen in both space and time, and the continuous operators $\partial/\partial t$ and A_i are replaced by discrete analogues, the complexity of which depends on the accuracy requested.

The system (1) is then replaced by a new set of evolution equations at the various points x_p with the initial conditions for $t = 0$ assumed to be known.

As an alternative, the fields F_i at any time can be considered as belonging to some vector space and can be expanded in terms of a complete set of functions (independent of time). If $\{e_m(x), m = 1, \infty\}$ forms a set, the fields F_i can be written

$$F_i(x, t) = \sum_{m=1}^{\infty} F_i^m(t) e_m(x). \tag{3}$$

The equivalent to selecting a finite number of points for a finite difference method is to project F_i on a finite-dimensional subset of our functional space (truncation procedure).

F_i is thus approximated by $\overline{F}_i$:

$$\overline{F}_i(x, t) = \sum_{m=1}^{M} \overline{F}_i^m(t) e_m(x). \tag{4}$$

In order to avoid unnecessary complications, the basic principles of the method can be better described for a single variable $F = F(x, t)$.

(1) becomes

$$\partial F/\partial t = A(F), \tag{5}$$

with the initial condition $F(x,0) = f(x)$. $\tag{6}$

If F is the truncated field

$$\bar{F}(x,t) = \sum_{m=1}^{M} F^m(t)e_m(x),\tag{7}$$

then we can define a residual R by

$$R(\bar{F}) = \frac{\partial \bar{F}}{\partial t} - A(\bar{F}) = \sum_{m=1}^{M} \frac{dF^m}{dt}e_m(x) - A\left(\sum_{m=1}^{M} F^m e_m(x)\right).\tag{8}$$

The dF^m/dt are then computed such as to minimize

$$J(\bar{F}) = \int_S \left\{R(\bar{F})\right\}^2 dx \tag{9}$$

(least square minimization). A necessary condition is

$$\frac{\partial J}{\partial(dF^m/dt)} = 0 \quad \text{for } m = 1,\dots,M.\tag{10}$$

(8) and (10) lead to

$$\int_S R(\bar{F})e_m(x)\, dx = 0 \quad \text{for } m = 1,\dots,M.\tag{11}$$

So the least square minimization is equivalent to a Galerkin procedure with the e_m as test functions. As a consequence,

$$\int_S R(\bar{F})\bar{F}\, dx = 0;\tag{12}$$

in other words, the residual is orthogonal to $\bar{F}$. (11) can also be written:

$$\sum_{n=1}^{M} I_{n,m}\frac{dF^n}{dt} = \int_S A(\bar{F})e_m(x)\, dx, \qquad m = 1,\dots,M,\tag{13}$$

with $I_{n,m} = \int_S e_m(x)e_n(x)\, dx$. If the e_m are chosen orthogonal to each other (i.e. $I_{n,m} = 0$ for $n \neq m$), and if they are normalised such that $I_{n,n} = 1$, the system (13) reduces to

$$\frac{dF^m}{dt} = \int_S A(\bar{F})e_m\, dx, \qquad m = 1,\dots,M.\tag{14}$$

We thus obtain a set of simple ordinary time differential equations, the right-hand side of which can be evaluated by some quadrature procedures.

In order to determine the initial values of the F^m, it seems logical to use a similar procedure, namely to minimize

$$E = \int_S \left\{F(x,0) - \bar{F}(x,0)\right\}^2 dx,\tag{15}$$

which is obtained for $\partial E/\partial F^m = 0$, $m = 1,\ldots,M$. This leads to:

$$\int_S (F - \bar{F})e_m(x)\, dx = 0, \qquad m = 1,\ldots,M. \qquad (16)$$

If the e_m are orthonormal, (16) is equivalent to the following system:

$$F^m(t = 0) = \int_S F(x, t = 0)e_m(x)\, dx, \qquad m = 1,\ldots,M. \qquad (17)$$

In practice, since the atmosphere is highly anisotropic, especially in the vertical, it is very often convenient to use different techniques for the horizontal and vertical parts of the equations. Most spectral models use a truncated expansion only in the horizontal, the vertical being treated with classical finite difference techniques. Only a few are combined with finite elements (such an approach is presently being investigated at ECMWF).

Therefore in the next section we shall restrict ourselves to the application of the spectral technique to the horizontal part of the equations.

3.3. *Choice of the expansion functions and basic properties.* The principles outlined in the previous section can apply to both spectral and finite element techniques. The distinction comes mainly from the choice of the expansion functions (and possibly of the minimization procedure): finite elements use sets of continuous functions with compact support and spectral methods use sets of globally differentiable functions.

It is desirable to choose expansion functions simplifying the operator A. As mentioned earlier A includes space differential operators (horizontal derivatives and Laplacian operator), nonlinear terms and possibly some vertical integrals. Such a simplification is achieved if the expansion functions are eigenfunctions of some operators of A. This is the case with *spherical harmonics*:

$$Y_n^m(\lambda, \theta) = e^{im\lambda}P_n^m(\theta), \qquad (18)$$

λ = longitude, θ = latitude. $P_n^m(\theta)$ is the associated Legendre function of first kind of order m and degree n (solution of the Legendre equations, see Annex B).

There are several analytical forms for the P_n^m. A classical one is given by the Rodrigues formula:

$$P_n^m(\mu) = \frac{\left(1 - \mu^2\right)^{|m|/2}}{2^n n!}\ \frac{d^{n+|m|}\left(1 - \mu^2\right)}{d\mu^{n+|m|}} \qquad \text{if } \mu = \sin\theta. \qquad (19)$$

The Y_n^m are eigenfunctions in particular of the 2-dimensional Laplacian on the sphere. This is one of their key properties for meteorological applications:

$$\Delta Y_n^m = -\frac{n(n+1)}{a^2} Y_n^m, \qquad (20)$$

a = radius of the earth. It is an obvious consequence of (B.5) in Annex B. They are also eigenfunctions of the zonal derivative operator

$$\partial Y_n^m / \partial \lambda = im Y_n^m. \tag{21}$$

Amongst the properties of the Legendre functions and spherical harmonics, we would like to mention the following ones, particularly useful for meteorologists:

Orthogonality properties.

$$\frac{1}{2} \int_{-1}^{1} P_n^m P_{n'}^m \, d\mu = \begin{cases} 1 & \text{if } n = n' \\ 0 & \text{if } n \neq n' \end{cases} \tag{22}$$

$$\Rightarrow \frac{1}{4\pi} \int_{-1}^{1} \int_{0}^{2\pi} Y_n^m Y_{n'}^{m'*} \, d\lambda \, d\mu = \begin{cases} 1 & \text{if } (m, n) = (m', n') \\ 0 & \text{if } (m, n) \neq (m', n') \end{cases} \tag{23}$$

Y^* represents the complex conjugate of Y.

Other properties.

$$P_n^m(\mu) = P_n^{-m}(\mu), \tag{24}$$

$$P_n^m(\mu) \equiv 0 \quad \text{if } |m| > n. \tag{25}$$

(24) and (25) are straight consequences of definition (19). As a result of (24),

$$Y_n^{-m} = Y_n^{m*}. \tag{26}$$

These two properties will prove important when selecting a truncation, as we shall see below.

$$P_n^m(-\mu) = (-1)^{n+|m|} P_n^m(\mu). \tag{27}$$

This property will be useful in order to improve the efficiency of the transform method (subsection 3.4).

Finally, the expression

$$(1 - \mu^2) \frac{dP_n^m}{d\mu} = a_n^m P_{n-1}^m(\mu) + b_n^m P_{n+1}^m(\mu) \tag{28}$$

will allow a way of computing the north-south derivatives more easily than starting from the complicated analytical form of the P_n^m.

The coefficients (spectral components) of the expansion are computed as described in subsection 3.2:

$$F_n^m = \frac{1}{4\pi} \int_{0}^{2\pi} \int_{-1}^{1} F(\lambda, \mu) Y_n^{m*}(\lambda, \mu) \, d\mu \, d\lambda, \tag{29}$$

and any meteorological field F can be written

$$F(\lambda, \theta, \eta, t) = \sum_{n,m} F_n^m(\eta, t) Y_n^m(\lambda, \theta), \tag{30}$$

where η corresponds to some vertical coordinate. The sum is done over an infinite number of n and m. However, one has to restrict to a finite expansion (truncation):

$$\bar{F} = \sum_{m=-M}^{M} \sum_{n=|m|}^{N(m)} F_n^m Y_n^m. \tag{31}$$

The sum over m goes from $-M$ to M, which ensures that F is real since

$$Y_n^{-m} = Y_n^{m*} \Rightarrow F_n^{-m} = F_n^{m*} \Rightarrow F_n^{-m}Y_n^{-m} + F_n^m Y_n^m \quad \text{real.}$$

The sum over n goes from $|m|$ to $N(m)$ as a consequence of (25).

It is worth noting that increasing m (and n) corresponds to features with decreasing horizontal scale. So when using spherical harmonic expansions we have a direct control on the scales we wish to neglect, similar to the one in grid point methods when selecting the size of the grid intervals.

The two most common truncations used in numerical prediction models are the so-called triangular and rhomboidal ones (see Figure 3).

Early spectral models employed mainly rhomboidal truncations (probably for programming convenience and efficiency). However, the tendency is now changing and most newly developed spectral models use triangular truncations (France, Japan, ECMWF).

A theoretical advantage of a triangular truncation is that it is isotropic. In other words, if $\bar{F}_1$ is an expansion of F truncated at wavenumber N in one spherical coordinate system (λ_1, θ_1) and $\bar{F}_2$ in another one (λ_2, θ_2), then $\bar{F}_1 = \bar{F}_2$. This results from the following property of the spherical harmonics:

$$Y_n^m(\lambda_1, \theta_1) = \sum_{m'=-n}^{n} C_n^{m,m'} Y_n^{m'}(\lambda_2, \theta_2).$$

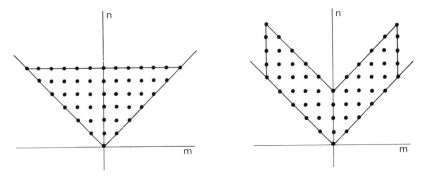

FIGURE 3. Triangular (left) and rhomboidal (right) truncations.

The resolution in a triangularly truncated system is therefore uniform on the sphere. Rhomboidal truncations, on the other hand, are only invariant by rotations around the earth's axis, and they thus correspond to a uniform resolution only in the east-west direction.

Some other considerations (e.g. spectra of kinetic energy, Baer [1972]) also plead in favour of triangular truncations. Very few comparisons have been carried out. Daley and Bourassa [1978] found no significant differences for short range weather forecasts. However, some experiments performed at ECMWF (not published) suggest some benefit from triangular truncations, especially in the stratosphere.

Before ending this section, we would like to mention another important property of the expansions in terms of spherical harmonics: the convergence of $\overline{F}$ (truncated field) toward F when $n, m \to \infty$ is extremely fast. In particular, if F is infinitely differentiable the truncation error $(\overline{F} - F)$ converges uniformly toward 0 faster than any finite power of $1/N$ (Orszag [1974]). It means that spherical harmonic truncated expansions are a very efficient way of concentrating the information for reasonably smooth fields. Another consequence is that spectral models can be expected to achieve an accuracy similar to grid point methods with a significantly smaller number of degrees of freedom.

3.4. *Transform methods.* Computation of linear terms is straightforward in a spectral model. Such is not the case for the nonlinear ones.

In practice, most nonlinear terms appear as products and in the following, without loss of generality, we shall restrict our presentation to quadratic terms $C = A \cdot B$. The conclusions can easily be extended to multiple products. Let $\overline{A} = \Sigma_{p,q} Q_q^p Y_q^p$ and $\overline{B} = \Sigma_{r,s} B_s^r Y_s^r$; if $\overline{C} = \Sigma_{m,n} C_n^m Y_n^m$, then the problem is, knowing the A_q^p and B_s^r, to compute the C_n^m:

$$\overline{A} \cdot \overline{B} = \sum_{p,q} \sum_{r,s} A_q^p B_s^r Y_q^p Y_s^r \Rightarrow C_n^m = \sum_{p,q} \sum_{r,s} I_{qsn}^{prm} A_q^p B_s^r, \tag{32}$$

with

$$I_{qsn}^{prm} = \frac{1}{4\pi} \int_0^{2\pi} \int_{-1}^1 Y_q^p Y_s^r Y_n^{m*} \, d\mu \, d\lambda. \tag{33}$$

The I_{qsn}^{prm} are called interaction coefficients since they describe the possible interactions between the various scales.

This method, however, cannot be used in practice, as soon as the truncation chosen is not very small: it leads to overwhelming storage and computing requirements. In a brave attempt to alleviate the problems several scientists elaborated "selection laws", showing that many of these interaction coefficients were 0. In other words only certain types of interaction are permitted. Despite some theoretical interest, it did not prove sufficient.

An alternative method was proposed simultaneously by Orszag [**1970**] and Eliasen et al. [**1970**]: the so-called transform method which allowed the subsequent rapid development of competitive spectral models. It is a good example of a lack of communication between mathematicians and meterologists since all the necessary ingredients were already available ten years earlier.

The basic idea is starting from A_n^m and B_n^m (with n, m belonging to the chosen truncation T) to compute $\overline{A}$ and $\overline{B}$ on a special grid, then to compute $\tilde{C} = \overline{A} \cdot \overline{B}$ and finally to recover C_n^m from $\tilde{C}$ by numerical quadrature. Note that we write $\tilde{C}$ and not C since

$$\tilde{C} = \sum_{r,s \in T} \sum_{p,q \in T} A_q^p B_s^r Y_q^p Y_s^r = \sum_{n,m \in T^P} C_n^m Y_n^m;$$

T^P is the "product" of truncation T by itself. It corresponds to all the Y_n^m necessary to represent the products of two Y_q^p belonging to T. We have always T included in T^P. The scheme is then:

Step 1.

$$A_n^m \rightarrow A_m(\mu) = \sum_n A_n^m P_n^m(\mu) \rightarrow \overline{A}(\lambda, \mu) = \sum_m A_m(\mu) e^{im\lambda},$$

$$B_n^m \rightarrow B_m(\mu) \rightarrow \overline{B}(\lambda, \mu).$$

Step 2.

$$\tilde{C}(\lambda, \mu) = \overline{A}(\lambda, \mu) \cdot \overline{B}(\lambda, \mu).$$

Step 3.

$$\tilde{C}(\lambda, \mu) \rightarrow C_m(\mu) = \frac{1}{2\pi} \int_0^{2\pi} \tilde{C}(\lambda, \mu) e^{-im\lambda} \, d\lambda, \tag{34}$$

$$C_m(\mu) \rightarrow C_n^m = \frac{1}{2} \int_{-1}^1 C_m(\mu) P_n^m(\mu) \, d\mu. \tag{35}$$

The integrals (34) and (35) can be computed exactly using quadrature formulae. (34) corresponds to classical Fourier transforms. Extremely fast Fourier transform programs are available for most computers. (35) is a so-called Legendre transform. It is easy to show that the integrand is a polynomial in μ. Several quadrature formulae have been proposed. The most efficient seems to be the Gaussian quadrature:

$$\frac{1}{2} \int_{-1}^1 g(\mu) \, d\mu = \sum_{j=1}^J w(\mu_j) g(\mu_j) \quad \text{if } g \text{ is a polynomial;} \tag{36}$$

the number of points J depends on the degree of g, the μ_j are the zeros of the Legendre polynomial $P_J^0(\mu)$, and the $w(\mu_j)$ are the Gaussian weights.

The grid corresponding to these Gaussian latitudes and to the longitudes needed for the Fourier transforms is called the Gaussian grid. Its size depends on the degree of the polynomials in the integrand and therefore on the polynomials in the integrand and therefore on the size of the truncation selected. If the grid is not large enough, the quadrature is no longer exact and the error made is called *aliasing error*.

In practice, most spectral models use a Gaussian grid providing alias-free calculations only for double products and linear terms. The triple products and the possible divisions are treated with some degree of aliasing (note that it would have been out of the question to compute such terms with the interaction coefficients method).

The advantage of the transform method over the interaction method is considerable in terms of storage and computations required: it no longer needs the storage of large interaction coefficients and the number of computations increases only as M^3 instead of M^5 when the cut-off number M (e.g. the maximum zonal wavenumber) is large enough. This is clearly illustrated in Figure 4.

Another important advantage is that grid point values are available at every time step, allowing an easy inclusion of processes such as precipitation, and convection which was not possible with the interaction coefficients method. However, despite many efforts, it has not been possible to design fast Legendre transforms similar to the fast Fourier transforms

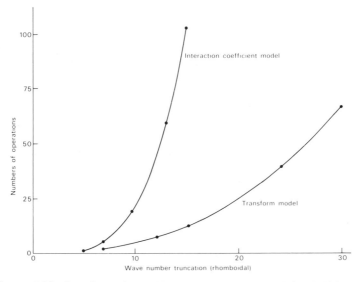

FIGURE 4. Number of operations (arbitrary units) as a function of rhomboidal truncation for an interaction coefficient model and a corresponding transform model (after Bourke [**1972**]).

(they typically correspond to 15% to 25% of the computations in a complete prediction model).

3.5. *Structure of spectral models and efficiency on modern computers.* In this subsection we shall describe only the structure of spectral models with a reasonably high truncation, using the transform method. They require the computation of values of the meteorological fields at the points of a latitude-longitude grid (e.g. Gaussian grid). However, on most computers such storage is not feasible for the complex multilevel spectral models; for example, with the resolution of the operational ECMWF model it would require more than two million words for one time level of grid point values.

In order to describe the possible solutions to this problem, it is convenient to use a simple advection equation

$$\frac{\partial F}{\partial t} = -\frac{u}{a(1-\mu^2)}\frac{\partial F}{\partial \lambda} - \frac{v}{a}\frac{\partial F}{\partial \mu} + R(F). \tag{37}$$

It contains most of the ingredients used in a numerical model: nonlinear terms (products and R), horizontal derivatives and time derivative. For simplicity let us choose a very simple time discretisation:

$$\frac{\partial F}{\partial t} \text{ is approximated by } \frac{F(t+\Delta t) - F(t-\Delta t)}{2\Delta t},$$

and all the other terms are evaluated at time level t. Let us adopt the following conventions:

$$x^+ = x(t+\Delta t), \qquad x^- = x(t-\Delta t), \qquad x = x(t).$$

(37) can be written

$$F^+ = F^- + 2\Delta t K \tag{38}$$

with

$$K = -\frac{u}{a(1-\mu^2)}\frac{\partial F}{\partial \lambda} - \frac{v}{a}\frac{\partial F}{\partial \mu} + R(F); \tag{39}$$

this corresponds to an explicit time scheme. The associated system for the spectral components equations is

$$F_n^{+m} = F_n^{-m} + 2\Delta t K_n^m. \tag{40}$$

The first solution used by early spectral models was to store three time levels of spectral components (F_n^m, F_n^{+m} and F_n^{-m}) and to scan once over the latitude lines. The corresponding structure can be schematized as follows: For each line of latitude (μ_i),

$$F_n^m \rightarrow F_m(\mu_i) = \sum_n F_n^m P_n^m(\mu_i) \rightarrow \left(\frac{\partial F}{\partial \lambda}\right)_m (\mu_i) = imF_m(\mu_i)$$

$$\rightarrow \left(\frac{\partial F}{\partial \mu}\right)_m (\mu_i) = \sum_n F_n^m \frac{\partial P_n^m}{\partial \mu}(\mu_i).$$

This yields Fourier components of F and of its horizontal derivatives. These Fourier components are then transformed into grid point values for the corresponding latitude line and the nonlinear terms are evaluated for this line, which gives $K(\lambda_j, \mu_i)$ and by Fourier transform $K_m(\mu_i)$; if we remember that

$$K_n^m = \sum_{i=1}^{I} K_m(\mu_i) P_n^m(\mu_i) w(\mu_i)$$

(e.g. by Gaussian quadrature, cf. subsection 3.4) we can compute the partial contribution of line μ_i to F_n^{+m} and accumulate it in F_n^{+m}:

$$F_n^{+m} := F_n^{+m} + 2\Delta t K_m(\mu_i) P_n^m(\mu_i) w(\mu_i).$$

When the scan is completed we can do

$$F_n^{+m} = F_n^{+m} + F_n^{-m}, \quad F_n^{-m} = F_n^{m}, \quad F_n^{m} = F_n^{+m},$$

and the full procedure can be applied again to the next time step. But this solution can also demand too much storage on many computers. Thus other solutions had to be looked for, aiming at reducing the number of spectral arrays requested (their size increases quadratically with resolution compared to a linear increase for global latitude-longitude grid point models using a scanning over latitude lines).

This leads to the *double scanning structure*: this approach was suggested by D. Burridge for the first version of the ECMWF spectral model (Baede et al. [1979]). Only one time level was requested for the spectral components of the prognostic variables, at the expense of some input/output (I/O) of grid point values and of the computation of some linear terms in grid points. It can be illustrated, again with the help of the advection equation (37). The principle is:

1. *First scan*.

$$F_n^m \rightarrow F, \frac{\partial F}{\partial \lambda}, \frac{\partial F}{\partial \mu}$$

in the way described for the one scan structure. These grid point values are then written (e.g. on disk) for every latitude line.

2. *Second scan*. For every line μ_i,
Read F, $\partial F/\partial \lambda$, $\partial F/\partial \mu$ and compute K.
Read F^-, add it to $2\Delta t K$, compute the corresponding Fourier components, and compute the partial contribution of line μ_i to F_n^{+m}, and accumulate into the array F_n^m which has been reset to zero prior to the start of the second scan:

$$F_n^m := F_n^m + \{ F^- + 2\Delta t K \}(\mu_i) P_n^m(\mu_i) w(\mu_i).$$

Compared to the one scan structure, the reduction in storage requirements is considerable and the amount of computation involved in each of the scans is generally sufficient to allow an overlapping of the input/output (I/O) by the computations.

Further extensions. Even these reduced requirements can prove too restrictive for the development on older and smaller computers of models to be used operationally on more modern and more powerful machines.

In view of that, we developed at ECMWF a multiple scan version of a spectral model where the memory requirements can be further reduced almost at will with an almost negligible extra cost.

Let us describe first the *three-scan* version.

First scan. This is identical to the first scan above, i.e.

$$F_n^m \to F, \ \frac{\partial F}{\partial \lambda}, \ \frac{\partial F}{\partial \mu},$$

and write these quantities.

Second scan. Read F, $\partial F/\partial \lambda$, $\partial F/\partial \mu$ and F^-, compute $F^- + 2\Delta tK$, and write it to disk.

Third scan. Read $F^- + 2\Delta tK$ and perform Fourier and Legendre transforms to get F_n^{+m}.

Compared to the two-scan version, an advantage of the three-scan version is that the spectral components are no longer needed in the second scan where the nonlinear terms are calculated, and in particular the parameterizations of physical processes, some of which are extremely demanding in terms of storage. The only extra cost lies in the increase of I/O. But for models with sufficiently sophisticated physical parameterizations, the amount of computation in the second scan should be more than enought to overlap the I/O.

It is further possible to separate the spectral form of the primitive equations used in NWP so that the storage of one time level of spectral components is no longer simultaneously, but sequentially, requested for the prognostic variables. For 4 main prognostic variables, as in the present ECMWF model, it reduces the spectral storage requirements by a factor of four.

For n prognostic variables it corresponds to a $(2n + 1)$-*scan structure*. Without going into details, let us just say that there is one scan to compute all the nonlinear terms and also some linear terms, and two scans for each prognostic variable (schematically, one to perform direct transforms and one for inverse transforms $F_n^m \to F$). Amongst other advantages, it makes the inclusion of new prognositc variables extremely easy and without substantial increase of storage requirements. All these structures can be made compatible: for example, the new ECMWF spectral model has been coded in a very flexible way and the choice between 2, 3 or 9 scans for its four prognostic variables is done through a simple switch. Furthermore, if the Fourier transforms are included in the same scan as the nonlinear computations the three-scan (or more) structure may be very well suited to multiprocessor machines. Let us just give the general principles.

In the scan corresponding to the grid point computations the lines of latitude can be distributed at will to the various processors. The interface with the other scans dealing with Legendre transforms can be done through I/O of Fourier components which can be divided in as many groups as processors. Each of these processors can then compute the corresponding spectral components and vice versa.

4. Comparison of spectral versus grid point models.

4.1. *Introduction.* In the previous chapter, we discussed the general application of spectral methods to numerical weather prediction models. The next problem is, of course, when designing a model, how to choose between grid point and spectral, or perhaps finite elements. The aim of a forecasting model is to get as good a forecast as possible. The quality of the forecast is the ultimate criterion, given, however, a certain amount of computer resources. This is again an insufficient specification since it may depend on such trivialities as programming skill or the available supporting software. Moreover, it must be specified precisely to what end the model will be used: short term regional forecasting, medium range global forecasting, climate research, or predictions.

Also, as mentioned in §2, the numerical method used to solve the model equations is but a small factor in today's very complex meteorological models. Many other factors may influence the results of a model's forecast at least as much as the numerical scheme. Therefore an intercomparison of numerical schemes within the complex environment of a full forecasting system is a very tricky business and should be conducted very carefully. Such a careful comparison has been carried out over the last few years at ECMWF between a spectral model and a grid point model. The purpose of this section will be to present the main results obtained.

As the main purpose of ECMWF is to produce forecasts of the atmospheric circulation for the "medium range", that is, up to ten days ahead, the models must first of all be global: the interhemispheric interactions, small at short range, may become crucially important beyond five or six days. This global nature is important for our discussion since spectral methods are not well suited to regional models. Second, in order to produce successful medium range forecasts, the models must represent a multitude of physical processes that may be left out for very short range (up to two days ahead) forecasts. In other words, the models compared are amongst the most sophisticated available, making our task more difficult.

4.2. *Theoretical differences between spectral and grid point models.* In this subsection, we do not claim to give an exhaustive list of all the differences between spectral and grid point techniques. We rather wish to insist on the differences which are considered the most relevant.

The pole problem. In a finite difference model on a latitude-longitude grid, there is a convergence of the meridians towards the poles. A consequence is the need to use a very short time step to avoid numerical instability (for a single advection equation the maximum time step permitted is roughly proportional to the grid interval).

Several solutions have been proposed to solve this problem. One is to use a modified grid where the size of the intervals does not reduce as rapidly near the poles (by having less and less grid points when approaching the poles). Another solution is to make a Fourier analysis for every latitude line and to chop the smallest scale components of the fields, or at least of their time tendencies when getting nearer the poles. It is aimed at simulating an isotropic resolution without the inconveniences of the first solution (e.g. complicated computations of north-south derivatives). A variant of this solution has been developed and implemented in the ECMWF grid point model.

Spectral models, on the other hand, have no pole problem. For example, as mentioned in §4, the triangular truncation is isotropic, which is sufficient to inhibit any pole problem.

Aliasing errors. Let us consider the simple advection equation on a circle:

$$\frac{\partial u}{\partial t} = -u\frac{\partial u}{\partial \lambda}. \tag{41}$$

This is a nonlinear equation. If we consider $2N$ points on the circle, that is

$$\lambda_i = \frac{2\pi}{2N}(i-1), \qquad i = 1,\dots,2N, \tag{42}$$

then the grid distance is $\Delta\lambda = \pi/N$; u, being periodic ($u(\lambda + 2\pi) = u(\lambda)$), can be expanded in Fourier series:

$$u(\lambda, t) = \sum_{m=-N}^{N} u_m(t)e^{im\lambda}, \tag{43}$$

with

$$u_m(t) = \frac{1}{2\pi}\int_0^{2\pi} u(\lambda, t)e^{-im\lambda}\,d\lambda. \tag{44}$$

If we assume the horizontal derivatives to be computed exactly,

$$\frac{\partial u}{\partial \lambda} = \sum_{n=-N}^{N} inu_n(t)e^{in\lambda}; \tag{45}$$

then

$$u\frac{\partial u}{\partial \lambda} = \sum_{n=-N}^{N} \sum_{m=-N}^{N} inu_n(t)u_m(t)e^{i(n+m)\lambda}. \tag{46}$$

At the points λ_i defined by (42) we have

$$u(\lambda_i, t)\frac{\partial u}{\partial \lambda}(\lambda_i, t) = \sum_{n=-N}^{N} \sum_{m=-N}^{N} inu_n(t)u_m(t)e^{i(n+m)\lambda_i}, \tag{47}$$

$n + m$ varies from $-2N$ to $2N$ and we have only $2N$ points. We can therefore only resolve

$$e^{ip\lambda}, \qquad p = -N,\ldots,N;$$

the consequence is that wavenumbers with $|p| > N$ are misrepresented (aliased) as longer waves since

$$e^{ip\lambda_i} = e^{i[p\lambda_i - k2\pi]}, \tag{48}$$

which implies

$$e^{ip\lambda_i} = e^{i(p-2N)\lambda_i}, \tag{49}$$

so p is aliased as $p - 2N$.

It can be a spurious source of energy in grid point models, and even if the time step is chosen as to avoid linear instability, it can lead to nonlinear instability (Phillips [1959]). There are again several solutions to prevent this instability.

The first one is to filter all the waves with $|p| > N$. It is equivalent to what is done in spectral models but it becomes more complicated for two-dimensional problems.

A second solution is to impose certain constraints on energy which inhibit any spurious growth of the amplitude of the smallest scales (e.g. Arakawa [1966]). It gets rid of the instability problem, but not of the aliasing error. This is the choice made by the ECMWF grid point model and most other grid point models.

Spectral models, as mentioned in subsection 3.4 do not suffer from aliasing error for quadratic terms, which also prevent the nonlinear instability mentioned above. They nevertheless present some aliasing for the triple (or more) products and for the quotients (as do grid point models). However, it is our belief that aliasing errors only play a marginal role in the differences observed between numerical forecasts made with spectral or grid point models.

Conservation laws. The continuous equations used to simulate the behaviour of the atmosphere present a number of conservation properties. In particular, the mass, the energy, the enstrophy and the angular momentum of the atmosphere are conserved. It seems desirable that their discretized equivalents have the same properties and most discretizations for grid point models have been selected on such considerations.

The problem is different for spectral models. There is nothing to choose here: either the truncated form of the equations conserves these quantities, or it does not. In practice, all conservation laws involving only linear or quadratic combinations of the spectrally truncated variables are fulfilled. The ones involving triple terms are not. For example, the ECMWF spectral model does not formally conserve energy, but in practice the deviation from energy conservation is very small; nevertheless, it remains to be checked whether it is important for long integrations (over several years, for example, in climatic simulations).

Linear phase error. We believe this to be one of the most relevant differences between spectral and grid point models. To illustrate the problem let us consider the simple linear advection equation

$$\frac{\partial u}{\partial t} = -\omega \frac{\partial u}{\partial \lambda};$$
(50)

$u = e^{i(m\lambda - ct)}$ is a solution of (50) if

$$c = \omega m.$$
(51)

A second order accurate discretization of $\partial u/\partial \lambda$ may be

$$\frac{\partial u}{\partial \lambda} \to \frac{u(\lambda + \Delta\lambda) - u(\lambda - \Delta\lambda)}{2\Delta\lambda},$$
(52)

which implies

$$c = \omega \frac{\sin m\Delta\lambda}{\Delta\lambda}.$$
(53)

Comparing (51) to (53) shows that the solution of the discretized equation has a phase speed c which is different from the exact one. Its absolute value is smaller, meaning that the waves move more slowly than in reality. In particular, the smallest wave representable ($m = N$ which implies $m\Delta\lambda = 2\pi$) is not advected at all ($c = 0$). The phase error is, moreover, a function of the wavelength leading to a spurious dispersion of meteorological systems corresponding to a combination of several waves. In a spectral model, within the initial truncation error there is no phase error of this type and therefore no spurious dispersion.

Coupling errors. Linear phase errors arise in grid point models as we have seen from the nonexact computation of derivatives in the linear terms. Coupling errors are a generalisation of this concept: they arise from the nonexact computation of derivatives in nonlinear terms. They tend to misrepresent the interactions between the various scales of motion and, in most cases, similarly to the linear phase error, to underestimate them. The linear phase error can be considered as a particular case of coupling errors between the mean part of the flow and the various other scales involved in advection processes.

Coupling errors will in practice act to reduce the effective resolution of grid point models, which can thus be expected to behave in some situations like spectral models with a significantly lower resolution. Some possible illustrations will be shown in subsection 4.5.

Initial representation of the fields. Grid point models do not have any problems here: their initial conditions coincide with the values of the fields at selected points. On the other hand, a minimization procedure (see §3) is necessary for spectral models; and since the associated grid has more degrees of freedom than the number of spectral components, it is not possible to have an exact coincidence with the values of the continuous initial field. The discrepancy will be larger for fields with a smaller scale structure, depending on the rate of convergence of the truncated expansion.

A good example is the orography. Figure 5 shows the orography soon to be used by the ECMWF spectral model (with a triangular truncation at wavenumber 63). Each point corresponds to a point of the associated Gaussian grid. Some ripples are obvious over the oceans and correspond to this initial truncation error. It can be seen even better in Figure 6 which displays a cross section of the orography along about 20° south latitude. The maximum amplitude of the ripples is seen near the steepest mountains (Andes) and their amplitude reduces considerably away from the steep slopes. Although their effect on the large scale simulation of the behaviour of the atmosphere is part of the truncation error, their local effect can be unaesthetic, and in some cases undesirable, for example, by inducing noisier precipitation fields.

4.3. *Design of the intercomparison.* The real difference between a spectral model and a grid point model being the technique used to discretize the adiabatic, horizontal part of the equations, it is essential to ensure that in such a comparison, all the other elements are kept identical as much as possible. This was achieved for the present comparison in the following way: the initial data were the same for both models, produced by the operational ECMWF analysis system, and they were initialized (see §2) by a nonlinear normal mode procedure (Temperton and Williamson [1979]) involving the operational grid point model. This may have produced a slight bias in favour of the grid point model, but it is likely to have been small compared to the differences seen. The vertical differencing schemes of both models are identical: a finite difference scheme

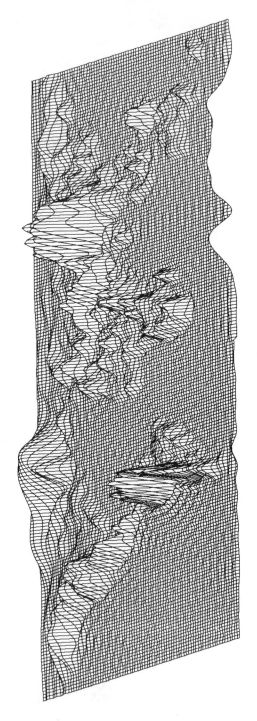

FIGURE 5. Global orography for the ECMWF spectral model.

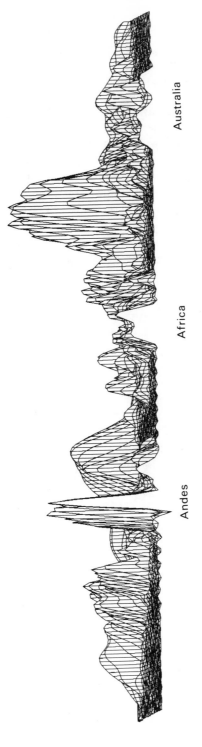

FIGURE 6. Cross section of the orography used by the ECMWF spectral model at about 20°S.

with 15 unequally spaced levels. The vertical coordinate is $\sigma = p/p_s$ in which p is the pressure and p_s the surface pressure. It has the advantage of being terrain-following near the surface and of providing thus a simple lower boundary condition.

The time differencing is almost identical in both models: it is a so-called semi-implicit leap-frog scheme. This means that the nonlinear terms are treated at the current time level and a time average over $t + \Delta t$ and $t - \Delta t$ levels is applied to the linear terms which are responsible for the fast moving gravity waves. This avoids the severe restrictions due to the CFL criterion associated with explicit schemes. The application of the semiexplicit scheme to our full system of equations requires the solution of a Helmholtz equation (i.e. the inversion of a Laplacian operator) which is trivial and costless in a spectral model and more cumbersome and expensive in a grid point model. The time step used in the grid point model was 15 minutes (18 minutes for the spectral model). The parameterization of "physical processes" (see §2) was identical in both models (Tiedtke et al. [1979]).

As already mentioned, the main difference between the models lies in the horizontal schemes. For a more detailed description, we refer to two technical reports of ECMWF: Burridge and Haseler [1977] for the grid point model, and Baede et al. [1979] for the spectral model. We shall here just emphasize the main characteristics.

The grid point model uses a latitude-longitude grid on the sphere with a regular spacing of 1.875 in both directions. It will be referred to in the following as N48 (there are 48 lines of latitude between pole and equator). In order to avoid nonlinear instability by aliasing, due to the nonlinear terms in the model equations (see subsection 4.2) a staggered grid (called Arakawa C-grid) is used in which different variables are defined on different grid points. It has the property of conserving some integral quantities in their finite difference analogue such as total energy and absolute potential enstrophy, thus inhibiting the nonlinear instability.

The spectral model operational at ECMWF since 21 April 1983 uses a triangular truncation at wavenumber 63. It will thereafter be referred to as T63. The associated grid (see subsection 3.4) used to compute the nonlinear terms, including the "physics", is almost identical to the N48 grid.

The amount of time required by both models to carry out a 10-day forecast was very similar (about 4 hours elapsed time). In order to cover as wide a range of situations as possible, the experiments were done regularly, once a week between September 1979 and August 1980, leading to one of the most comprehensive of such comparisons. Further tests were during the winter of 1982/83 which basically confirmed, or even enhanced, the conclusions drawn from the original set of experiments. The evidence obtained led to the introduction of the spectral model for operational forecasting in April 1983 as a replacement for the grid point model used so far.

4.4. *Objective evaluation.* In this subsection, we present some of the results of the evaluation of the 53 parallel forecasts. At this stage we want to warn the nonmeteorologists among our readers: the differences seen between the models are small, often smaller then the forecast errors themselves. Nevertheless, the observed improvement in favour of the spectral model was found to be significant, especially in view of the general trend of improvement for numerical weather predictions (on average, a gain of 1 day in predictability has been achieved every 4th year in the last 15 to 20 years).

When dealing with objective evaluation, one has to make use of some skill scores, preferably as few as possible, but concentrating most of the information. Such a score for medium range forecasts was found to be the anomaly correlation of the height field. Weather maps are usually displayed in the form of isolines of height on a given pressure surface. The forecast pattern can then be compared with the observed one by computing their correlation coefficient. It is better though to compute the correlation coefficient between the forecast and observed deviations from the normal climatological height patterns, since this is the important feature to predict for medium range forecasts. Experience shows that this anomaly correlation (AC) has to be larger than 60% for the forecast to be of any value. The time needed to reach this level is often taken as a measure of the predictive skill of the model.

In Figure 7 we compare the anomaly correlations (AC) from day 3 up to day 10 of all forecasts (53) made with both models, averaged over the troposphere (1000–200 mb) and most over the northern hemisphere (20°–82.5°N). For days 3–5 the clouds of points lie slightly above the diagonal on average indicating a slight superiority of T63 over N48. Note also a stronger dispersion, along the diagonal rather than across it, which corresponds to a larger variability from case to case than between models. By $D + 6$, most forecasts have already lost any predictive skill according to the criterion above.

A comparison of the same score at two selected levels, 1000 mb (near the surface) and 500 mb (middle troposphere) (Figure 8), shows that the improvement by T63 is larger at 1000 mb than at 500 mb. Note that most very big differences (when at least one of the models has predictive skill) are in favour of the spectral model.

Figure 9 shows the scores for two different groups of zonal wavenumbers, and it is clear that the improvement by T63 comes mainly from the longest (1-3) waves. The improvement was found to increase from day 1

A.C. Z1000-200

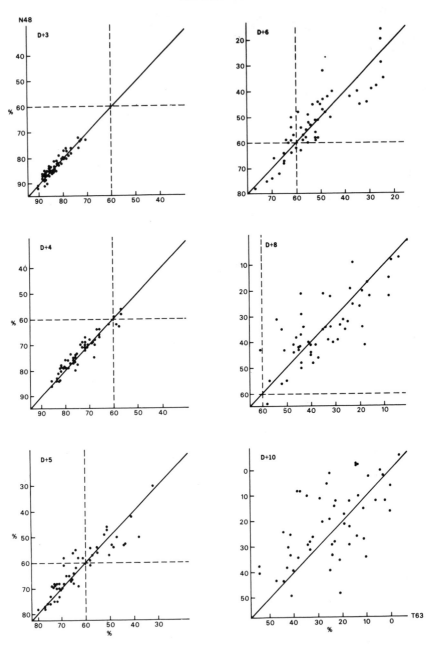

FIGURE 7. Scatter diagrams showing the anomaly correlation of height (AC) for several forecast days, averaged horizontally from 20°N to 82.5°N and vertically from 1000 mb to 200 mb comparing 53 forecasts by T63 (horizontal axis) and N48 (vertical axis).

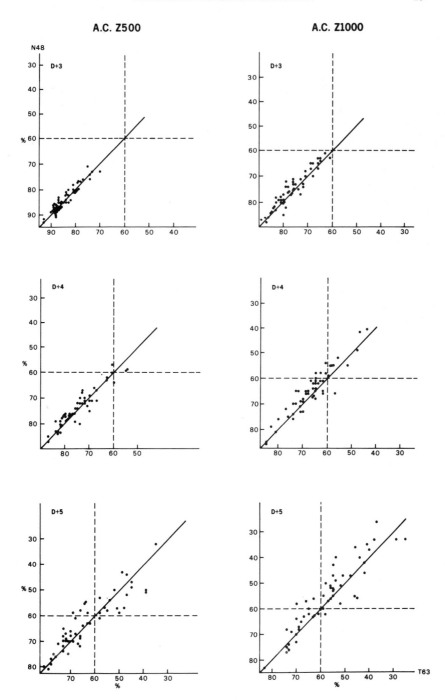

FIGURE 8. As Figure 7 but for 500 mb and 1000 mb height separately.

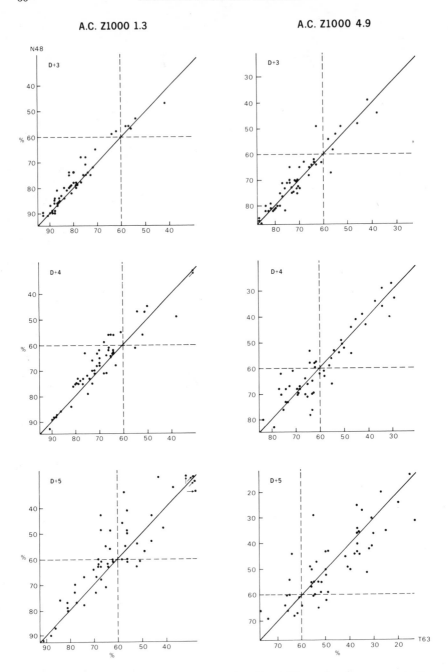

FIGURE 9. As Figure 7 but for long (1-3) and medium (4-9) zonal wave components at 1000 mb separately.

to day 5 and then to drop and vanish around day 7, suggesting that other sources of errors were becoming predominant. It is important to recall that this experiment was carried out in 1979–80. Since then the ECMWF forecasting system has been improved, and new comparisons made since tend to show that the advantage of T63 over N48 is now larger, increases up to day 6 or day 7 and vanishes only at about day 10.

Let us finally look at the improvement from the point of view of predictability (P). Figure 10 shows the percentage of cases where the difference in predictability exceeds some thresholds $(3, 6, \ldots, 24$ hours) at 1000 mb and 500 mb levels for the height field. For example, at 1000 mb, T63 improved predictability over N48 by more than half a day in 25% of the cases, the reverse being true in only 5% of the cases. In 12% of the cases T63 proved useful more than a day longer, the reverse never being true. These gains are substantial in view of the general historical trend mentioned earlier and they can be of considerable benefit to users.

Summarizing, we may say that the superiority of T63 over N48 has been confirmed by the objective evaluation. It is largest near the surface and for the longest waves. Four times a month a predictability improvement of more than 24 hours at 1000 mb (and three times a month at 500 mb) can be expected from T63. A comparison with new experiments made since also suggests that the improvement is now even larger and stands out more clearly when other error sources become smaller.

4.5. *Subjective evaluation and relation between theoretical and observed differences.* In this paper we do not wish to bother the nonmeteorologists among our readers with too many maps. We shall rather insist on typical

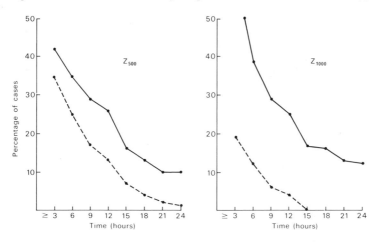

FIGURE 10. Predictability improvement of one model over the other. Full (resp. dashed) lines indicate the percentage of cases when the predictability difference between both models exceeds the time indicated, in favour of T63 (resp. N48).

M. JARRAUD AND A. P. M. BAEDE

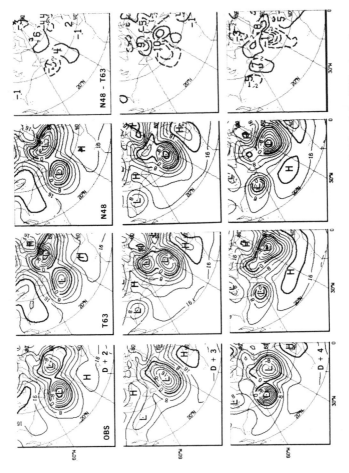

FIGURE 11. A comparison of two model forecasts of the height of the 1000 mb surface over the North Atlantic, based on the analysis of February 7, 1980. Top row: day 2 forecast; middle row: day 3 forecast; bottom row: day 4 forecast. Columns from left to right: verifying analysis; spectral (T63) forecast; grid-point (N48) forecast; difference between both forecasts.

differences. It is fair to say that the most commonly observed and easy to classify difference was linked with the position of so-called depressions (low pressure systems): they normally move eastwards (or north-east-wards) over the nothern Pacific and Atlantic Oceans. In most cases models tend to move them too slowly, but this proved to be worse for the grid point model than for the spectral model.

A typical example is given in Figure 11; it displays the observed pattern over the North Atlantic on February 9–11, 1980, as well as the corresponding 2-day, 3-day and 4-day forecasts by the spectral and grid point model, based on the analysis of February 7, 12 GMT. The difference maps between N48 and T63 are additionally provided to help evaluate the main features. One can clearly see how the main low lags behind in the N48 forecast (note the typical $-/+$ cycles in the difference maps). A second low on $D + 4$ is treated in the same way. Besides this phase error, a track difference is also visible: N48 tends to move the lows on a more southern track, not curving the trajectories northwards sufficiently.

Some statistics show that for fast moving lows the differences amount to about 120 km per day which is far from negligible. These phase differences can safely be related to the "linear" phase error described in subsection 4.2. In practice, the linear analysis is only valid over time and space intervals where the flow is more or less constant. They are preferably observed along the main cyclone tracks. The track differences are believed to be related to the anisotropy of the grid in the N48 model inducing a larger north-south than east-west phase error at middle and high latitudes.

In some other situations, very large differences were observed which could not be explained by linear phase differences. In most cases they were in favour of T63 and they are most likely the coupling errors described in subsection 4.2. These act to reduce the effective resolution of the grid point model, which was found sometimes to behave more like a spectral model with a coarser resolution. Some differences were also observed in the behaviour of some systems near the pole, in favour of T63, but it has not been possible to relate them safely to a pole problem (see subsection 4.2) for N48.

It must also be stressed that most differences between T63 and N48 originate from the transient systems, the stationary part of the flow being handled (or mishandled) in a similar way. This is illustrated in Figure 12 which displays the mean errors on day 4 averaged over 13 winter cases for both T63 and N48 at both 1000 and 500 mb. Most notable are the large systematic errors at 500 mb over the west coasts of the American and European continents. Ironically, the ECMWF model has the largest systematic errors over Europe! The errors originate from the failure of

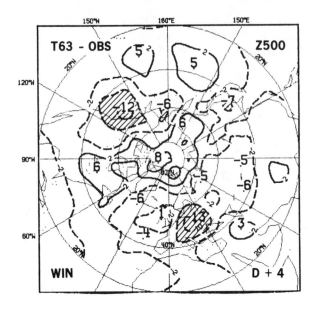

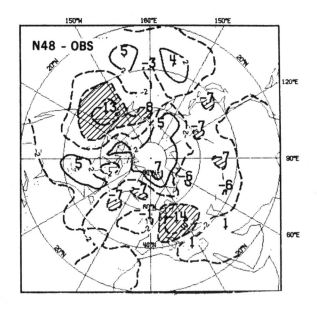

FIGURE 12. Mean errors of height at 500 and 1000 mb on day 4 of the forecasts (December 1979 to February 1980).

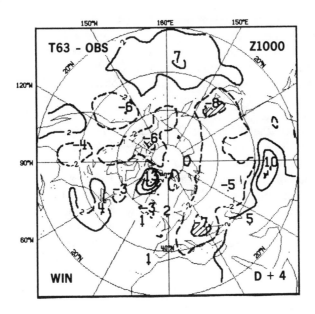

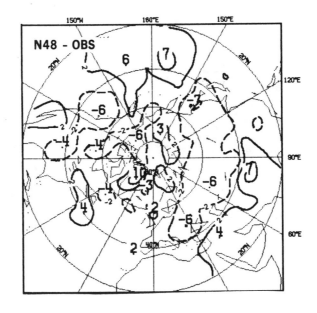

FIGURE 12. (continued)

the models to describe properly the decay of low pressure systems approaching the continents. It is worth mentioning that these errors have been substantially reduced in the present ECMWF model by a better representation of the mountain forcing.

In conclusion, we can say that the synoptic evaluation which is the ultimate criterion for our errors confirmed the significance of the differences found in the objective comparison. In some cases it was possible to trace back the differences between the two models to well-known theoretical differences, but in many more situations this proved impossible due to the very complicated nature of the models and to the importance of nonlinear interactions which can make the forecasts diverge through positive feedbacks. It is also important to mention that, unlike mathematicians, we do not know the initial state perfectly well (data, observation errors), and that can sometimes lead to a better forecast by the coarser model. Thus we had to rely on the size of our sample to get some statistical evidence of the better behaviour of the spectral model.

5. Annexes.

Annex A. *The primitive equations used by the ECMWF operational spectral model.* They can be written for a moist atmosphere as:

Momentum equations (in their divergence and vorticity forms).

$$\frac{\partial \xi}{\partial t} = \frac{1}{a(1 - \mu^2)} \frac{\partial}{\partial \lambda} (F_v + P_v) - \frac{1}{a} \frac{\partial}{\partial \mu} (F_u + P_u), \quad (A.1)$$

$$\frac{\partial D}{\partial t} = \frac{1}{a(1 - \mu^2)} \frac{\partial}{\partial \lambda} (F_u + P_u) + \frac{1}{a} \frac{\partial}{\partial \mu} (F_v + P_v) - \nabla^2 G. \quad (A.2)$$

Hydrostatic equation.

$$\frac{\partial \phi}{\partial \eta} = - \frac{R_d T_v}{p} \frac{\partial p}{\partial \eta}. \quad (A.3)$$

Thermodynamic equation.

$$\frac{dT}{dt} = \frac{x T_v \omega}{(1 + \delta q) p} + P_T. \quad (A.4)$$

Continuity equation.

$$\frac{\partial}{\partial \eta} \left(\frac{\partial p}{\partial t} \right) + \nabla \cdot \left(\mathbf{V} \frac{\partial p}{\partial \eta} \right) + \frac{\partial}{\partial \eta} \left(\dot{\eta} \frac{\partial p}{\partial \eta} \right) = 0. \quad (A.5)$$

Moisture equation.

$$\frac{dq}{dt} = P_q. \quad (A.6)$$

TABLE 1. Description of the main quantities and variables used in the primitive equations.

Coordinates

t = time

λ = longitude

θ = latitude, $\quad \mu = \sin \theta$

η = general vertical coordinate varying from 0 to 1.

Main variables

T = temperature

q = specific humidity

p = pressure (p_s = pressure at the surface of the earth)

$\mathbf{V}$ = horizontal wind vector ($u, v, 0$)

$U = u \cos \theta, \qquad V = v \cos \theta$

$\xi = \dfrac{1}{a} \{ \dfrac{1}{1 - \mu^2} \dfrac{\partial V}{\partial \lambda} - \dfrac{\partial U}{\partial \mu} \}$ = relative vorticity

$D = \dfrac{1}{a} \{ \dfrac{1}{1 - \mu^2} \dfrac{\partial U}{\partial \lambda} + \dfrac{\partial V}{\partial \mu} \}$ = divergence

$\dot{\eta} = \dfrac{d\eta}{dt}$ = vertical velocity in η coordinate

$\omega = \dfrac{dp}{dt}$ = vertical velocity in pressure coordinate

ϕ = geopotential = gz, g being the acceleration due to gravity

Operators

∇ = horizontal gradient operator

∇^2 = horizontal Laplacian operator in spherical coordinates

$\dfrac{d}{dt}$ = material derivative = $\dfrac{\partial}{\partial t} + \mathbf{V} \cdot \nabla + \dot{\eta} \dfrac{\partial}{\partial \eta}$

Physical quantities

R_d = gas constant for dry air (R_v = idem for water vapour)

C_{pd} = specific heat at constant pressure for dry air

(C_{pv} = idem for water vapour).

$\delta = \dfrac{C_{pd}}{C_{pv}} - 1, \qquad \varepsilon = \dfrac{R_d}{R_v} - 1, \qquad \chi = \dfrac{R_d}{C_{pd}}$

a = radius of the earth

f = Coriolis parameter

Derived quantities

$T_v = T(1 + \varepsilon q)$ = virtual temperature

$F_u = (f + \xi)V - \dot{\eta} \dfrac{\partial U}{\partial \eta} - \dfrac{R_d T_v}{a} \dfrac{\partial \ln p}{\partial \lambda}$

$F_v = -(f + \xi)U - \dot{\eta} \dfrac{\partial V}{\partial \eta} - \dfrac{R_d T_v}{a}(1 - \mu^2) \dfrac{\partial \ln p}{\partial \mu}$

$G = \phi + \dfrac{U^2 + V^2}{2(1 - \mu^2)}$

The meaning of the various quantities is described in Table 1. (Only the main quantities are described.) The natural boundary conditions with the general vertical coordinate selected at ECMWF are $\dot\eta = 0$ for $\eta = 0$ and $\eta = 1$, that is at the top and the bottom of the atmosphere.

Using these and integrating the continuity equation leads to a prognostic equation for the surface pressure

$$\frac{\partial p_s}{\partial t} = -\int_0^1 \nabla \cdot \left(\mathbf{V} \frac{\partial p}{\partial \eta} \right) d\eta, \qquad (A.7)$$

and a diagnostic equation for $\dot\eta$

$$\dot\eta \frac{\partial p}{\partial \eta} = -\frac{\partial p}{\partial t} - \int_0^\eta \nabla \cdot \left(\mathbf{V} \frac{\partial p}{\partial \eta} \right) d\eta, \qquad (A.8)$$

which in turn leads to a diagnostic equation for the vertical velocity in pressure coordinate (ω)

$$\omega \equiv \frac{dp}{dt} = -\int_0^\eta \nabla \cdot \left(\mathbf{V} \frac{\partial p}{\partial \eta} \right) d\eta + \mathbf{V} \cdot \nabla p. \qquad (A.9)$$

The terms P_x correspond to source or sink terms for the prognostic equations:

P_u, P_v: correspond to horizontal forces other than the pressure force, acting on an air parcel (e.g. friction);

P_T: sources and sinks of heat (e.g. radiative processes or release of latent heat due to condensation, etc.);

P_q: sources and sinks of humidity (evaporation, precipitation, etc.).

Annex B. Definition of the spherical harmonics. By definition a spherical harmonic of degree n is an homogeneous function $u(x, y, z)$ of degree n which is a solution of the Laplace equation

$$\Delta u = 0. \qquad (B.1)$$

The homogeneity condition in spherical coordinates (λ, θ, r) gives

$$r \frac{\partial u}{\partial r} = nu. \qquad (B.2)$$

(B.2) implies that

$$u(\lambda, \theta, r) = r^n S_n(\lambda, \theta) \qquad (B.3)$$

when S is an arbitrary function of λ and σ. Δu can be written in spherical coordinates:

$$\Delta u = \frac{1}{r^2} \left\{ \frac{1}{\cos \theta} \frac{\partial}{\partial \theta} \left(\cos \theta \frac{\partial u}{\partial \theta} \right) + \frac{1}{\cos^2 \theta} \frac{\partial^2 u}{\partial \lambda^2} + \frac{\partial}{\partial r} \left(r^2 \frac{\partial u}{\partial r} \right) \right\}; \qquad \text{(B.4)}$$

inserting (B.3) into (B.4) gives

$$\frac{1}{\cos \theta} \frac{\partial}{\partial \theta} \left(\cos \theta \frac{\partial S_n}{\partial \theta} \right) + \frac{1}{\cos^2 \theta} \frac{\partial^2 S_n}{\partial \lambda^2} + n(n+1) S_n = 0. \qquad \text{(B.5)}$$

$S_n(\lambda, \theta)$, solution of (B.5), is also called spherical harmonic of degree n. It can be obtained by a classical separation method. We look for a solution of the form

$$S_n(\lambda, \theta) = A(\theta) \cdot B(\lambda). \qquad \text{(B.6)}$$

(B.5) then gives

$$\frac{B(\lambda)}{\cos \theta} \frac{d}{d\theta} \left(\cos \theta \frac{dA}{d\theta} \right) + \frac{A(\theta)}{\cos^2 \theta} \frac{d^2 B}{d\lambda^2} + n(n+1) A \cdot B = 0. \qquad \text{(B.7)}$$

This leads to

$$\alpha(\theta) \frac{d^2 B}{d\lambda^2} + \beta(\theta) B = 0. \qquad \text{(B.8)}$$

The solutions are

$$B(\lambda) = a e^{im\lambda}. \qquad \text{(B.9)}$$

Inserting (B.9) back into (B.7) leads to

$$\frac{1}{\cos \theta} \frac{d}{d\theta} \left(\cos \theta \frac{dA}{d\theta} \right) + \left\{ n(n+1) - \frac{m^2}{\cos^2 \theta} \right\} A = 0. \qquad \text{(B.10)}$$

This is the so-called *Legendre equation* and it can be shown that the solutions are the $P_n^m(\theta)$ or *associated Legendre function of first kind* of order m and degree n.

ACKNOWLEDGEMENTS. We want to acknowledge C. Girard for having played an important role in the comparison between spectral and grid point models and for his contribution to the design of the new ECMWF spectral model. We are also grateful to D. Burridge and A. Simmons for many helpful discussions and suggestions, and to U. Cubasch for his help in the preparation of the experiments.

REFERENCES

General references on dynamical meteorology and numerical methods.

GARP Publ. No. 17 (1979). *Numerical methods used in atmospheric models*, Vols. 1, 2, WMO, Geneva, Switzerland.

Holton, J. R. (1972). *An introduction to dynamic meteorology*, Internat. Geophys. Ser., Vol. 16, Academic Press, New York.

Proceedings of the 1983 *ECMWF Seminar on Numerical Methods for Weather Prediction*, ECMWF, Shinfield Park, Reading, Berkshire, U.K.

References on the ECMWF models.

Baede, A. P. M., M. Jarraud and U. Cubasch (1979). *Adiabatic formulation and organisation of ECMWF's spectral model*, ECMWF Technical Report No. 15.

Burridge, D. M. and J. Haseler (1977). *A model for medium range weather forecasting, adiabatic formulation*, ECMWF Technical Report No. 4.

Temperton, C. and D. L. Williamson (1979). *Normal mode initialisation for a multi-level grid point model*, ECMWF Technical Report No. 11.

Tiedtke, M., J.-F. Geleyn, A. Hollingsworth and J.-F. Louis (1979). *ECMWF model parameterization of sub-grid scale processes*, ECMWF Technical Report No. 10.

Other references.

Arakawa, A. (1966). *Computational design for long term integrations of equations of fluid motion*, J. Comput. Phys. **1**, 119–143.

Baer, F. (1972). *An alternate scale representation of atmospheric energy spectra*, J. Atmos. Sci. **29**, 649–664.

Bourke, W. (1972). *An efficient one-level primitive equation spectral model*, Mon.Wea.Rev. **100**, 683–689.

————. (1974). *A multi-level spectral model. Formulation and hemispheric integrations*, Mon.Wea.Rev. **102**, 687–701.

Daley, R. and Y. Bourassa (1978). *Rhomboidal versus triangular spherical harmonic truncation: some verification statistics*, Atmosphere–Ocean **16**, 187–196.

Daley, R., C. Girard, J. Henderson and I. Simmonds (1978). *Short term forecasting with a multi-level spectral primitive equation model*, Atmosphere **14**, 98–134.

Eliasen, E., B. Machenhauer and E. Rasmussen (1970). *On a numerical method for integration of the hydro-dynamical equations with a spectral representation of the horizontal fields*, Report No. 2, Institut for Teoretisk Meteorologi, Københavens Universitet, Copenhagen, Denmark.

Girard, C. and M. Jarraud (1982). *Short and medium range forecast differences between a spectral and grid point model. An extensive quasi-operational comparison*, ECMWF Technical Report No. 32.

Gottlieb, D. and S. A. Orszag (1977). *Numerical analysis of spectral methods: theory and application*, Soc. Indust. Appl. Math., Philadelphia, Pa.

Haurwitz, B. and R. A. Craig (1952). *Atmospheric flow patterns and their representation by spherical surface harmonics*, A.F.C.R.L. Geophysical Research Paper No. 14.

Hoskins, B. J. and A. J. Simmons (1975). *A multi-layer spectral model and the semi-implicit method*, Quart. J. Roy. Meteor. Soc. **101**, 637–655.

Jarraud, M., C. Girard and U. Cubasch (1981). *Comparison of medium range forecasts made with models using spectral or finite difference techniques in the horizontal*, ECMWF Technical Report No. 23.

Machenhauer, B. (1979). *The spectral method*, Vol. 2, WMO/GARP Publ. Ser. No. 17, pp. 121–275.

Machenhauer, B. and R. Daley (1972). *A baroclinic primitive equation model with a spectral representation in 3-dimensions*, Report No. 4, Institut for Teoretisk Meteorologi, Københavns Universitet, Copenhagen, Denmark.

Machenhauer, B. and E. Rasmussen (1972). *On the integration of the spectral hydro-dynamical equations by a transform method*, Report No. 3, Institut for Teoretisk Meteorologi, Københavns Universitet, Copenhagen, Denmark.

Orszag, S. A. (1970). *Transform method for calculation of vector coupled sums. Application to the spectral form of the vorticity equation*, J. Atmos. Sci. **27**, 890–895.

_____. (1974). *Fourier series on spheres*, Mon.Wea.Rev. **102**, 56–75.

Phillips, N. A. (1959), *An example of nonlinear computational instability. The atmosphere and the sea in motion*, Rossby Memorial Volume Rockefeller Inst. Press, New York, pp. 501–504.

Platzman, G. (1979). *The ENIAC computations of 1950. Gateway to numerical weather prediction*, Bull. Amer. Meteor. Soc. **60**, 302–321.

Silberman, I. (1954). *Planetary waves in the atmosphere*, J. Meteor. **11**, 27–34.

EUROPEAN CENTRE FOR MEDIUM RANGE WEATHER FORECASTS, SHINFIELD PARK, READING, BERKSHIRE RG2 9AX, UNITED KINGDOM

ROYAL NETHERLANDS METEOROLOGICAL INSTITUTE (KNMI), P.O. BOX 201, 3730 AE DE BILT, THE NETHERLANDS

Lectures in Applied Mathematics
Volume **22**, 1985

Improved Flux Calculations
for Viscous Incompressible Flow
by the Variable Penalty Method

Haroon Kheshgi[1] and Mitchell Luskin[2]

ABSTRACT. The classical penalty method for viscous, incompressible flow replaces the continuity equation $\nabla \cdot \mathbf{u} = 0$ by the approximation $\nabla \cdot \mathbf{u} = -\varepsilon p$ where $\varepsilon > 0$ is a small parameter. This procedure gives an error of size $O(\varepsilon)$ for flux calculations. The variable penalty method replaces the continuity equation by $\nabla \cdot \mathbf{u} = \varepsilon \varphi_h p$, where φ_h is chosen to locally conserve mass. We prove in this paper that the variable penalty method gives an error of size $O(\varepsilon^2 + \varepsilon h^2)$ for flux calculations. Numerical experiments are given which show this dramatic improvement for flux calculations.

1. Introduction. The Navier-Stokes system for viscous, incompressible flow is given by

$$R(\mathbf{u} \cdot \nabla)\mathbf{u} + \nabla p = 2\nabla \cdot \underline{D}(\mathbf{u}) + \mathbf{f}, \qquad x \in \Omega, \tag{1.1}$$

$$\nabla \cdot \mathbf{u} = 0, \qquad x \in \Omega,$$

where $\mathbf{u} = (u_1, u_2)$ is velocity, p is pressure, $\mathbf{f}$ is a given body force, R is the Reynolds number, Ω is a bounded domain in $\mathbf{R}^2$,

$$D_{ij}(\mathbf{u}) = \frac{1}{2}\left(\frac{\partial u_i}{\partial x_j} + \frac{\partial u_j}{\partial x_i} \right), \qquad i, j = 1, 2,$$

$$\left(\nabla \cdot \underline{D}(\mathbf{u}) \right)_i = \sum_{j=1}^{2} \frac{\partial}{\partial x_j} D_{ij}, \qquad i = 1, 2.$$

1980 *Mathematics Subject Classification.* Primary 65N30, 75D05.
[1] Supported by a Grant-in-Aid from Kodak Research Laboratories.
[2] Supported by the NSF, Grant MCS 810-1631.

The classical penalty approximation of (1.1) is to replace the contiuity equation

$$\nabla \cdot \mathbf{u} = 0$$

by the perturbed continuity equation

$$\nabla \cdot \mathbf{u} = -\varepsilon p, \qquad x \in \Omega, \tag{1.2}$$

where $\varepsilon > 0$ is a small parameter [1, 8]. The purpose of introducing the approximation (1.2) is that it allows the pressure variable to be eliminated from (1.1) to give the system of equations for the approximate velocity $\mathbf{u}_\varepsilon$,

$$R(\mathbf{u}_\varepsilon \cdot \nabla)\mathbf{u}_\varepsilon - \nabla\left(\frac{1}{\varepsilon}\nabla \cdot \mathbf{u}_\varepsilon\right) = 2\nabla \cdot \underline{D}(\mathbf{u}_\varepsilon) + \mathbf{f}, \qquad x \in \Omega. \tag{1.3}$$

The penalty approximation is often applied to numerical discretizations of (1.1) since it gives a reduction in the size and band-width of the system of equations. The penalty approximation is also easier to program than (1.1) since the elminination of the pressure simplifies the data structures.

It has been shown that the difference between the soltuion to (1.1) (with appropriate boundary conditions) and the solution $\mathbf{u}_\varepsilon$, p_ε to the penalty approximation

$$R(\mathbf{u}_\varepsilon \cdot \nabla)\mathbf{u}_\varepsilon + \nabla p_\varepsilon = 2\nabla \cdot \underline{D}(\mathbf{u}_\varepsilon) = \mathbf{f}, \qquad x \in \Omega,$$
$$\nabla \cdot \mathbf{u}_\varepsilon = -\varepsilon p_\varepsilon, \qquad x \in \Omega, \tag{1.4}$$

is $O(\varepsilon)$ [1, 8]. Thus, ε must be sufficiently small to insure an accurate approximation. However, if ε is too small, then the penalty approximation becomes unstable to round-off error [2–6]. This is most easily understood by observing that the condition number of numerical discretizations of (1.3) is $O(\varepsilon^{-1}h^{-2})$, where h is the grid size. Hence, it is desirable that our penalty approximation be as accurate as possible for a given ε. This is the motivation for the variable penalty method.

The variable penalty method [2–7] replaces the continuity equation by

$$\nabla \cdot \mathbf{u}_{\varepsilon,h} = -\varepsilon\varphi_h p_{\varepsilon,h}, \qquad x \in \Omega, \tag{1.5}$$

where φ_h is normalized so that

$$\|\varphi_h\|_{L^\infty(\Omega)} = 1, \tag{1.6}$$

and φ_h satisfies the estimate

$$\left|\int_\Omega \varphi_h \zeta \, dx\right| \leqslant ch^2, \tag{1.7}$$

for smooth functions ζ with two continuous derivatives (the constant, c, in (1.7) is dependent on ζ, but independent of h). For example, on an irregular rectangular grid, it is convenient to take

$$\varphi_h(x) = \pm A_i/A, \qquad x \in Q_i,$$

where Q_i is the ith rectangle, A_i is the area of Q_i, A is the area of the largest rectangle, and the sign alternates with a checkerboard pattern (see Figure 1). Here h denotes the maximum side length of the rectangular grid. The variable penalty method also allows the pressure variable to be eliminated to give the following system of equations for the approximate velocity $\mathbf{u}_{\varepsilon,h}$:

$$R(\mathbf{u}_{\varepsilon,h} \cdot \nabla)\mathbf{u}_{\varepsilon,h} - \nabla\left(\frac{\varphi_h}{\varepsilon}\nabla \cdot \mathbf{u}_{\varepsilon,h}\right) = \underline{D}(\mathbf{u}_{\varepsilon,h}) + \mathbf{f}, \qquad x \in \Omega.$$

Error estimates for the variable penalty approximation have been presented by the authors in [3, 4]. These estimates show that the variable penalty approximation of the velocity is of higher order than the classical penalty approximation. It is also shown that a higher order method for the pressure can be obtained if the computed pressure is appropriately smoothed (or post-processed). The purpose of this paper is to show that the variable penalty method gives a higher order approximation to the total flux. An error estimate is given in §2 and numerical experiments are presented in §3.

2. Error estimates. In this section, we give error estimates for the calculation of the total flux. We note that these estimates are for the error due the approximation of the incompressibility constraint by the variable penalty approximation only. Our results here can be extended to a complete error analysis including discretization error by using our results in [3].

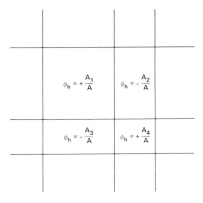

FIGURE 1. A patch of four elements.

It was demonstrated in [4] that (with appropriate boundary conditions)

$$\|p - p_\varepsilon\|_{L^2(\Omega)} \leqslant c\varepsilon, \tag{2.1}$$

and

$$\|p - p_{\varepsilon,h}\|_{L^2(\Omega)} \leqslant c\varepsilon. \tag{2.2}$$

The total flux for (1.1) is given by

$$\int_{\partial\Omega} \mathbf{u} \cdot \mathbf{n} \, ds$$

where $\mathbf{n}$ is the unit exterior normal. The approximate total flux is given by

$$\int_{\partial\Omega} \mathbf{u}_\varepsilon \cdot \mathbf{n} \, ds, \tag{2.3}$$

for the penalty method and

$$\int_{\partial\Omega} \mathbf{u}_{\varepsilon,h} \cdot \mathbf{n} \, ds \tag{2.4}$$

for the variable penalty method. We given now the following error estimate for the approximation of the total flux by the variable penalty method.

THEOREM. *Suppose p has two continuous derivatives and suppose that (2.2) is valid. Then we have that*

$$\left| \int_{\partial\Omega} \mathbf{u} \cdot \mathbf{n} \, ds - \int_{\partial\Omega} \mathbf{u}_{\varepsilon,h} \cdot \mathbf{n} \, ds \right| \leqslant c(\varepsilon^2 + \varepsilon h^2). \tag{2.5}$$

PROOF. By the divergence theorem,

$$\int_{\partial\Omega} \mathbf{u} \cdot \mathbf{n} \, ds = \int_\Omega \nabla \cdot \mathbf{u} \, dx = 0,$$

and

$$\int_{\partial\Omega} \mathbf{u}_{\varepsilon,h} \cdot \mathbf{n} \, ds = \int_\Omega \nabla \cdot \mathbf{u}_{\varepsilon,h} \, dx = -\varepsilon \int_\Omega \varphi_h p_{\varepsilon,h} \, dx$$
$$= \varepsilon \int_\Omega \varphi_h(p - p_{\varepsilon,h}) \, dx + \varepsilon \int_\Omega \varphi_h p \, dx. \tag{2.6}$$

Now by (2.2)

$$\left| \int_\Omega \varphi_h(p - p_{\varepsilon,h}) \, dx \right| \leqslant c\varepsilon,$$

and by (1.7)

$$\left| \int_{\Omega} \varphi_h p \, dx \right| \leqslant ch^2.$$

It then follows from the above that

$$\left| \int_{\partial\Omega} \mathbf{u} \cdot \mathbf{n} \, ds - \int_{\partial\Omega} \mathbf{u}_{\varepsilon,h} \cdot \mathbf{n} \, ds \right| \leqslant c(\varepsilon^2 + \varepsilon h^2). \qquad \square$$

We note that the optimal estimate for the classical penalty approximation of the total flux is

$$\left| \int_{\partial\Omega} \mathbf{u} \cdot \mathbf{n} \, ds - \int_{\partial\Omega} \mathbf{u}_{\varepsilon} \cdot \mathbf{n} \, ds \right| \leqslant c\varepsilon \qquad (2.7)$$

since

$$\int_{\partial\Omega} \mathbf{u} \cdot \mathbf{n} \, ds - \int_{\partial\Omega} \mathbf{u}_{\varepsilon} \cdot \mathbf{n} \, ds = \varepsilon \int_{\Omega} (p - p_{\varepsilon}) \, dx + \varepsilon \int_{\Omega} p \, dx$$

$$= O(\varepsilon^2) + \varepsilon \int_{\Omega} p \, dx.$$

3. Numerical experiments. We now describe the results of two numerical experiments which illustrate the error estimates given in §2. First, we consider plane Poiseuille flow in a half-channel ($R = 0$)

$$\begin{aligned} \nabla p &= 2\nabla \cdot \underline{D}(\mathbf{u}), & 0 \leqslant x_1, x_2 \leqslant 1, \\ \nabla \cdot \mathbf{u} &= 0, & 0 \leqslant x_1, x_2 \leqslant 1, \end{aligned} \qquad (3.1)$$

with boundary conditions

$$\begin{aligned} u_1(x_1, 0) = u_2(x_1, 0) &= 0, & 0 \leqslant x_1 \leqslant 1, \\ u_1(0, x_2) = 2x_2 - x_2^2, \quad u_2(0, x_2) &= 0, & 0 \leqslant x_2 \leqslant 1, \\ D_{12}(\mathbf{u}(x_1, 1)) = 0, \quad u_2(x_1, 1) &= 0, & 0 \leqslant x_1 \leqslant 1, \\ 2D_{11}(\mathbf{u}(1, x_2)) - p(1, x_2) = 0, \quad u_2(1, x_2) &= 0, & 0 \leqslant x_2 \leqslant 1. \end{aligned}$$

It is well known that the solution to (3.1) is given by

$$u_1(x_1, x_2) = 2x_2 - x_2^2, \quad u_2(x_1, x_2) = 0, \quad p(x_1, x_2) = 2(1 - x_1).$$

The variable penalty finite element approximation to the Poiseuille flow problem is given in [2, 6]. Bilinear, continuous basis functions were used for the velocity and piecewise constant functions were used for the pressure on a uniform 10×10 square grid (Figure 2). We note that the finite element solution for the velocity (with $\varepsilon = 0$) is exact at the nodes and that the finite element solution for the pressure (with $\varepsilon = 0$) is exact at the midpoint of the grid squares for this problem so we can neglect discretization error. Also, since the pressure is linear, $\int_\Omega \varphi_h\, p\, dx = 0$, and since the boundary conditions specify that the flux out of three sides is given, the estimate (2.5) can be improved to

$$\left| \int_0^1 u_1(1, x_2)\, dx_2 - \int_0^1 u_{1,h,\varepsilon}(1, x_2)\, dx_2 \right| < c\varepsilon^2. \qquad (3.2)$$

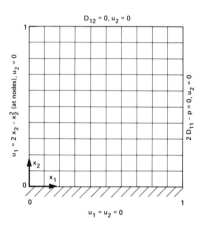

FIGURE 2. Finite element mesh for Poiseuille flow in a half-channel.

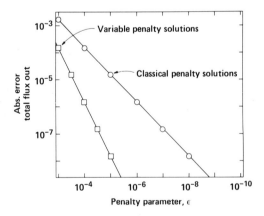

FIGURE 3. Error in total flux for Poiseuille flow problem.

The following numerical result clearly shows that the error in the total flux is $O(\varepsilon)$ for the classical penalty method and $O(\varepsilon^2)$ for the variable penalty method (Figure 3).

The next problem we consider is flow through a rotating channel (Figure 4). The dimensionless momentum equations in the rotating frame are

$$\mathscr{R}(\mathbf{u} \cdot \nabla)u + 2\boldsymbol{\omega} \times \mathbf{u} + \nabla p = E\Delta\mathbf{u}, \qquad 0 \leqslant x_1, x_2 \leqslant 1, -\infty < x_2 < \infty,$$
$$\nabla \cdot \mathbf{u} = 0, \qquad 0 \leqslant x_1, x_2 \leqslant 1, -\infty < x_3 < \infty,$$

$$(3.3)$$

where $\mathbf{u} = (u_1, u_2, u_3)$ is the velocity which is independent of x_3, i.e., $\mathbf{u} = \mathbf{u}(x_1, x_2)$, $p(x_1, x_2, x_3) = x_3 + p(x_1, x_2)$ is the gyrostatic pressure, $\boldsymbol{\omega} = (0, 1, 0)$, $\mathscr{R}$ is the Rossby number, and E is the Ekman number. No slip boundary conditions are given for the velocity. Numerical solutions to this problem by penalty finite element methods for various Rossby and Ekman numbers are given in [2, 6, 7]. Again, bilinear, continuous basis functions were used for the velocity and piecewise constant functions were used for the pressure on a rectangular grid. However, this time a nonuniform mesh was used to better approximate the thin boundary

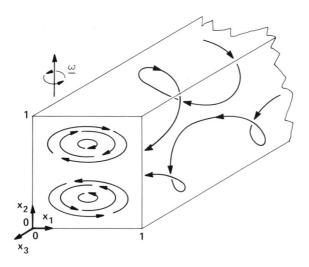

FIGURE 4. Flow through a rotating channel.

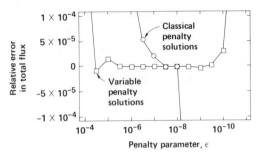

FIGURE 5. Error in total flux for flow through a rotating channel ($\mathscr{R} = 0$, $E = 10^{-6}$).

layer when the Ekman number is small. The relative error in the total flux,

$$\frac{\int_0^1 \int_0^1 u_3(x_1, x_2)\, dx_1\, dx_2 - \int_0^1 \int_0^1 u_{3,h,\varepsilon}(x_1, x_2)\, dx_1\, dx_2}{\int_0^1 \int_0^1 u_3(x_1, x_2)\, dx_1\, dx_2}, \qquad (3.4)$$

is given in Figure 5 for the case of $\mathscr{R} = 0$ and $E = 10^{-6}$. It is clear that the variable penalty method gives an accurate solution for a much wider range of ε than the classical penalty method. The increased accuracy of the solution at large ε is due to the improved stability of the variable penalty method [2–7] to round-off error for problems with a nonuniform mesh.

BIBLIOGRAPHY

1. V. Girault and P.-A. Raviart, *Finite element approximation of the Navier-Stokes equations*, Lecture Notes in Math., vol. 749, Springer-Verlag, Berlin, 1979.

2. H. Kheshgi, *The motion of viscous liquid films*, Ph.D. Thesis, University of Minnesota, Minneapolis, 1983.

3. H. Kheshgi and M. Luskin, *Analysis of the finite element variable penalty method for Stokes equations*, University of Minnesota Mathematics Report 82-168, 1982.

4. _____, *On the variable sign penalty approximation of the Navier-Stokes equation*, Nonlinear Partial Differential Equations (Joel Smoller, ed.), Contemporary Math., Vol. 17, Amer. Math. Soc., Providence, R. I., 1983.

5. H. Kheshgi and L. Scriven, *Finite element analysis of incompressible viscous flow by a variable penalty function method*, Penalty-Finite Element Methods in Mechanics (J. N. Reddy, ed.), AMD-Vol 51, Amer. Soc. Mech. Engr., New York, 1982.

6. _____, *Variable penalty method for finite element analysis of incompressible flow*. Internat. J. Numer. Methods Fluids, submitted.

7. _____, *Viscous flow through a rotating square channel*, Phys. Fluids, submitted.

8. R. Temam, *Navier-Stokes equations*, Second Ed., North-Holland, Amsterdam, 1979.

DEPARTMENT OF CHEMICAL ENGINEERING AND MATERIALS SCIENCE, UNIVERSITY OF MINNESOTA, MINNEAPOLIS, MINNESOTA 55455

SCHOOL OF MATHEMATICS, UNIVERSITY OF MINNESOTA, MINNEAPOLIS, MINNESOTA 55455

Current address (H. Kheshgi): Lawrence Livermore National Laboratory, Livermore, California 94550

Lectures in Applied Mathematics
Volume **22**, 1985

TVD Schemes in One
and Two Space Dimensions

Randall J. LeVeque[1] and Jonathan B. Goodman[2]

ABSTRACT. We briefly review the theory of total variation diminishing (TVD) schemes for scalar conservation laws and our recent work in this area. In two (or more) space dimensions TVD schemes are at most first order accurate. In 1D we present a TVD implementation of a second order MUSCL-type scheme. We also discuss the spreading of rarefaction waves and one approach to obtain bounds on the rate of spreading in numerical solutions.

1. Introduction. At one time numerical computations of discontinuous solutions of conservation laws were plagued by one of two difficulties [**14**]. Either the method was first order accurate on smooth flows and the discontinuities were excessively smeared, or else spurious oscillations were introduced which polluted the solution and sometimes lead to nonlinear instability. It has now become possible to avoid these difficulties on many problems with the recent development of schemes which are second order accurate in smooth regions and give sharp discontinuities with no oscillation.

Unfortunately, imposing this last requirement for systems of conservation laws is notoriously difficult. Consequently, much of the effort has gone into devising and studying methods for scalar conservation laws, where precise statements can be made. The hope is that good methods for

[1] Supported in part by a National Science Foundation Postdoctoral Fellowship and by the Courant Institute of Mathematical Sciences.
[2] Supported in part by National Science Foundation Grant NSF-MCS-82-01599.
1980 *Mathematics Subject Classification*. Primary 65M05, 65M10.

scalar problems can be generalized to obtain good methods for systems. This hope is borne out surprisingly well. The process is discussed, for example, by Roe [15] and by Harten, Lax and van Leer [8]. Here we restrict our attention to scalar conservation laws.

The main theoretical advance was made by Harten [7], who introduced the concept of a total variation diminishing (TVD) scheme and first constructed schemes that are simultaneously TVD and second order accurate on smooth solutions. For the scalar conservation law

$$u_t + f(u)_x = 0, \tag{1.1}$$

any weak solution $u(x, t)$ has diminishing total variation [9]

$$\mathrm{TV}(u(\cdot, t)) \leqslant \mathrm{TV}(u(\cdot, 0)), \tag{1.2}$$

where

$$\mathrm{TV}(u(\cdot, t)) = \int_{-\infty}^{\infty} |u_x(x, t)| \, dx$$

in the distribution sense. A TVD scheme is one for which the resulting numerical approximation has the same property. Let $U_j^n \approx u(x_j, t_n)$ denote the approximate solution where $h = \Delta x$ and $k = \Delta t$ are the mesh width and time step, respectively. Since we are interested in the effect of a single time step, we will drop the superscripts and use $U_j = U_j^n$ and $\overline{U}_j = U_j^{n+1}$. Then the discrete total variation is

$$\mathrm{TV}(U) = \sum_j |U_{j+1} - U_j|,$$

and a difference scheme is TVD if

$$\mathrm{TV}(\overline{U}) \leqslant \mathrm{TV}(U) \tag{1.3}$$

for all data U. This property is important in theory, since it leads to convergence proofs [2, 7], and in practice, since it guarantees freedom from spurious oscillations.

A property stronger than (1.3) is monotonicity. A scheme is monotone if $U \geqslant V \Rightarrow \overline{U} \geqslant \overline{V}$, where $U \geqslant V$ means $U_j \geqslant V_j$ for all j. Although monotone schemes are TVD [2, 7], they are at most first order accurate [2]. Hence they are unsuitable for practical computations in more than one space dimension where higher order accuracy is generally required to obtain good approximations with a reasonable number of meshpoints.

If a scheme is in conservation form

$$\overline{U}_j = U_j - \frac{k}{h} [F(U; j) - F(U; j - 1)], \tag{1.4}$$

then monotonicity is equivalent to L_1 contraction, which requires that

$$\|\overline{U} - \overline{V}\|_1 \leqslant \|U - V\|_1$$

for all sets of data $\{U_j\}$ and $\{V_j\}$ such that

$$\|U - V\|_1 \equiv \sum_j |U_j - V_j| < \infty.$$

We use this fact, which follows from a lemma of Crandall and Tarter [3] (see also [4]), in the next section.

In this paper we summarize some of our recent theoretical work in this area. In §2 we sketch our proof that TVD schemes in two or more space dimensions are at most first order accurate. We discuss the practical implications of this theorem and some possible ways around it towards an analysis of accurate multidimensional schemes. §3 contains an analysis of a simple second order Godunov-type scheme similar to the MUSCL scheme of van Leer [12]. We show that if sufficient care is taken in implementation, such a scheme can be made TVD. These results give some theoretical explanation for the impressive resolution of well-engineered Godunov-type schemes, e.g. [1].

Monotone schemes have the advantage that they always converge to the correct (i.e. entropy-satisfying) weak solution of (1.1) [2]. Higher order schemes may lack this property and give an entropy-violating "rarefaction shock" in place of a physically correct rarefaction wave [7]. More subtly, a rarefaction wave may spread too slowly causing a "dogleg" in the numerical approximation [17]. In §4 we briefly discuss this phenomenon and describe one approach to obtain estimates for the rate of spreading.

2. TVD schemes in 2 space dimensions. While in one space dimension there are many ways to derive second order accurate TVD schemes [6, 12 15, 18] in 2D no such schemes exist, at least not with the obvious definition of total variation. For the scalar problem

$$u_t + f(u)_x + g(u)_y = 0, \qquad (2.1)$$

in 2D and initial data having compact support the total variation can be defined as

$$\mathrm{TV}(u) = \int \int \|\nabla u\| \, dx \, dy. \qquad (2.2)$$

If we use the L_1 norm $\|\nabla u\| = |u_x| + |u_y|$, then the discrete version of this is

$$\mathrm{TV}(U) = \mathrm{TV}_x(U) + \mathrm{TV}_y(U), \qquad (2.3)$$

where

$$\mathrm{TV}_x(U) = \Delta y \sum_{j,k} |U_{j+1,k} - U_{j,k}|, \qquad (2.4)$$

and

$$\text{TV}_y(U) = \Delta x \sum_{j,k} |U_{j,k+1} - U_{j,k}| \qquad (2.5a)$$

$$= \sum_k \|U_{\cdot,k+1} - U_{\cdot,k}\|_1. \qquad (2.5b)$$

Here $\| \cdot \|_1$ is the one-dimensional discrete L_1 norm. Using this definition we have the following result.

THEOREM. *In 2 space dimensions any TVD scheme is at most first order accurate.*

The proof can be found in Goodman and LeVeque [4]. The main idea is that by using (2.5b) we can relate total variation diminishing in 2 dimensions to L_1-contracting in 1 dimension. Note that any two-dimensional scheme defines a corresponding one-dimensional scheme simply by applying it to data which is constant in one direction. Given two sets of one-dimensional data $\{V_j\}$ and $\{W_j\}$ we construct two-dimensional data $\{U_{ij}\}$ such that the 2D scheme is TVD on U only if the corresponding 1D scheme is L_1-contracting on V and W ($\|\overline{V} - \overline{W}\|_1 \leq \|V - W\|_1$). Since any L_1-contracting scheme is at most first order accurate [2], it follows that the 2D scheme is also at most first order accurate.

It should not be inferred from this result that there is no hope of obtaining second order accurate results in two dimensions with sharp discontinuities. On the contrary, results we have seen which were obtained using 1D second order TVD schemes and dimensional splitting look very good indeed, with no visible oscillations. Therefore we point out some possible ways around this theorem. One obvious possibility is that a different norm should be used in (2.2). The total variations based on $\|\nabla u\| = (u_x^2 + u_y^2)^{1/2}$, or on $\|\nabla u\| = \max(|u_x|, |u_y|)$, are also nonincreasing for the exact solution. Perhaps there exist higher order accurate difference schemes that are TVD in one of these senses. Secondly, it may be possible (although it seems difficult) to obtain useful bounds for the total variation of a solution obtained with a scheme which is not TVD. Finally, it may be that an entirely different measure of stability is required in two dimensions.

3. A second order accurate TVD scheme in 1D. One approach to defining a second order accurate TVD scheme is to generalize Godunov's method, a standard first order accurate TVD scheme. In Godunov's method the data $\{U_j\}$ at time t_n is viewed as a piecewise constant function taking the value U_j in the interval $I_j = (x_{j-1/2}, x_{j+1/2})$. We denote this piecewise constant function by $v^n(x, t_n)$ and solve the problem (1.1) exactly up to time t_{n+1} with this initial data to obtain $v^n(x, t_{n+1})$.

This can easily be done if k is sufficiently small. Then $\{\overline{U}_j\}$, the approximation at time t_{n+1}, is obtained by averaging $v^n(x, t_{n+1})$:

$$\overline{U}_j = \frac{1}{h} \int_{x_{j-1/2}}^{x_{j+1/2}} v^n(x, t_{n+1}) \, dx. \qquad (3.1)$$

In the future we drop the superscript n on the function $v(x, t)$.

We can rewrite Godunov's method in "conservation form" by first integrating (1.1) over the rectangle $I_j \times [t_n, t_{n+1}]$ to obtain

$$0 = \int_{t_n}^{t_{n+1}} \int_{x_{j-1/2}}^{x_{j+1/2}} v_t(x, t) + f(v(x, t))_x \, dx \, dt$$

$$= \int_{x_{j-1/2}}^{x_{j+1/2}} \left(v(x, t_{n+1}) - v(x, t_n) \right) dt$$

$$+ \int_{t_n}^{t_{n+1}} \left(f(v(x_{j+1/2}, t)) - f(v(x_{j-1/2}, t)) \right) dt,$$

or, using (3.1),

$$\overline{U}_j = U_j - \frac{k}{h} \left[F(U; j) - F(U; j - 1) \right], \qquad (3.2)$$

where

$$F(U; i) = \frac{1}{k} \int_{t_n}^{t_{n+1}} f(v(x_{i+1/2}, t)) \, dt \qquad (3.3)$$

is the "numerical flux" across $x_{i+1/2}$. This can be simplified due to the simple structure of the solution v for a scalar conservation law. If we assume that $f'(u) \neq 0$ for all u between U_i and U_{i+1} (i.e. we are away from sonic point u_0 where $f'(u_0) = 0$), then $v(x_{i+1/2}, t)$ is equal to either U_i or U_{i+1} for $t_n \leqslant t \leqslant t_{n+1}$ depending on whether $f'(u)$ is positive or negative, and

$$F(U; i) = \begin{cases} f(U_i) & \text{if } f'(u) > 0 \; \forall u \in \text{int}[U_i, U_{i+1}], \\ f(U_{i+1}) & \text{if } f'(u) < 0 \; \forall u \in \text{int}[U_i, U_{i+1}]. \end{cases} \qquad (3.4)$$

Van Leer, with his MUSCL scheme [12], pioneered the effort to obtain second order accuracy by using a piecewise linear function $v(x, t_n)$ rather than a piecewise constant. The resulting scheme is conservative providing v is of the form

$$v(x, t_n) = U_j + s_j(x - x_j) \quad \text{for } x \in I_j. \qquad (3.5)$$

There are various ways to pick the slopes s_j. One simple choice which leads to a TVD scheme is

$$s_j = \begin{cases} 0 & \text{if } (U_{j+1} - U_j)(U_j - U_{j-1}) \leqslant 0, \\ \text{sgn}(U_{j+1} - U_j) \min\left\{ \left| \dfrac{U_{j+1} - U_j}{h} \right|, \left| \dfrac{U_j - U_{j-1}}{h} \right| \right\} & \text{otherwise,} \end{cases}$$
(3.6)

as illustrated in Figure 1. It is easy to show that this choice of slope guarantees that

$$\text{TV}(v(\cdot, t_n)) \leqslant \text{TV}(U).$$
(3.7)

If we now solve (1.1) for $v(x, t_{n+1})$ and obtain $\overline{U}_j$ by (3.1) as in Godunov's method, then $\text{TV}(\overline{U}) \leqslant \text{TV}(U)$. This follows from (3.7) and the fact that the exact solution operator taking $v(\cdot, t_n)$ to $v(\cdot, t_{n+1})$ and the averaging process (3.1) are both TVD. By considering the truncation error introduced by the approximation (3.5) it can be shown that this scheme is second order accurate in smooth regions of the flow.

The problem with this method is that in general it cannot be efficiently implemented as just described, since it requires solving the conservation law (1.1) exactly with piecewise linear initial data. This is not nearly so trivial as solving the problem with piecewise constant data. Instead of computing the exact solution, various approximations can be used. Most such approximations do not, however, retain the TVD property. In [5] we have introduced an approximation which does give a TVD, second order accurate numerical scheme which is easy to implement.

The idea is to approximate not only the initial data but also the flux function f occurring in (1.1) by a piecewise linear function. If this approximation is chosen correctly, then the problem of computing the flux across any $x_{j+1/2}$ reduces, locally, to solving a linear problem with piecewise linear initial data—a simple task.

The sonic point again causes some difficulty and so for simplicity we begin by assuming that $f'(v(x, t_n)) \neq 0$ for $x \in I_j \cup I_{j+1}$. The formulas for the sonic case will be presented afterwards. We always assume that the flux function f is convex: $f'' > 0$.

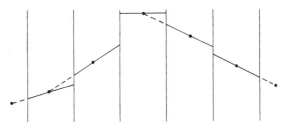

FIGURE 1. The piecewise linear function $v''(x, t_n)$ with slopes s_j given by (3.6). Dots denote the U_j.

To compute our numerical flux across $x_{j+1/2}$, we first set

$$U_i^\pm = U_i \pm \tfrac{1}{2}hs_i.\tag{3.8}$$

The points U_j^-, U_j^+, U_{j+1}^-, U_{j+1}^+ are monotonically ordered (though two or more may coincide) by virtue of (3.6). Let $g(u)$ be the piecewise linear function connecting the values $f(U_j^-), f(U_j^+), f(U_{j+1}^-)$ and $f(U_{j+1}^+)$, and set

$$g_i' = \begin{cases} \left(\dfrac{f(U_i^+) - f(U_i^-)}{U_i^+ - U_i^-} \right) & \text{if } s_i \neq 0, \\[2mm] f'(U_i) & \text{if } s_i = 0 \end{cases}\tag{3.9}$$

for $i = j, j + 1$ (see Figure 2).

For the problem $v_t + g(v)_x = 0$ with piecewise linear initial data v we can easily compute the flux $G(U; j)$ across $x_{j+1/2}$ during $[t_n, t_{n+1}]$. We find that

$$v(x_{j+1/2}, t) = \begin{cases} U_j^+ - (t - t_n)s_j g_j' & \text{if } f' > 0, \\[2mm] U_{j+1}^- - (t - t_n)s_{j+1} g_{j+1}' & \text{if } f' < 0, \end{cases}\tag{3.10}$$

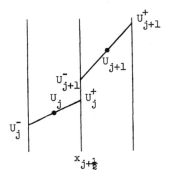

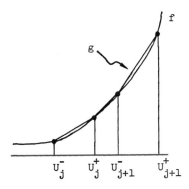

FIGURE 2. The piecewise linear functions $v^n(x, t_n)$ and $g(u)$ over two cells I_j and I_{j+1}.

and since

$$g(u) = \begin{cases} f(U_j^+) + (u - U_j^+)g_j' & \text{for } u \in \text{int}[U_j^-, U_j^+], \\ f(U_{j+1}^-) + (u - U_{j+1}^-)g_{j+1}' & \text{for } u \in \text{int}[U_{j+1}^-, U_{j+1}^+], \end{cases}$$

$$(3.11)$$

the flux is

$$G(U; j) = \frac{1}{k} \int_{t_n}^{t_{n+1}} g(v(x_{j+1/2}, t)) \, dt$$

$$= \begin{cases} f(U_j^+) - \frac{1}{2}ks_j(g_j')^2 & \text{if } f' > 0, \quad (3.12a) \\ f(U_{j+1}^-) - \frac{1}{2}ks_{j+1}(g_{j+1}')^2 & \text{if } f' < 0. \quad (3.12b) \end{cases}$$

Note that if we set $s_j \equiv 0$ for all j, i.e. if we use piecewise constant initial data, we recover the flux (3.4) of Godunov's method. Also note the similarity of both (3.12a) and (3.12b) to the flux of the Lax-Wendroff method,

$$F_{LW}(U; j) = \frac{1}{2}(f(U_j) + f(U_{j+1}))$$

$$- \frac{1}{2}k\left(\frac{U_{j+1} - U_j}{h}\right)\left(\frac{f(U_{j+1}) - f(U_j)}{U_{j+1} - U_j}\right)^2. \quad (3.13)$$

In fact,

$$G(U; j) = F_{LW}(U; j) + \mathcal{O}(h^2)$$

with the $\mathcal{O}(h^2)$ term a smooth function of j (except where $s_j = 0$). In computing

$$\overline{U}_j = U_j - \frac{k}{h}[G(U; j) - G(U; j - 1)], \quad (3.14)$$

the $\mathcal{O}(h^2)$ terms cancel to $\mathcal{O}(h^3)$, showing that our scheme agrees with Lax-Wendroff to $\mathcal{O}(h^3)$ locally, and hence is second order accurate (in smooth regions, except at extrema of U).

The use of the limited values s_j and g_j' in (3.12a) and (3.12b), rather than the corresponding expressions in (3.13), has the great advantage that our scheme, unlike Lax-Wendroff, is TVD [5].

The sonic case is discussed in [5]. Here we simply present the formulas. First, set

$$g_{j+1/2}' = \begin{cases} \left(\dfrac{f(U_{j+1}^-) - f(U_j^+)}{U_{j+1}^- - U_j^+}\right) & \text{if } U_{j+1}^- \neq U_j^+, \\ f'(U_j^+) & \text{if } U_{j+1}^- = U_j^+, \end{cases}$$

$$(3.15)$$

$$v_0 = \min\left(\max\left(U_j^+, u_0\right), U_{j+1}^-\right),$$

$$G_j = f\left(U_j^+\right) - \tfrac{1}{2}ks_j\left(g_j'\right)^2,$$

$$G_{j+1} = f\left(U_{j+1}^-\right) - \tfrac{1}{2}ks_{j+1}\left(g'_{j+1}\right)^2.$$

Then the flux is given by

1) if $g_j' > 0$, $g_{j+1/2}'(U_{j+1}^- - U_j^+) = 0$, and $g_{j+1}' < 0$, then

$$G(U; j) = \begin{cases} G_j & \text{if } \left(g_j'\right)^2 s_j \geqslant \left(g_{j+1}'\right)^2 s_{j+1}, \\ G_{j+1} & \text{otherwise;} \end{cases} \qquad (3.16)$$

2) otherwise,

$$G(U; j) = \begin{cases} G_j & \text{if } g_j' \geqslant 0, \, g_{j+1/2}' \geqslant 0, \\ G_{j+1} & \text{if } g_{j+1/2}' \leqslant 0, \, g_{j+1}' \leqslant 0, \\ f(v_0) & \text{if } g_j' < 0, \, g_{j+1}' > 0. \end{cases} \qquad (3.17)$$

These formulas cover all cases, including the nonsonic cases (3.12a) and (3.12b).

4. The entropy condition and spreading of rarefaction waves. Although much attention has been given to developing schemes with sharp discontinuities, the proper resolution of rarefaction waves has also proved difficult [7]. In general, the weak solution of a conservation law is not unique and some additional condition, such as an entropy condition, must be used to select the physically correct solution [7, 11]. For scalar conservation laws this amounts to ensuring that an initial discontinuity with characteristics spreading apart resolves into an expansion fan (rarefaction wave) rather than persisting as a sharp "entropy-violating" discontinuity. If the flux function $f(u)$ in (1.1) is strictly convex,

$$f''(u) \geqslant \alpha > 0, \qquad (4.1)$$

then an initial jump

$$u(x,0) = \begin{cases} u_l & \text{for } x \leqslant 0, \\ u_r & \text{for } x > 0, \end{cases}$$

should resolve into a rarefaction fan if $u_l < u_r$. One can show [13] that $u(x, t)$ then satisfies the spreading estimate

$$u_x(x, t) \leqslant 1/\alpha t \quad \text{for } t > 0, \text{ all } x. \qquad (4.2)$$

More generally,

$$u(x, t) - u(y, t) \leqslant \frac{x - y}{\alpha t} \quad \text{for } x > y, \, t > 0, \qquad (4.3)$$

is valid for arbitrary (bounded measurable) initial data. Note that (4.3) rules out entropy-violating shocks but not the physically correct shocks for which $u_l > u_r$. Oleinik [13] has shown uniqueness of solutions satisfying (4.3).

It is clearly desirable that a numerical approximation U_j^n satisfy (4.3) in the form

$$U_{j+1}^n - U_j^n \leqslant 1/cn \qquad (4.4)$$

for some constant $c > 0$. A TVD scheme in conservation form satisfying (4.4) will converge to the correct entropy-satisfying solution of (1.1). Moreover, if (4.4) holds with $c = \alpha k/h$, then rarefaction waves spread at the correct rate. Not all TVD schemes have this property and consequently one sometimes observes "doglegs" [17, 18], particularly near the sonic point. These are regions where the rarefaction is spreading too slowly and consequently a kink appears in what should be a smooth rarefaction wave (see Figure 3).

To prove (4.2) one shows that

$$\frac{d}{dt} D(t) = -\alpha D^2(t),$$

where $D(t) = \max_x u_x(x, t)$, and then applies Gronwall's inequality. Analogously, we can show (4.4) for the numerical approximation by induction if

$$D_{n+1} \leqslant D_n - cD_n^2, \qquad (4.5)$$

where $D_n = \max_j (U_{j+1}^n - U_j^n)$.

We will briefly sketch our proof of (4.5) for Godunov's method to demonstrate the main ideas used in our approach. The possibility of proving estimates like (4.4) was independently noticed by Tadmor [19] who used a different approach to obtain (4.4) for the Lax-Friedrichs scheme.

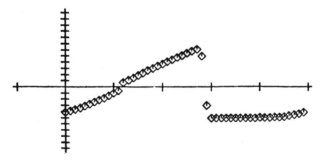

FIGURE 3. A kinked rarefaction wave (dogleg) obtained with Godunov's method on Burgers' equation, $f(u) = \frac{1}{2}u^2$. The sonic point is at $u_0 = 0$. Computations and picture courtesy of P. K. Sweby.

Consider a single meshpoint and suppose we wish to show that

$$U_{j+1}^{n+1} - U_j^{n+1} \leqslant D_n - cD_n^2. \tag{4.6}$$

For simplicity in this discussion we assume $f'(u) > 0$ everywhere. We use two main ideas. First, it suffices to consider the extremal case where $U_{j+1}^n - U_j^n = U_j^n - U_{j-1}^n = D_n$. It is straightforward to check that the actual value of $U_{j+1}^{n+1} - U_j^{n+1}$ will be bounded above by the bound obtained for the extremal case. Second, we begin by considering the linearized equation

$$v_t + f'(U_j^n)v_x = 0. \tag{4.7}$$

Solving this equation with Godunov's method gives values V^{n+1} which are easily seen to satisfy

$$V_{j+1}^{n+1} - V_j^{n+1} \leqslant D_n. \tag{4.8}$$

The bound (4.6) follows by comparing the true solution of (1.1) to the true solution of (4.7). The strict convexity of f implies the spreading estimate (4.2). Since there is no spreading in the linear problem, a comparison of U^{n+1} and V^{n+1} yields

$$U_j^{n+1} - V_j^{n+1} \geqslant \tfrac{1}{2}cD_n^2 \quad \text{and} \quad U_{j+1}^{n+1} - V_{j+1}^{n+1} \leqslant -\tfrac{1}{2}cD_n^2.$$

Subtracting and using (4.8) gives (4.6). Details of these arguments may be found in [5].

For Godunov's method, we find that the constant c in (4.6), and hence (4.4), is equal to $\alpha k/h$ except near the sonic point where $c = \alpha k/2h$ is the best we can obtain in general. Thus the part of a rarefaction spanning the sonic point may spread at only half the correct rate. This explains the dogleg phenomenon for Godunov's method.

For the second order method described in §3 we cannot obtain (4.5) since the linearized scheme may not satisfy (4.8). However, it is possible to enforce (4.8) for linear problems and (4.6) for nonlinear problems by adding extra smoothing.

REFERENCES

1. P. Collela and P. Woodward, *The piecewise-parabolic method (PPM) for gas dynamical simulations*, J. Comput. Phys. **54** (1984), 174–201.

2. M. Crandall and A. Majda, *Monotone difference approximations for scalar conservation laws*, Math. Comp. **34** (1980), 1–21.

3. M. Crandall and L. Tartar, *Some relations between nonexpansive and order preserving mappings*, Proc. Amer. Math. Soc. **78** (1980), 385–390.

4. J. B. Goodman and R. J. LeVeque, *On the accuracy of TVD schemes in two space dimensions*, Math. Comp. (to appear).

5. _____, *A geometric approach to high resolution TVD schemes*, ICASE Report No. 84-55, 1984.

6. A. Harten, *High resolution schemes for hyperbolic conservation laws*, J. Comput. Phys. **49** (1983), 357–393.

7. A. Harten, J. M. Hyman and P. D. Lax, *On finite difference approximations and entropy conditions for shocks*, Comm. Pure Appl. Math. **29** (1976), 297–322.

8. A. Harten, P. D. Lax and B. van Leer, *On upstream differencing and Godunov-type schemes for hyperbolic conservation laws*, SIAM Review **25** (1983), 35–61.

9. B. Keyfitz, Appendix to [9].

10. P. D. Lax, *The formation and decay of shock waves*, Amer. Math. Monthly **79** (1972), 227–241.

11. _____ , *Hyperbolic systems of conservation laws and the mathematical theory of shock waves*, SIAM Regional Conf. Ser. Appl. Math., No. 11, Soc. Indust. Appl. Math., Philadelphia, Pa., 1973.

12. B. van Leer, *Towards the ultimate conservative difference scheme*. III, J. Comput. Phys. **23** (1977), 263–275.

13. O. A. Oleinik, *Construction of a generalized solution of the Cauchy problem for a quasi-linear equation of first order by the introduction of "vanishing viscosity"*, Amer. Math. Soc. Transl. (2) **33** (1963), 277–283.

14. R. D. Richtmyer and K. W. Morton, *Difference methods for initial value problems*, Interscience, New York, 1967.

15. P. L. Roe, *Approximate Riemann solvers, parameter vectors, and difference schemes*, J. Comput. Phys. **43** (1981), 357–372.

16. _____ , *Some contributions to the modeling of discontinuous flows*, these PROCEEDINGS.

17. P. K. Sweby, *High resolution schemes using flux limiters for hyperbolic conservation laws*, SIAM J. Numer. Anal. **21** (1984), 995–1011.

18. _____ , *High resolution TVD schemes using flux limiters*, these PROCEEDINGS.

19. E. Tadmor, *The large-time behavior of the scalar, genuinely nonlinear Lax-Friedrichs scheme*, NASA Contractor Report 172138, ICASE, 1983.

DEPARTMENT OF MATHEMATICS, UNIVERSITY OF CALIFORNIA, LOS ANGELES, CALIFORNIA 90024

COURANT INSTITUTE OF MATHEMATICAL SCIENCES, NEW YORK UNIVERSITY, NEW YORK, NEW YORK 10012

Lectures in Applied Mathematics
Volume **22**, 1985

Computational Methods for Discontinuities in Fluids

Oliver A. McBryan[1,2]

ABSTRACT. A wide range of physical phenomena involve shocks or discontinuous solutions. Examples include oil reservoir simulation, gas dynamics, combustion, aerodynamics, laser fusion physics and crystal growth. Physics-independent numerical methods for such problems are currently being developed, based on data-structures for representing multivalued data and topologically complex discontinuity surfaces. This is in contrast with standard numerical methods which are usually based on rectangular arrays. Applications to a variety of physical systems are under way.

We describe these methods and some of the applications, and discuss in more detail the efficient solution of elliptic equations with discontinuous coefficients. In particular we discuss solution methods for elliptic equations using high-order finite elements, with solution costs proportional to the number of unknowns. A high-order ADI pre-conditioner for conjugate gradient iteration is presented, as well as a multigrid scheme, both of which have been applied successfully to the solution of discontinuous coefficient problems.

1. Introduction. We will describe a range of computational methods for fluid flows in which shocks or discontinuities occur. Collectively, these methods constitute the method of front tracking. Most of this work has

---

1980 *Mathematics Subject Classification*. Primary 65M60, 35A40.

[1] Supported in part by Department of Energy grant DEACO278ER03077, and in part by NSF Grant DMS-83-12229.

[2] A. P. Sloan Foundation Fellow.

been developed at the Courant Institute by Jim Glimm, O. McBryan, and coworkers, including Bradley Plohr, Brent Lindquist, Sara Yaniv, Eli Isaacson, I.-L. Chern, and by David Sharp and Ralph Menikoff at Los Alamos.

Front tracking refers to certain computational methods in which special degrees of freedom are placed at fronts or discontinuities in order to obtain increased resolution in those areas. Unlike mesh-refinement schemes, where a computational grid is refined in the neighborhood of a front or singularity, in shock tracking an extra lower-dimensional grid is introduced, geometrically fitting the shocks. For flow problems in two space dimensions, this lower-dimensional grid is a system of curves, which we will collectively refer to as an *interface*. Typically, the interface is represented as a series of points distributed along the curves. In addition to spatial location, each point stores associated multivalued physical state variables, corresponding to the discontinuities present in the physical variables at the shock. At most such points, double-valued state variables suffice to describe the physics, but at intersection points of curves, higher degrees of multivaluedness are encountered.

The interface is propagated in time dynamically according to the appropriate physics. For example, for hyperbolic conservation laws the Rankine-Hugoniot jump conditions provide the correct dynamics for propagating interface points, and the interface contains precisely the data needed to compute the propagation speed—namely the multivalued state variables. In addition to moving the interface points, propagation of the interface involves updating the multivalued states stored along it. Because of the lower-dimensionality of the interface in the computational space, one can afford to compute these state updates very carefully without affecting the overall efficiency of the computational scheme. Again taking the hyperbolic conservation law as the example, it is worth the expense of solving a Riemann problem in the normal direction to obtain the correct evolution of the front states. In this way maximal analytic information about the behavior at shocks can be used in these critical regions.

In some problems, new shocks or discontinuities may appear dynamically, which require algorithms to detect their occurrence and introduce corresponding curves in the interface. Frequently the topology of the interface may change in time due to collisions between curves, requiring algorithms to detect intersections and perform surgery.

Away from shocks or discontinuities the physical states are reasonably smooth, and standard numerical methods may be employed. In particular, a rectangular grid is used to describe the state variables in these *interior* regions. Since we have not developed new methods for interior schemes we will say little about them in this paper. However we note that at all times boundary data for these interior schemes are obtained only from the branch of the multivalued shock states which corresponds to the same side of the shock as the interior region under consideration.

For three dimensions, a moving two-dimensional grid, i.e. a set of surfaces, would be used in addition to a standard three-dimensional grid. All of the work reported here relates to two-dimensional studies, but we believe these methods extend in a natural way to three dimensions.

Shock tracking is appropriate only for those problems where there are discontinuities and other singularities concentrated on surfaces (or curves in two dimensions). Examples include shock waves, contact discontinuities, material interfaces, phase boundaries, slip lines, chemical reaction fronts, crystallization and melting surfaces. In some cases, the surfaces or curves are unstable in time, developing fingers or droplets for example. Eventually the topology of the interface becomes so convoluted that shock tracking is no longer appropriate and a different representation of the physics is required.

Mathematical discontinuities are not usually found in physical fluid flows. However in many situations extremely sharp gradients are observed and the physical equations have obvious idealized mathematical approximations which have true discontinuous solutions. For example, the Euler equations are approximations to the Navier-Stokes equations and permit discontinuous solutions. The missing physics in this approximation is related to viscosity effects, and the viscosity parameter defines a length scale that determines the width of a boundary layer. Within a zone of this width surrounding the idealized discontinuity, the quantity in question changes very rapidly. For many flow problems this length is much smaller than realistic computational mesh lengths. If the physics at subgrid levels can be ignored then shock tracking based on the idealized equations is adequate for modeling the physics. If subgrid physics is important, then some other procedure such as local mesh refinement, or perhaps a hybrid computational scheme combining different representations in the boundary and interior regions, will be needed to resolve the boundary layer.

To summarize, the methods we describe here are appropriate to that range of problems containing near-discontinuities and shocks, where subgrid physics is not significant, and where the topology of the interface is such that it is still appropriate to consider the fluid as a finite number of connected components separated by shocks.

The first published shock fitting discussions are in Richtmeyer and Morton [1]. These ideas were implemented and further developed by Moretti [2]. These computations were of high quality but appear to be limited to somewhat simplified situations. Our goal has been to develop a flexible, modular, physics-independent computational package suitable for solving a wide range of partial differential equations in which shocks or discontinuities appear, and giving high resolution solutions at shocks even on coarse grids. By supplying various physics-dependent routines and parameters, the package may be tailored to specific physical problems. We are currently studying oil reservoir simulation, gas dynamics and incompressible multiphase inertial flows using the same package in each case.

2. Discontinuities in fluid physics. As indicated above, shocks occur in a wide variety of physical phenomena and reliable methods for computing them are of great importance. As an example, the processes occurring in secondary and tertiary oil recovery involve shocks which separate the oil from an injected fluid. The oil industry has invested in the development of numerical simulators to study these processes because experimental measurements are too expensive. Similar remarks apply to gas dynamics where the study of shocks is vital, for example, to aerodynamics and to combustion physics. The equations of transonic flow over an airfoil are of mixed type–hyperbolic in one area and elliptic in another. The motion of the ablation front in a laser fusion pellet may also be described by a shock—in fact, recent experiments use a space and time-focused sequence of shocks of varying velocities to induce ignition. For reviews of the equations encountered in secondary oil recovery, see the books of Peacemen [3] and Scheidegger [4], and the paper of Douglas, Dupont and Rachford [5], for an introduction to shocks in gas dynamics see the book of Courant and Friedrichs [6] and for a discussion of shock phenomena in laser fusion physics see the article of K. Brueckner [7].

A particularly interesting class of problems involving shocks is those where the shockfront is unstable. In these cases, small ripples in an otherwise smooth front rapidly develop into fingers. Typically such phenomena are observed when a "lighter" medium presses on a "heavier" one. Thus for two-phase flow in a porous medium, the shockfront is unstable if the material of lower viscosity is flowing into the one of higher viscosity. For example, this is the normal situation in oil reservoir problems where water or CO_2 gas is injected in order to displace residual oil. Similarly the classical Rayleigh-Taylor instability in an accelerated two-phase fluid occurs when the direction of acceleration of the medium is from the lighter to the heavier material. For a review of the Taylor problem see the papers of Birkhoff [8] and of Sharp [9]. This instability is responsible for the flow of water out of an inverted glass—elementary physics would suggest the water should not fall since the upward air pressure on its surface can support about 30 feet of water. A related shock instability, the Kelvin-Helmholtz instability, is responsible for the roll-up and mushrooming of fingers.

While shocks are most familiar from hyperbolic equations, many hyperbolic flow problems have an associated elliptic equation. Incompressible flows also give rise to elliptic equations. For example, in two-phase incompressible flows and in porous media flows, the fluid pressure is determined by an elliptic equation whose coefficients are functions of the density of the fluid or of the concentration of one of the fluids in the flow, respectively. Since the density or concentration is discontinuous at a shockfront, the pressure equation requires careful treatment—for example, if the fluids are air and water, the discontinuity in density is of order 600, leading to a corresponding discontinuity in the coefficient of the elliptic pressure equation. In particular, for an unstable shockfront the associated elliptic equation may have to be solved with very high resolution near the shock, since such discontinuities will be encountered across the edges of all fingers or droplets.

3. Numerical computations. The nature of discontinuous solutions of one-dimensional hyperbolic and elliptic equations is reasonably well understood analytically. For higher-dimensional studies it is almost impossible to make any progress analytically, and even existence proofs are very rare. Thus numerical computations are essential, in order both to understand the underlying physics and to gain mathematical insight into the behavior of the equations.

Discontinuous solutions of hyperbolic equations have generally proved difficult to compute. In particular, most standard methods such as finite differences or finite elements, effectively smear the shockfront leading to reduced resolution (numerical diffusion). In some cases this smearing may actually lead to a qualitatively incorrect solution of the problem, for example, if the correct solution is a highly fingered shock or if physical diffusion is small. For elliptic equations, discontinuous coefficients also make it difficult to compute solutions, unless very carefully tailored grids are utilized. In particular, standard differencing schemes are not appropriate as differencing across a discontinuity interface is not acceptable. Frequently enhancements, such as local mesh refinement, or the use of dynamically evolving grids, have been used in order to reduce the errors at shocks. While these techniques certainly can improve the quality of solutions, they are no substitute for a correct discretization of the differential equations.

4. Shock tracking methods. Recently, J. Glimm, E. Isaacson, D. Marchesin, B. Lindquist, O. McBryan, B. Plohr and S. Yaniv have developed a shock-tracking method for two-dimensional fluid flow problems. For detailed exposition I refer to our papers [**10–23**]. For an early discussion of shock fitting see Richtmyer and Morton [**1**], and for an application to two-phase flow problems see the paper of Gardner,

Peaceman and Pozzi [24]. In the shock-tracking method, degrees of freedom are assigned to the shockfront itself. The front is approximated by a piecewise-linear (or parabolic) curve and the points of intersection of the line segments are propagated along characteristics for short time intervals. To avoid pile-ups the number and location of points along the interface is modified dynamically. Updating of degrees of freedom in the regions away from the front is then performed using a standard method. Interactions across the shock are computed by solving one-dimensional Riemann problems in the normal direction. A Riemann problem is a one-dimensional hyperbolic equation with a step function as initial data. This allows the maximum amount of analytic information to be used in the solution in cases where such information is available. This scheme has proven to be very satisfactory even for unstable flow problems [17, 22]. An important aspect of the method is its ability to track even very narrow fingers with high resolution.

One difficulty of this method is that in order to move the shock points along characteristics, the fluid velocity is required at points on the front. In some problems this may require the accurate solution of an associated elliptic equation. Such an equation will frequently have discontinuous coefficients at the shockfront, precisely where the solution is required. We have developed an elliptic solver specifically designed for equations with coefficients discontinuous across a given interface, see McBryan [15, 16]. The method uses finite elements on a suitable triangulation of the domain. The core of the solver is an algorithm which, given a discontinuity interface in a rectangular domain, triangulates the domain in such a way that no triangle is intersected by the interface. Thus within each triangle, the elliptic coefficients are continuous. A further feature of the solver is that away from the interface the triangles are obtained by bisecting rectangles. In fact the grid of finite element nodes is a topological deformation of a regular rectangular mesh which is rectangular away from the interface but quadrilateral at the interface. The resulting algebraic equations may be effectively solved using the conjugate gradient or multigrid methods; see below.

The shock-tracking method for hyperbolic problems and the elliptic grid construction method for elliptic problems have been extended to handle an arbitrary number of discontinuity curves with complex topology; see Glimm and McBryan [25]. Thus it is now possible to track various shock fronts of interest in a physical problem. The topology of the interfaces may change as a result of shock interactions or due to spontaneous generation of shocks. Similarly an elliptic solver has been developed for problems with discontinuities across such general interfaces and provides a triangulation of the domain in which each triangle lies entirely within a single connected component; see McBryan [26]. The resulting codes are structured in a very physics-independent way, allowing for example for arbitrary numbers of physical degrees of freedom at each point. In a particular physical application, one then supplies a small number of physics-dependent routines which describe the number, types and initial values of physical variables, the exact form of equation coefficients and various boundary data.

Some problems, for example the Rayleigh-Taylor instability study, require very accurate evaluation of solutions because multiple derivatives of the solution are needed on the interface. To handle these cases, McBryan [26] has extended the discontinuous elliptic solver to handle higher-order elements, up to bicubics. The increased accuracy of the solutions allows coarser grids to be used—which has one disadvantage. The resolution of the interface as represented in the triangulation decreases. To overcome this problem, elements with curved edges (isoparametrics) are allowed. Solving the large systems of linear equations resulting from such higher-order finite element discretizations requires development of special purpose preconditioning operators. We have explored several preconditioning schemes [27] which typically lead to numbers of iterations independent of grid size. Recently a multigrid version of this algorithm has been developed, using a chain of successively finer triangulations, and also gives computation costs proportional to the number of unknowns [28]. We discuss these elliptic solution topics in more detail below.

5. Current applications of tracking. These methods are currently being tested on three problems with markedly different physics: oil reservoir recovery, gas dynamics and inertial flow problems.

J. Glimm, B. Lindquist, O. McBryan and S. Yaniv are currently working on a range of problems related to oil recovery. In one problem, we simulate a gas-oil-water region in which there are two discontinuity surfaces one betweeen oil and gas, the other between oil and water in the neighborhood of a drill shaft; see [18, 23]. The coning problem frequently referenced in the petroleum literature involves understanding the flow of oil and gas when the drill shaft is opened in the oil region. Opening the shaft reduces the oil pressure dramatically in the vicinity, forcing dissolved gas to come out of solution. The resulting gas cap may seriously impede oil production. By tracking the oil-gas front and following the gas bubble-point curve, the percolation of the gas can be studied. In the well region this may likely require either mesh refinement or implicit solution of the fluid equations. A second project studies the statistics of unstable fingering in porous media. We have also performed simulations with large numbers of randomly spaced wells, as a test of the robustness of the code [17]. This is an important phenomenon in many secondary oil recovery situations especially those with a high mobility ratio of water to oil. One goal is to find physical quantities which are more or less independent of fingering, and attempt to quantify the fingering in terms of growth rates, influence of geometrical factors and other criteria.

The gas dynamics tests involve flow in a tube with an obstacle. We have studied supersonic flow past a wedge and regular reflections. One or more shocks or contact discontinuities may be present in the initial data or time-dependent boundary data. The code solves for the flow in the tube, simultaneously tracking one or more waves, allowing for interactions between waves and the changing topology of the wavefront. We have also studied circularly symmetric shocks and contact discontinuities where comparison to a one-dimensional solution is feasible. Finally we have studied the onset of Kelvin-Helmholtz instability in flows with a tangential shear interface. Current work in this area involves development of methods to handle automatic changes in front topology, and applications to combustion physics. Our work on shock-tracking in gas dynamics is reported in the papers by Glimm, McBryan and Plohr [19] and Glimm, McBryan, Plohr and Yaniv. [29].

Particularly for hyperbolic problems, interaction of tracked waves poses interesting problems, both physical and mathematical. At the intersection point, the dynamics is no longer governed by a one-dimensional Riemann problem. In many cases this requires deeper understanding of the physics. There is no difficulty in incorporating bifurcations of the topology of the front into the calculation once the physics is understood—the current method supports general front topologies. Interactions between tracked and untracked waves are a simpler matter and are already handled in a general way.

The classical Taylor instability has been studied both numerically and theoretically for many years, and is still far from resolved. As examples, see the recent work of Orszag and coworkers [30] and of Menikoff and Zemack [31]. Over the last two years, J. Glimm and O. McBryan, in collaboration with R. Menikoff and D. Sharp at Los Alamos, have developed an application of the shock tracking method to the study of Rayleigh-Taylor instability. They study the dynamics of an air-water interface under gravity. Even in the case of incompressible fluids, the interface becomes a vorticity layer due to velocity shear. The velocities are computed using a stream-function formulation with the vorticity layer as a line source. The velocity field is then used to compute the pressure, from an elliptic equation which has both highly-discontinuous coefficients and a line source. Finally, Euler's equations provide time propagation of the vorticity. Because of the central role of the interface in this problem the elliptic solver described previously plays a very important role by providing a well-defined finite-element formulation even in the presence of the irregular interface. Since the derivatives of the velocity solution are used as source terms for the pressure solution, the stream-function must be solved for with particularly high accuracy. This work [9, 22, 32] has progressed to the point of initial validation against other methods (boundary integral techniques, conformal transformations) and scientific studies are now being performed. Of particular interest is the study of nonlinear modal interactions, the statistical analysis of fingering and droplet formation in the approach to turbulence, the inclusion of surface tension and vorticity diffusion effects and the role of both small and large-scale heterogeneities and random perturbations. Because of the generality of the front-tracking approach, there is no restriction to constant densities or irrotational flows inherent in the methods. The related development of Helmholtz instabilities is also being studied.

6. Efficient elliptic solution methods. In recent years there has been considerable progress in solving efficiently discrete systems of equations of the type encountered in discretizing partial differential equations. As an example, consider the familiar five-point matrix obtained by discretizing Poisson's equation on a regular rectangular grid in two dimensions. To solve the resulting linear equations using Gaussian elimination, taking into account the band structure of the equation matrix, $O(n^2)$ operations would be required where n is the number of unknowns (i.e. grid points). In addition, Gaussian elimination would require $O(n^{3/2})$ storage. Some improvement is possible if an optimal reordering scheme is used to minimize fill-in during elimination. These time and space requirements make elimination methods unacceptable on even the simplest equations for large grid sizes. Other direct solution methods such as ones based on the fast Fourier transform or cyclic reduction, allow such equations to be solved in $O(n \log n)$ operations, where n is the number of unknowns. Such methods are collectively described as *fast solvers*. However these methods are only available for very special problems (constant coefficients), with severe restrictions on the grids (rectangular), domains (rectangles) or boundary conditions.

For more general equations or domains, iterative solution methods are used due to the prohibitive computer time and storage requirements of Gaussian elmination solvers. In many cases, the number of iterations required is enormously reduced by using a fast solver to *precondition* the equations. Suppose that **B** is a matrix for which a fast solver has been developed. Then solving

$$\mathbf{Au} = \mathbf{F} \tag{1}$$

is equivalent to solving the equations

$$\mathbf{AB}^{-1}\mathbf{v} = \mathbf{F}, \qquad \mathbf{u} = \mathbf{B}^{-1}\mathbf{v}. \tag{2}$$

If the matrix $\mathbf{AB}^{-1}$ is close to the identity matrix $\mathbf{I}$ then the number of iterations required to solve the latter equation will be less than to solve the original one. The cost per iteration will have increased, of course, by the extra cost of an application of $\mathbf{B}^{-1}$, but since this is $O(n \log n)$ it is comparable with the cost of applying $\mathbf{A}$ itself. Thus provided the number of iterations decrease by a substantial factor, there will be a net improvement.

An effective iterative method for positive-definite systems is the *conjugate gradient method* of Hestenes and Stiefel [33], in conjunction with preconditioning as above. Typically the number of iterations required is independent of grid size if a suitable preconditioner is available, so that the computational cost is essentially $O(n)$. Another efficient solution scheme is the class of *multigrid* methods which distribute the iterative solution over a number of grids of different sizes. Here the strategy is to perform as much of the work as possible on coarse grids where the computational cost is minimal. These methods also lead to accurate solutions in approximately $O(n)$ operations. In the folloiwng sections we describe an effective finite-element formulation for discontinuities, and then describe efficient solution schemes using a high-order ADI fast solver as preconditioner and using a multigrid method.

7. Finite elements and discontinuities. As an illustration of the use of the methods discussed in the previous section, I will describe their use to solve a typical discontinuous coefficient elliptic equation:

$$-\nabla \cdot (k(x, y)\nabla u)(x, y) = f(x, y),$$

where $k(x, y)$ is a function which has a discontinuity across a curve in the domain of interest, a rectangle for example. The solutions of this equation are the minima of the energy functional:

$$E(u) = \int dx \, dy \frac{1}{2} k(x, y) \nabla u \cdot \nabla u - \int dx \, dy f(x, y) u.$$

The finite-element method involves the approximate minimization of the energy functional—the minimum is taken over a finite-dimensional space of piecewise polynomial functions. Typically the domain is triangulated in some way and the trial functions are taken to be the space of continuous functions which are polynomials of a specified order on each triangle. Minimization of the energy functional over this space leads to a finite-dimensional matrix equation for the polynomial coefficients.

In order to obtain a sensible discretization in the presence of discontinuities, we have developed a specialized triangulation grid generator; see McBryan [15, 16]. This generator considers the location of specified discontinuity curves, and fits a triangulation to them in such a way that triangles are not intersected by these curves. Thus the coefficient function $k(x, y)$ is smooth within each triangle, which leads to an accurate discretization scheme even in the presence of discontinuities. Furthermore, the triangle vertices form a logically rectangular grid, and in general almost all the vertices coincide with vertices of a rectangular grid, with a small number of vertices in shifted locations so as to place them on an adjacent curve. In fact, away from curves the triangulation is chosen to be that obtained by bisecting rectangles of a rectangular grid. It is advantageous to use a tensor product Lagrange element on such rectangles (e.g., a bicubic) rather than treating the pair of triangles separately (e.g., with cubics). The resulting matrix equation is then very similar to that obtained by finite-element discretization on a rectangular gird, except that there are perturbations at matrix locations corresponding to those grid points which have been shifted.

In addtion to handling discontinuous coefficients, the method described above allows line sources in the source function $f(x, y)$ above to be treated accurately. The curves which represent the support of such line sources are simply added to the list of discontinuity curves as input to the grid generator. In the resulting triangulation, the line sources lie only along the edges of triangules, ensuring an accurate discretization. Point sources are also easily incorporated.

8. High-order solutions by ADI pre-conditioning. I will now describe an effective approach to solving the resulting discrete equations efficiently, for the case of bilinear, biquadratic or bicubic finite elements. The matrices involved are banded, having at most 49 nonzero diagonals (in the case of bicubic elements). Assuming that $k(x, y)$ varies slowly away from the interface, the matrix should first be scaled by multiplying on each side by the square root of its diagonal. This results in a matrix which is close to that for constant coefficients, except at the interface. The resulting equations are next solved by conjugate gradient iteration using a special fast solver, described below, as preconditioner. Since the matrix after scaling is close to that for the Poisson equation with the same order discretization, a fast solver for higher-order discretization of the Poisson equation provides a very effective preconditioner. The fact that the grid is irregular at the curves affects the number of iterations in a mild way, while the fact that the scaling is not perfect at the discontinuity has a more substantial effect. In practice, the number of conjugate gradient iterations is independent of grid size, though it grows as the magnitude of the discontinuity is increased.

We have developed a fast solver for rectangular grid higher-order finite-element discretizations of the Poisson equation [27], based on ADI ideas. The method is based on the formal decomposition of the Laplacian as a sum of one-dimensional operators in the x and y directions:

$$\Delta u = \frac{d^2 u}{dx^2} + \frac{d^2 u}{dy^2}.$$

Suppose that we wish to solve the equation

$$Lu = f, \qquad L = L_1 + L_2,$$

with L_1 and L_2 commuting operators, where each of L_1, L_2 is itself a product of simpler operators:

$$L_1 = L_{1x} \cdot L_{1y}, \qquad L_2 = L_{2x} \cdot L_{2y},$$

and where the operators

$$L_{1x} + a \cdot L_{2x}, \qquad L_{1y} + a \cdot L_{2y},$$

are cheaply invertible for positive scalars a. As suggested by the notation, this can occur naturally if L_1 and L_2 are tensor products of derivatives along coordinate directions. The iteration step consists of repeating the basic step:

$$x_{n+1} = (L_{1x} + a \cdot L_{2x})^{-1}(L_{1y} + a \cdot L_{2y})^{-1}$$
$$\cdot \{(L_{1x} - a \cdot L_{2x})(L_{1y} - a \cdot L_{2y})x_n + 2a \cdot f\},$$

starting with an initial guess $u_0 = u0$. The parameter a is allowed to vary from step to step. This formula is an identity if u_n, u_{n+1} are replaced by the exact solution u of $Lu = f$, and yields a convergent iteration scheme if a, L_{1x}, L_{2x}, L_{2y} are all positive.

The application of this ADI approach to obtain a fast solver is based on the observation that the discrete Laplacian matrix obtained from a tensor product finite-element basis on a rectangular grid decomposes into a direct sum of tensor products of one-dimensional matrices. Specifically, let $K_{i,j}$ be the energy matrix for the Poisson equation with Lagrangian tensor product finite elements, where i denotes a standard rectangular indexing in two dimensions of the finite-element nodes. We will use the notation $c(i)$, $r(i)$ to denote the column and row containing the grid point i. Then

$$K_{i,j} = kx_{c(i),c(j)} my_{r(i),r(j)} + mx_{c(i),c(j)} ky_{r(i),r(j)},$$

where kx, ky and mx, my are the corresponding energy and mass matrices for one-dimensional finite elements of the same order and grid dimensions. Thus the 2d energy matrix has been decomposed into a sum of tensor products of 1d operators. This is exactly the requirement for the development of an ADI solver as above. A single ADI step utilizes a range of ADI parameters a that span the spectrum of the operator

$mx^{-1/2} \cdot kx \cdot mx^{-1/2}$. In the conjugate gradient context it actually suffices to use a single such compound step as the preconditioner, since all that is required is an approximate inverse of the Laplacian. This is much cheaper than allowing the ADI steps to converge completely at every conjugate gradient iteration.

An even simpler preconditioning is also quite effective for moderate size problems, though the number of iterations increases steadily with grid size. Consider a logically rectangular $(N + 1) \cdot (M + 1)$ grid on which say bicubic Lagrange elements have been used on rectangles away from the interface, and cubic elements are used on the triangles near the front obtained by bisecting quadrilaterals. The resulting grid of finite-element nodal points is still logically rectangular, indeed rectangular almost everywhere with only a small number of nodes in shifted locations. These nodal points constitute a $(3N + 1) \cdot (3M + 1)$ grid. The resulting discretization matrix has 49 nonzero diagonals. As a preconditioning matrix one could try the regular 5-point Laplace matrix for a $(3N + 1) \cdot (3M + 1)$ grid. While this is not so close to the operator we are dealing with as the ADI fast solver, it compensates by the fact that there are very fast 5-point solvers available. In fact on moderate sized grids, this solution strategy is often faster than using the ADI fast solver, although the iteration count may be substantially higher.

9. Multigrid solutions. Recently, the multigrid method developed by Brandt [**34**], Hackbusch [**35**], Stüben-Trottenberg [**36**], and others provides efficient solutions of a wide range of algebraic discretizations to a suitable accuracy in $O(n)$ time. This has already been demonstrated for equations such as the Laplace equation where multigrid solvers are competitive with standard fast solvers on special geometries, but it works equally well with irregular domains or nonconstant coefficients where the standard methods do not apply. Unlike other solution methods, multigrid methods operate with a chain of increasingly finer discretizations of the given equation, the finest grid being the one desired. The principle idea behind the multigrid method is to exploit the fact that certain simple iterative relaxation methods (e.g. Gauss-Seidel) reduce the high-frequency error of an approximate solution very efficiently, while having little effect on low-frequency errors. Note that the highest frequency part of a solution is very local, and appears only on the finest grid. Thus the high-frequency errors are reduced by a few iterations on a fine grid, after which the equation and approximate solution are projected onto a coarser grid. Some of the low-frequency components for the fine grid are actually high-frequency components for the coarse grid, and thus may be reduced effectively using a few coarse-grid iterations. The remaining low-frequency error is again transferred and solved for on still coarser grids. The advantage is that most of the solution work is performed on very coarse grids where the cost per iteration is minuscule relative to the cost on the finest grid. Effectively one is "integrating out" the low momentum part of the solution by using a type of block-spin description on coarse grids. The remaining high-frequency part of the solution is easily found using only a local iteration operator such as Gauss-Seidel.

We have developed such a multigrid solver for elliptic equations containing discontinuities of the type described above discretized with high-order finite elements, see McBryan [28]. The interface-fitted triangulations described earlier provide a natural interpolation operator between the coarse and fine grids, in particular, an operator that never interpolates across a discontinuity. At every grid level we obtain an optimal triangulation to resolve the discontinuity on that level. The use of isoparametric (curved) elements further enhances the quality of the low-grid approximations. As an alternate to using coarse grid information, one can also use lower-order solutions from the same grid. This solver is substantially faster than what we had available previously. It has been used extensively in the study of the Rayleigh-Taylor instability, where large grids and higher-order finite elements are needed to provide adequate resolution. Further applications of these ideas to singular differential equations look promising, particularly to cases where mesh-refined grids or variable finite-element orders lead to highly irregular finite-element nodes.

10. Interfaces and discontinuities. As an example of modular code design, as required in large-scale scientific computation, I will describe an interface-handling package that is one component of our shock-tracking codes. We will describe some of the capabilities of this library here to illustrate the interesting computing problems encountered in developing a tracking program. Applications of the package to fluid flow problems have been described elsewhere.

Many physical systems involve several physical materials separated by interfaces. We have developed a library of subroutines for specifying and manipulating such interfaces in the plane [25]. These routines are purely geometrical and topological, knowing nothing of the physics of a problem. However they have been an essential tool in the development of a successful shock-tracking method [12–15].

Consider a set of points (nodes) in the plane, joined in some way by noninteresecting directed curves. We will use the word *interface* to describe such a system. In another terminology, an interface is a planar graph each of whose edges is a curve. Note that given an interface in the plane, the plane is decomposed into disjoint components whose boundaries are composed of interface curves. We can associate an integer (color) with each such component, and then each curve has associated left and right component values. We do not require that every component have a distinct component value—indeed, all components may be assigned the same value if desired. However, the requirement that curves be nonintersecting (except at their endpoints), ensures that a curve separates at most two components. An important operation is one that determines the component number at an arbitrary point x, y in the plane.

As a simple example a circle is an interface containing one node (any point on the circle) and one curve (joining the node to itself counterclockwise say). A circle divides the plane into its interior and exterior regions, which we might label with integers 1 and 2 respectively. The curve then has 1 and 2 as its associated left- and right-hand components.

The package of subroutines provides a mechanism for defining, and manipulating planar interfaces of arbitrary complexity in an efficient way. Thus addition or deletion of a point is an effectively $O(1)$ operation. Similarly, the operation to determine the connected component containing a point x, y is effectively $O(1)$ on the average. Considerable internal preprocessing of data is involved to achieve these efficiencies. The routines provide for automatic storage allocation, garbage collection and error handling. In addition, the user may tailor the package to his own needs by adding entries to the basic data-structures describing interfaces and by supplying subroutines which are called automatically after each interface operation.

REFERENCES

1. R. Richtmeyer and K. Morton, *Difference methods for initial value problems*, Interscience, New York, 1952.

2. G. Moretti, *Thoughts and afterthoughts about shock computations*, Report No. PIBAL-72-37, Polytechnic Institute of Brooklyn, 1972.

3. D. W. Peaceman, *Fundamentals of numerical reservoir simulation*, Elsevier, Amsterdam, 1978.

4. A. Scheidegger, *The physics of flow through porous media*, The University of Toronto Press, Toronto, 1974.

5. J. Douglas, T. Dupont and H. Rachford, *The application of variational methods to waterflooding problems*, Proc. 20th Annual Technical Meeting of the Petroleum Society of CIM (Edmonton, May 1969).

6. R. Courant and K. Friedrichs, *Supersonic flow and shock waves*, Interscience, New York, 1948.

7. K. Brueckner, *Laser-plasma interaction*, 1980 Les Houches Lecture Notes, North-Holland, Amsterdam, 1982.

8. G. Birkhoff, *Taylor instability and laminar mixing*, LA-1862, Los Alamos Report, 1954.

9. D. H. Sharp, *An overview of Rayleigh-Taylor instability*, Los Alamos preprint LA-UR-83-2130, 1982.

10. J. Glimm, D. Marchesin and O. McBryan, *Subgrid resolution of fluid discontinuities. II*, J. Comput. Phys. **37** (1980), 336–354.

11. _____, *Statistical fluid dynamics: Unstable fingers*, Comm. Math. Phys. **74** (1980), 1–13.

12. J. Glimm, E. Isaacson, D. Marchesin and O. McBryan, *A shock tracking method for hyperbolic systems*, Proc. 1980 Army Numerical Analysis and Computers Conf. (September 1980).

13. _____ , *Front tracking for hyperbolic systems*, Adv. in Appl. Math. **2** (1981), 91–119.

14. J. Glimm and O. McBryan, *Front tracking for hyperbolic conservation laws*, Proc. 1981 Army Numerical Analysis and Computers Conf.

15. O. McBryan, *Elliptic and hyperbolic interface refinement in two phase flow*, Boundary and Interior Layers (J. J.H. Miller, ed.), Boole Press, Dublin, 1980.

16. _____ , *Shock tracking for 2d flows*, Computational and Asymptotic Methods for Boundary Layer Problems (J. J. Miller, ed.), Boole Press, Dublin, 1982.

17. J. Glimm, E. Isaacson, B. Lindquist, O. McBryan and S. Yaniv, *Statistical fluid dynamics: The influence of geometry on surface instability*, Frontiers in Applied Mathematics, Vol. 1, Soc. Indust. Appl. Math., Philadelphia, Pa., 1983.

18. J. Glimm, B Lindquist, O. McBryan and L. Padmanhaban, *A front tracking reservoir simulator: 5-spot validation studies and the water coning problem*, Frontiers in Applied Mathematics, Vol. 1, Soc. Indust. Appl. Math., Philadelphia, Pa., 1983.

19. J. Glimm, O. McBryan and B. Plohr, *Applications of front tracking to two-dimensional gas dynamics calculations*, Lecture Notes in Engrg., Springer-Verlag, New York, 1983.

20. J. Glimm, O. McBryan, B. Plohr and S. Yaniv, *Shock tracking for gas dynamics,*, New York Univ., preprint, 1984.

21. O. McBryan, *Fluids, discontinuities and renormalization group methods*, Mathematical Physics. VII (Brittin, Gustafson and Wyss, eds.), North-Holland, Amsterdam, 1984, pp. 481–494.

22. J. Glimm, O. McBryan, D. Sharp and R. Menikoff, *Front tracking applied to Rayleigh-Taylor instability*, Los Alamos preprint LA-UR-84-941, Los Alamos, 1984.

23. J. Glimm, B. Lindquist, O. McBryan, B. Plohr and S. Yaniv, *Front tracking for petroleum reservoir simulation*, Proc. SPE Meeting on Numerical Methods (San Francisco, November 1983).

24. A. O. Gardner, D. W. Peaceman and A. L. Pozzi, *Numerical calculation of multi-dimensional miscible displacements by the method of characteristics*, SPE Reprint Series, No. 8, Soc. Petroleum Engrs. of AIME, Dallas.

25. J. Glimm and O. McBryan, *Planar interfaces*, Courant Institute, preprint, 1984.

26. O. McBryan, *Shock tracking in fluid flows*, Proc. Ninth Nat. Congr. of Appl. Mech. (Cornell, June 1982).

27. _____ , *Preconditioning operators for higher order finite elements on rectangular grids*, Courant Institute, preprint, 1984.

28. _____ , *A multi-grid finite element solution of discontinuous elliptic equations*, Proc. Internat. Multigrid Conf. (Copper Mountain, Colorado, 1983) Elsevier (to appear).

29. J. Glimm, O. McBryan, B. Plohr and S. Yaniv, *The coupling of tracked and interior waves in a front tracking scheme*, Proc. First Army Conf. on Appl. Math. and Computing (1983).

30. R. Baker, R. McCrory, D. Meiron and S. Orszag, *Rayleigh-Taylor instability of ideal fluid layers*, Conf. on Computational Methods for Instabilities (Los Alamos, June 1982).

31. R. Menikoff and C. Zemack, *Rayleigh-Taylor instability and the use of conformal maps for ideal fluid flows*, Los Alamos preprint LA-UR-83-1964, 1982.

32. O. McBryan, *Computing discontinuous flows*, Presentation, Meeting on Fronts, Patterns and Interfaces (Los Alamos, 1983).

33. M. R. Hestenes and E. Stiefel, *Method of conjugate gradients for solving linear systems*, J. Res. Nat. Bur. Standards **49** (1952), 409–436.

34. A. Brandt, *Multi-level adaptive solutions to boundary-value problems*, Math. Comp. **31** (1977), 333–390.

35. W. Hackbusch, *Convergence of multi-grid iterations applied to difference equations*, Math. Comp. **34** (1980), 425–440.

36. K. Stüben and U. Trottenberg, *On the construction of fast solvers for elliptic equations*, Computational Fluid Dynamics, Rhode-Saint-Genese, 1982.

DEPARTMENT OF MATHEMATICS, COURANT INSTITUTE, NEW YORK UNIVERSITY, NEW YORK, NEW YORK 10012

Lectures in Applied Mathematics
Volume **22**, 1985

Problems and Numerical Methods of the Incorporation of Mountains in Atmospheric Models

Fedor Mesinger[1] and Zaviša I. Janjić

1. Introduction. A rather specific set of problems appears in numerical modeling of the stratified atmospheric flows as a consequence of the existence of mountains. These problems take a different form depending on the vertical coordinate chosen for the model. With the coordinate chosen most frequently, the major problem is that very large pressure gradient force errors are possible above steep mountain slopes. In one form or another, however, these numerical problems have a long history of being incompletely understood; moreover, some of the error analyses along traditional lines may even have increased the confusion. Few, if any, of the schemes presently used are free of the difficulties.

Within this review lecture we

• describe the problem of the calculation of the pressure gradient force in finite-difference atmospheric models, and give an outline of various approaches proposed to circumvent, hopefully, the major difficulty;

• give an analysis of the mechanism responsible for this error problem; and

• discuss a number of techniques, with most attention paid to a blocking technique, available for the reduction of the error.

1980 *Mathematics Subject Classification*. Primary 39-02, 65M05, 76N10, 86A10.
[1] Lecture presented by F. Mesinger.

The mentioned major problem is related to the hydrostatic approximation, made for reasons of computational economy in all large-scale
atmospheric prediction/simulation models. It consists of the replacement
of the vertical momentum equation by the hydrostatic equation, which
may be written as

$$\frac{\partial \phi}{\partial p} = -\frac{RT}{p}. \tag{1.1}$$

Here ϕ is geopotential, that is, the potential energy per unit mass in the
field of gravity; p is pressure, R is the gas constant and T the temperature. The problem itself is the calculation of the pressure gradient force,
appearing in the remaining horizontal momentum equation. With pressure, as in (1.1), used for the vertical coordinate, the horizontal component of the pressure gradient force takes a very simple form, $-\nabla_p \phi$, the
subscript of the del operator denoting the variable held constant under
the differentiation process. Since the pressure gradient force will thus
involve horizontal differencing of the geopotential, discretization of the
pressure gradient force will be related to the discretization of the hydrostatic equation.

Another term of the governing equations strongly coupled with the
pressure gradient force is the "omega-alpha" term of the thermodynamic
equation, $\kappa T \omega / p$. Here κ is R/c_p, where c_p is the specific heat at constant
pressure, and $\omega \equiv dp/dt$ is the "vertical velocity" in the pressure system,
d/dt being the material time derivative. It is perhaps generally accepted
as very important that this term be defined so as to guarantee that there
will be no false energy generation in transformations between the kinetic
and the total potential energy. Thus, along with (1.1), we shall in this
lecture mainly be concerned with the terms/equations

$$\frac{\partial \mathbf{v}}{\partial t} = \cdots - \nabla_p \phi + \cdots, \tag{1.2}$$

$$\frac{dT}{dt} = \frac{\kappa T \omega}{p} + \cdots. \tag{1.3}$$

Here $\mathbf{v}$ is the horizontal velocity component, $\mathbf{v} = (u, v, 0)$; and additional
terms in (1.3) describe radiative and other nonadiabatic heating effects.

To avoid difficulties with the lower boundary condition, in almost all
comprehensive atmospheric models terrain-following coordinates are
used. This, however, results in the appearance of two terms in the
expression for the pressure gradient force. For example, with the original
"sigma" coordinate of Phillips [1957]

$$\sigma \equiv p/p_S, \tag{1.4}$$

where the subscript S stands for surface values, one obtains

$$-\nabla_p \phi = -\nabla_\sigma \phi - RT \nabla \ln p_S. \tag{1.5}$$

Over sloping terrain the two terms on the right-hand side of (1.5) tend to be large in absolute values and to have opposite signs. If, say, they are individually ten times greater than their sum, a 1% error in temperature (2–3°C) will result in a 10% error in the pressure gradient force (Sundqvist [1975]).

It has not proven possible to construct a scheme which would be free of this error, and, for example, give no false pressure gradient force in the simple case of an atmosphere in hydrostatic equilibrium. However, a number of techniques have been designed aimed at keeping the error within hopefully tolerable limits. In addition, schemes have been constructed which maintain an integral property of the pressure gradient force (Arakawa [1972]). This feature was apparently also largely directed at controlling the effects of the error, representing a constraint on the error not to result in spurious sources or sinks of the vertically integrated vorticity (Arakawa and Suarez [1983]). We shall in the remainder of this section briefly review these various approaches.

A technique practiced as an early response to the recognition of the problem was the vertical interpolation of geopotential from sigma back to constant pressure surfaces (Smagorinsky et al. [1967], Kurihara [1968], Tomine and Abe [1982]). Problems remain near the ground, however, where extrapolation to obtain subterranean geopotentials is then needed. In addition, resulting schemes may be relatively time consuming. Finally, with this approach it is not obvious how to construct the scheme for the omega-alpha term in such a way as to prevent false energy generation in the energy conversion process.

The Arakawa scheme ([1972], Arakawa and Lamb [1977]) maintained the property of the pressure gradient force to generate no circulation of vertically integrated momentum along a contour of the surface topography. Namely, we have, with g denoting gravity, and subscript T values at the top pressure surface of the model,

$$-\frac{1}{g}\int_{p_T}^{p_S}\nabla_p\phi\,dp = -\frac{1}{g}\left[\nabla\int_{p_T}^{p_S}(\phi-\phi_S)\,dp + (p_S-p_T)\nabla\phi_S\right]. \quad (1.6)$$

Thus, the integral of the left-hand side of (1.6) along any closed curve following a contour of the surface topography will be zero, and no circulation of the vertically integrated momentum along such a curve will be generated by the pressure gradient force. The Arakawa scheme [1972] has been further developed to satisfy additional requirements (Phillips [1974], Tokioka [1978]), and is being extensively used in general circulation and weather prediction models (e.g., Stackpole et al. [1980]).

A scheme maintaining the one-dimensional analog of the integral of (1.6), necessary from the point of view of the angular momentum conservation, has been developed by Simmons and Burridge [**1981**]. The Simmons and Burridge scheme was, however, designed in terms of a very general "hybrid" vertical coordinate; and for a "local" hydrostatic equation, thereby avoiding the physically unrealistic dependence of geopotential of the lowest level on the temperature of all model levels that occurs in the Arakawa scheme [**1972**]. A most concise formulation of the vertical differencing requirement in sigma coordinates resulting in schemes that satisfy an analog of (1.6) has recently been achieved by Arakawa and Suarez [**1983**]. The Arakawa and Suarez formulation permits an arbitrary choice of the hydrostatic equation.

For a given atmosphere in hydrostatic equilibrium, Corby et al. [**1972**] have defined the difference analog of the pressure gradient (second) term of the pressure gradient force, so that it was exactly balancing the geopotential gradient (first) term. It was expected that with no error for this chosen, typical, atmosphere, large errors would be avoided in more general cases. Subsequently, this technique has been used and/or generalized by a number of authors: Nakamura [**1978**], Sadourny et al. [**1981**], Simmons and Burridge [**1981**], and Arakawa and Suarez [**1983**].

Janjić [**1977**] addresses the generation of the error in the general case. The basic idea is present also in an earlier paper by Rousseau and Pham [**1971**]. Note that, say, with the original definition of the sigma coordinate

$$-\nabla_p \phi = -\nabla_\sigma \phi - RT \nabla \ln p_S = -\nabla_\sigma \phi + \left(\nabla_\sigma \phi - \nabla_p \phi \right)$$

$$= -\nabla_\sigma \phi + \nabla \left(\phi_\sigma - \phi_p \right).$$

Thus, the pressure gradient term of the pressure gradient force can be considered to represent a hydrostatic correction to the geopotential gradient term. In a difference scheme, it corresponds to a vertical extrapolation/interpolation of geopotential from the sigma surface, or

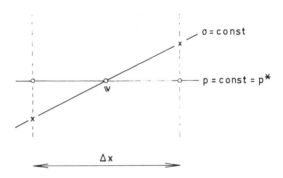

FIGURE 1. Illustration of the hydrostatic consistency problem of the sigma coordinate schemes.

surfaces, back to the constant pressure surface of the considered velocity point, say the surface p^* in Figure 1. It was pointed out by Rousseau and Pham that for minimization of the error, the formulation of that second term should be "coherent" with that of the first term and with the formulation of the hydrostatic equation. In the paper by Janjić, apparently the same requirement is put forth as that for a "hydrostatic consistency" of the scheme. The geopotential gradient term implies a certain difference scheme relating vertical increments of geopotential from pressure to the sigma surface, increments from ○ to x points in Figure 1, to some grid point values of temperature. If a different scheme and/or different grid point values are used to evaluate the *same* increments by the pressure gradient term, an erroneous remainder can be produced, large compared to the change of geopotential along the constant pressure surface.

Another technique introduced by Janjić [1977], and recently proposed also by Arakawa and Suarez [1983], is directed at the accuracy of vertical differencing in the hydrostatic equation. Note that the geopotential of a pressure surface is given by

$$\phi = \phi_S - \int_{p_S}^{p} RT d\ln p. \qquad (1.7)$$

Thus, the geopotential of a constant pressure surface depends only on temperatures *at and below* the considered surface. If this property of the continuous equations is to be maintained by the finite-difference scheme, it will not be possible to use for the numerical integration of the hydrostatic equation space-centered schemes of a high order of accuracy. Indeed, with almost no exception, simple divided differences are used. In this situation, it was pointed out by Janjić that the actual accuracy of the vertical differencing and of the resulting pressure gradient force can be increased if the variable used for the vertical differencing is judiciously chosen so as to minimize the error. In other words, (1.2) can be replaced by

$$\frac{\partial \phi}{\partial \zeta} = -\frac{RT}{p \, d\zeta/dp}, \qquad (1.8)$$

where

$$\zeta = \zeta(p), \qquad (1.9)$$

is a monotonic function of pressure, not necessarily equal to the vertical coordinate of the model. Instead, it is chosen so as to optimize the error properties of the scheme. This can be done with the idea of, once again, eliminating the error completely for a specific atmosphere. However, the technique can also be aimed at minimizing, in some average sense, the error of the difference approximation to $\partial\phi/\partial\zeta$ for a variety of the expected profiles.

An earlier suggestion of Phillips [1973] and Gary [1973] was to formulate the pressure gradient force in terms of deviations from a suitably chosen reference state. The effect of this technique is, in fact, equivalent to that of the Janjić and Arakawa-Suarez method. In recent experiments of Johnson and Uccellini [1983] these two methods have indeed given extremely similar results.

Most of these points we shall discuss in more detail in the continuation of this lecture.

2. A general form of the pressure gradient force. We shall need an expression for the pressure gradient force enabling use of the separate hydrostatic equation coordinate (1.9), in addition to a general vertical coordinate

$$\eta = \eta(p, p_S, z). \tag{2.1}$$

Eta is also assumed to be a monotonic function of pressure (and/or geometric height, z); it can be a terrain following (sigma), but also a more general type of coordinate. The choice of variables displayed here is motivated by an actual definition of eta used for a specific purpose later on; note that a still wider selection of variables would not, in fact, change the considerations to follow.

To derive the general formula for the pressure gradient force in such an eta system, consider the situation schematically represented in Figure 2. Namely, using the notation introduced in the figure, we may write

$$-\frac{\phi_2 - \phi_1}{\Delta s} = -\frac{\phi_3 - \phi_1}{\Delta s} - \frac{\phi_2 - \phi_3}{\Delta \zeta}\frac{\Delta \zeta}{\Delta s}.$$

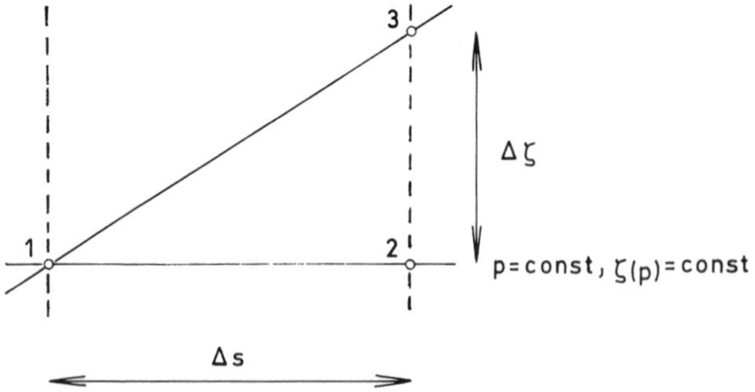

FIGURE 2. Stencil and notation used to derive the general formula for the pressure gradient force in the eta (e.g., sigma) system.

If Δs is oriented in the direction of the largest variation of geopotential along the pressure surface, in the limit as Δs tends to zero this expression tends to

$$-\nabla_p \phi = -\nabla_\eta \phi + \frac{\partial \phi}{\partial \zeta} \nabla_\eta \zeta. \qquad (2.2)$$

Here

$$\frac{\partial}{\partial \zeta} \equiv \frac{1}{\partial \zeta / \partial \eta} \frac{\partial}{\partial \eta}, \qquad (2.3)$$

that is, eta is the vertical coordinate actually used.

The formula (2.2) represents the general form of the pressure gradient force in the eta (e.g., sigma) system in the sense that any of the commonly used expressions can be derived from it by a particular choice of one or both of the functions ζ and η. For instance, if we choose

$$\zeta = \ln p, \qquad (2.4)$$

formula (2.2) takes the form

$$-\nabla_p \phi = -\nabla_\eta \phi - RT \nabla_\eta \ln p, \qquad (2.5)$$

(e.g., Simmons and Burridge [1981]). As an example where both of these functions are specified, consider the choice

$$\zeta = p; \quad \eta = \sigma = \frac{p - p_T}{\pi}, \quad \pi \equiv p_S - p_T. \qquad (2.6)$$

It results in

$$-\nabla_p \phi = -\nabla_\sigma \phi + \frac{\sigma}{\pi} \frac{\partial \phi}{\partial \sigma} \nabla_\sigma \pi, \qquad (2.7)$$

which is another frequently used expression (e.g., Arakawa and Lamb [1977]).

One form of the pressure gradient force not resulting from (2.2) in a direct way is the "flux form" of Johnson (Johnson [1980] and Uccellini [1983])

$$\frac{1}{\pi} \left[-\frac{\partial}{\partial \sigma} (p \nabla_\sigma \phi) + \nabla_\sigma \left(p \frac{\partial \phi}{\partial \sigma} \right) \right]. \qquad (2.8)$$

This form can, however, be obtained by a manipulation of the right-hand side of (2.7). A straightforward discretization of the first term in the bracket of (2.8) again leads to a geopotential gradient term, same as that obtained by a discretization of (2.7).

3. Discretization of the pressure gradient force and calculation of geopotential by the hydrostatic equation. Following Janjić [1979], it will be convenient to consider the construction of the pressure gradient force scheme as consisting of three steps. In the first step the geopotential is calculated at constant eta surfaces integrating the hydrostatic equation. As the second step, starting from known values at the neighboring eta surfaces, the geopotential is extrapolated or interpolated to a constant pressure surface, again via the hydrostatic equation. Finally, in the third step, the values of geopotential obtained in this way are used to calculate the finite-difference pressure gradient force approximation.

In order to apply this procedure, a specific distribution of variables over grid points in the vertical is needed. Almost all models carry temperatures (or potential temperatures) at the levels of horizontal velocity components, and the vertical velocities in between (the "Lorenz distribution" according to Arakawa). Among these models, differences exist in schemes used to calculate geopotential. Possibly all schemes can in this respect be classified as being either schemes which calculate geopotentials at the same levels as horizontal velocities, or schemes which calculate geopotentials at levels in between horizontal velocities. We shall call these schemes "nonstaggered" (or "level"), and "staggered" (or "layer") schemes, respectively. These two possibilities are schematically represented in Figure 3. The symbol k appearing in the figure represents the vertical index, and $\delta_\zeta \phi$ is the finite-difference approximation to $\partial \phi / \partial \zeta$. It is defined by the particular form of the hydrostatic equation used.

One should, however, note that a finite-difference hydrostatic equation for calculation of, for example, "full level" geopotentials, when it is supplied with a definition of "half level" (interface) geopotentials, can be rewritten so as to obtain a hydrostatic equation in terms of half level geopotentials, and a definition of full level geopotentials. The same kind of reformulation can be done starting with this latter set of equations, a

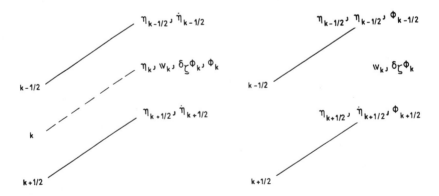

FIGURE 3. "Nonstaggered", or "level" (left panel) and "staggered", or "layer" (right panel) schemes for calculation of geopotential by the hydrostatic equation.

hydrostatic equation for half level geopotentials and a definitions of layer geopotentials. Thus, for a given scheme, it may not be obvious what should be considered as levels at which geopotentials are calculated by the finite-difference hydrostatic equation.

Given such a choice, we shall assume that the discrete values of $\phi(p)$ are actually defined at the points between which, with the given finite-difference hydrostatic equation, we obtain the thicknesses of the layers which, in case of arbitrary spacing of vertical levels, do not depend on the temperatures outside of the layers, or depend on them to the minimum extent possible. Thus, as examples of nonstaggered schemes the Arakawa [1972] and the Corby et al. [1972] schemes can be given; staggered schemes are those of Janjić [1977], Burridge and Haseler [1977], and the Arakawa-Suarez [1983] "local hydrostatic equation" schemes.

Let us first consider the perhaps more frequently used, nonstaggered scheme. For simplicity, we shall restrict ourselves to, say, the x component of the pressure gradient force. Let the x component of the pressure gradient force be calculated at the pressure level p^*, and let the values of ζ and geopotential at this level be denoted by ζ^* and ϕ^* respectively. Furthermore, we shall assume that the first step has been completed and concentrate on the second and third step only. Using the notation introduced in Figure 4 we may write

$$
\begin{aligned}
\phi_1^* &= \phi_1^k + \delta_\zeta \phi_1^k \big(\zeta^* - \zeta_1^k \big), \\
\phi_2^* &= \phi_2^k + \delta_\zeta \phi_2^k \big(\zeta^* - \zeta_2^k \big),
\end{aligned}
\tag{3.1}
$$

where the subscripts denote the grid points in the horizontal, and the superscript k indicates the eta level at which the variables are defined.

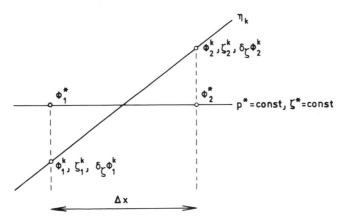

FIGURE 4. Stencil and notation used to calculate pressure gradient force approximation in case of the nonstaggered scheme for calculation of geopotential by the hydrostatic equation.

We have now completed the second step of our procedure for calculating the pressure gradient force. The third step yields

$$-\frac{\phi_2^* - \phi_1^*}{\Delta x} = -\frac{\phi_2^k - \phi_1^k}{\Delta x} + \frac{1}{2}\left(\delta_\zeta \phi_1^k + \delta_\zeta \phi_2^k\right)\frac{\zeta_2^k - \zeta_1^k}{\Delta x}$$

$$-\left[\zeta^* - \frac{1}{2}\left(\zeta_1^k + \zeta_2^k\right)\right]\frac{\delta_\zeta \phi_2^k - \delta_\zeta \phi_1^k}{\Delta x},$$

or in a more compact notation

$$-\delta_x \phi^* = -\delta_x \phi^k + \overline{\delta_\zeta \phi}^{\,x}\delta_x \zeta^k - \left(\zeta^* - \overline{\zeta^k}^{\,x}\right)\delta_x\left(\delta_\zeta \phi^k\right). \qquad (3.2)$$

If we define

$$\zeta^* = \overline{\zeta^k}^{\,x} \qquad\qquad (3.3)$$

we obtain the usual form of the pressure gradient force approximation with the nonstaggered scheme

$$-\delta_x \phi^* = -\delta_x \phi^k + \overline{\delta_\zeta \phi}^{\,x}\delta_x \zeta^k. \qquad (3.4)$$

As we can see from (3.3), unless ζ is a linear function of η, the approximation (3.4) is not defined at the level η_k as it is usually believed.

Let us now turn our attention to the staggered scheme. In the second step of our procedure for calculation of the pressure gradient force, we shall here use linear interpolation to obtain the geopotential at the pressure level p^* corresponding to $\zeta^* = \zeta(p^*)$. Thus, with the notation

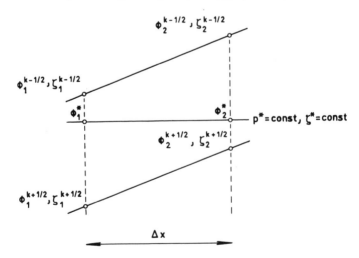

FIGURE 5. Stencil and notation used to calculate pressure gradient force approximation in case of the staggered scheme for calculation of geopotential by the hydrostatic equation.

introduced in Figure 5, we may write

$$\phi_1^* = \phi_1^{k-1/2} + \frac{\phi_1^{k+1/2} - \phi_1^{k-1/2}}{\zeta_1^{k+1/2} - \zeta_1^{k-1/2}}\left(\zeta^* - \zeta_1^{k-1/2}\right),$$

$$\phi_2^* = \phi_2^{k-1/2} + \frac{\phi_2^{k+1/2} - \phi_2^{k-1/2}}{\zeta_2^{k+1/2} - \zeta_2^{k-1/2}}\left(\zeta^* - \zeta_2^{k-1/2}\right).$$

(3.5)

Defining

$$\overline{A}^{\eta} \equiv \frac{1}{2}\left(A^{k-1/2} + A^{k+1/2}\right), \qquad \delta_\zeta A \equiv \frac{A^{k+1/2} - A^{k-1/2}}{\zeta^{k+1/2} - \zeta^{k-1/2}},$$

these relations, after rearrangement, can be written as

$$\phi_1^* = \overline{\phi_1}^{\eta} + \delta_\zeta\phi_1\left(\zeta^* - \overline{\zeta_1}^{\eta}\right),$$

$$\phi_2^* = \overline{\phi_2}^{\eta} + \delta_\zeta\phi_2\left(\zeta^* - \overline{\zeta_2}^{\eta}\right).$$

Proceeding with the third step, we find that

$$-\delta_x\phi^* = -\delta_x\overline{\phi}^{\eta} + \overline{\delta_\zeta\phi}^{x}\delta_x\overline{\zeta}^{\eta} - \left(\zeta^* - \overline{\zeta}^{-x}_{\eta}\right)\delta_x\left(\delta_\zeta\phi\right).$$

(3.6)

If we define the level at which the pressure gradient force is calculated by

$$\zeta^* = \overline{\zeta}^{-x}_{\eta},$$

(3.7)

we finally obtain

$$-\delta_x\phi^* = -\delta_x\overline{\phi}^{\eta} + \overline{\delta_\zeta\phi}^{x}\delta_x\overline{\zeta}^{\eta}.$$

(3.8)

Again we see that a relation (3.7), analogous to (3.3), was necessary in order to arrive at the result usually assumed.

In constructing the pressure gradient force schemes various authors, of course, typically did not follow the three-step procedure outlined here. Usually, a difference analog to the hydrostatic equation would be chosen, and, subsequently, a difference approximation made to one of the differential expressions for the pressure gradient force. It is possible, however, to demonstrate that, in spite of the difference in approach, each of the commonly used schemes does in fact implicitly consist of the three-step procedure described here. We shall have a look at some specific schemes, from that point of view, in later sections.

and, integrating, obtain

$$\phi(p) = \phi_0 + \tfrac{1}{2}RA\left(\ln^2 p_0 - \ln^2 p\right) + RB(\ln p_0 - \ln p). \quad (4.7)$$

Thus, ϕ_S and p_S are seen to be related by

$$\phi_S = \phi_0 + \tfrac{1}{2}RA\left(\ln^2 p_0 - \ln^2 p_S\right) + RB(\ln p_0 - \ln p_S). \quad (4.8)$$

Corby et al. now evaluate (4.4) by inserting geopotentials given by (4.3) and (4.8), and sigma level temperatures obtained from (4.5), that is

$$T_n = A\ln(\sigma_n p_S) + B. \quad (4.9)$$

After some algebra, they arrive at

$$\delta_x \overline{\phi_k}^{\,x} = -R\,\overline{\overline{T_k}^{\,x}\delta_x\ln p_S}^{\,x}. \quad (4.10)$$

Accordingly, to eliminate completely the error in the case of the atmosphere which is being considered, Corby et al. choose

$$-R\,\overline{\overline{T_k}^{\,x}\delta_x\ln p_S}^{\,x} \quad (4.11)$$

as the difference analog of the x component of the pressure gradient term.

For further simplification, let us consider the version of the Corby et al. scheme obtained using a horizontally staggered grid, with velocity,

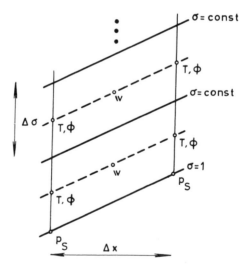

FIGURE 8. A horizontally staggered arrangement of the variables needed by the Corby et al. scheme. Note the difference in Δx as compared to the nonstaggered arrangement shown in Figure 7.

(Reprinted by permission of Gordon and Breach Science Publishers from F. Mesinger, *On the convergence and error problems of the calculation of the pressure gradient force in sigma coordinate models*, Geophys. Astrophys. Fluid Dynamics **19** (1982), 105–117.)

and temperature and geopotential defined at alternate grid points (Figure 8). Then, following the same procedure, one arrives at the analog

$$-\delta_x \phi_k - R \overline{T_k}^x \delta_x \ln p_S. \tag{4.12}$$

For a given value of Δx, (4.12) has a smaller truncation error than the original scheme.

To obtain now the scheme (4.12) using our three-step procedure, we shall, as the second step, need the extrapolation scheme

$$\phi_1^* = \phi_{1,k} - RT_{1,k}(\ln p^* - \ln p_{1,k}),$$
$$\phi_2^* = \phi_{2,k} - RT_{2,k}(\ln p^* - \ln p_{2,k}). \tag{4.13}$$

As could be expected in view of the second term of (4.12), only two grid point values of temperature appear in (4.13). As the third step, the expression

$$-\frac{\phi_2^* - \phi_1^*}{\Delta x} \tag{4.14}$$

is evaluated. Inserting (4.13), assuming that

$$\ln p^* \equiv \tfrac{1}{2}(\ln p_{1,k} + \ln p_{2,k}), \tag{4.15}$$

and, finally, taking into account that

$$\delta_x \ln p = \delta_x \ln p_S,$$

one indeed arrives at (4.12).

The values ϕ^* used in (4.13) are, however, not on the geopotential profile (4.3) defined by the finite-difference hydrostatic equation (4.2), and, thus, the scheme is hydrostatically inconsistent. The finite-difference hydrostatic equation implied by (4.13), involving single sigma level temperatures only, is different from the hydrostatic equation (4.2), involving two-point vertical averages of temperature. In addition, in case of steep sigma surfaces and/or thin sigma layers, the pressure surface $p^* = $ const can intersect additional sigma levels in between the grid points 1 and 2 used to form the analog (4.14).

Thus, rather than the values (4.13), a method consistent with the geopotential profile defined by (4.3) would have to use the values of geopotential located on that profile. Denoting these values by ϕ^c, and assuming that the pressure p^* is at the grid points 1 and 2 located in between the levels $k - r$ and $k - r - 1$, and $k + r$ and $k + r + 1$, respectively, we have

$$\phi_1^c = \phi_{1,k-r} - \tfrac{1}{2}R(T_{1,k-r} + T_{1,k-r-1})(\ln p^* - \ln p_{1,k-r}),$$
$$\phi_2^c = \phi_{2,k+r} - \tfrac{1}{2}R(T_{2,k+r} + T_{2,k+r+1})(\ln p^* - \ln p_{2,k+r}). \tag{4.16}$$

4. Hydrostatic consistency. As we have seen, the first and the second step of our procedure for calculating pressure gradient force involve the integration of the hydrostatic equation in order to obtain the values of geopotential at constant η and constant pressure surfaces, respectively. In the first step, integration of the finite-difference hydrostatic equation will at each point of the horizontal grid define a certain vertical profile of geopotential, as a function of the integration variable ζ. Typically, as stated in the introductory section, this integration is done by approximating $\partial\phi/\partial\zeta$ in between discrete η_k levels by a constant, so that the resulting vertical profile of geopotential will be a piecewise linear function of ζ, as schematically represented in Figure 6.

As proposed by Janjić, we can now define a scheme to be hydrostatically consistent if the procedure used in the second step yields a value of geopotential which lies on the geopotential profile defined by the finite-difference hydrostatic equation of the first step. For example, this could be the value represented by the plus sign in Figure 6. For hydrostatic consistency, in the second step, explictly or implictly, the same finite-difference hydrostatic equation has to be used, and the same grid point values, as in the first step.

Let us now, from the point of view of preceding considerations, have a look at some specific schemes for the hydrostatic equation and the pressure gradient force. We shall first consider possibly the simplest scheme with the nonstaggered calculation of geopotential by the hydrostatic equation, that of Corby et al. [**1972**] (also Gilchrist [**1975**] and Corby et al. [**1977**]). They have used the original definition of the sigma coordinate, (1.4), and a nonstaggered arrangement of horizontal velocities, temperatures and geopotentials both in horizontal and in the vertical direction (Figure 7). The geopotential of the lowest level ($k = K$) they have defined as

$$\phi_K = \phi_S + RT_K \ln\left(1/\sigma_K\right), \tag{4.1}$$

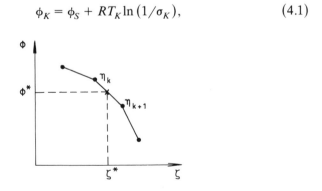

FIGURE 6. Schematic representation of vertical profile of geopotential defined by the finite-difference hydrostatic equation.

and the difference analog of the hydrostatic equation above that level as

$$\Delta_\sigma \phi = -R \overline{T}^\sigma \Delta_\sigma \ln \sigma. \tag{4.2}$$

Thus,

$$\phi_k = \phi_S + \sum_{n=k}^{K-1} \frac{1}{2} R(T_n + T_{n+1}) \ln \frac{\sigma_{n+1}}{\sigma_n} + RT_K \ln \frac{1}{\sigma_K} \tag{4.3}$$

defines the vertical profile of geopotential prescribed by the finite-difference hydrostatic equation, that is, by the first step of our procedure.

To choose the analog of the pressure gradient force, Corby et al. adopt

$$\delta_x \overline{\phi_k}^x \tag{4.4}$$

as the difference approximation to $(\partial \phi / \partial x)_\sigma$, and seek to find a consistent approximation to the pressure gradient term so that combined the two terms give a zero pressure gradient in case of the resting atmosphere defined by

$$T(p) = A \ln p + B. \tag{4.5}$$

The profile (4.5) they have taken as a reasonable average profile for the troposphere. To this end, they substitute (4.5) into the hydrostatic equation

$$d\phi = -RT \, d\ln p, \tag{4.6}$$

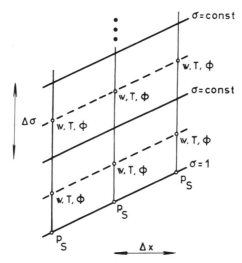

FIGURE 7. The vertical grid of the Corby et al. scheme [1972], with a nonstaggered arrangement of the time-dependent variables p_S, v and T, and locations where ϕ is defined by the finite-difference scheme.
(Reprinted by permission of Gordon and Breach Science Publishers from F. Mesinger, *On the convergence and error problems of the calculation of the pressure gradient force in sigma coordinate models*, Geophys. Astrophys. Fluid Dynamics **19** (1982), 105–117.)

with the values $\phi_{1,k-r}$ and $\phi_{2,k+r}$ here given by (4.3). In this way, with the extrapolation prescribed by (4.16) performed along the piecewise linear profile of geopotential defined by hydrostatic equation scheme, evaluation of

$$-\frac{\phi_2^c - \phi_1^c}{\Delta x} \tag{4.17}$$

gives the pressure gradient force with the vertical discretization as accurate as it can be achieved within the accuracy limits of the hydrostatic equation scheme.

In the simple case $r = 0$, when the slope of the sigma surface is not too great and/or the considered sigma layers are not too thin, so that the surface p^* intersects no additional sigma levels in between the grid poins 1 and 2, (4.17) gives

$$-\delta_x \phi_k - \tfrac{1}{4} R \left(T_{j-1,k-1} + T_{j-1,k} + T_{j+1,k} + T_{j+1,k+1} \right) \delta_x \ln p_S. \tag{4.18}$$

Here, of course, again (4.15) had to be assumed, and j represents the grid point index along the x direction.

The difference between the Corby et al. scheme (4.12) and the hydrostatically consistent analog (4.18)

$$\tfrac{1}{4} R \left(T_{j+1,k+1} - T_{j+1,k} - T_{j-1,k} + T_{j-1,k-1} \right) \delta_x \ln p_S, \tag{4.19}$$

can be considered to represent the error of (4.12) due to its hydrostatic inconsistency. It can be written as

$$\frac{1}{4} R \left[\left(\frac{\Delta_\sigma T}{\overline{\Delta \sigma}^\sigma} \overline{\Delta \sigma}^\sigma \right)_{j+1,k+1/2} - \left(\frac{\Delta_\sigma T}{\overline{\Delta \sigma}^\sigma} \overline{\Delta \sigma}^\sigma \right)_{j-1,k-1/2} \right] \delta_x \ln p_S, \tag{4.20}$$

and is thus, formally, a small quantity. Nevertheless, the presence of (4.19) seems to be a rather undesirable feature of the scheme. Grid point values of temperature are known to exhibit large grid point to grid point variations due to the effects of the so-called physical processes in atmospheric models, and the factor $\delta_x \ln p_S$ increases in magnitude with the slope of model mountains. Scheme (4.18), free of this error, appears from that point of view certainly much more appealing. It has, however, the problem of being asymmetric, and thus possibly time consuming. In addition, it is not applicable at the uppermost as well as at the lowermost level of the model.

The staggered scheme for calculation of geopotential permits a straightforward construction of hydrostatically consistent pressure gradient force schemes without these disadvantages. As an example, consider the scheme of Burridge and Haseler [**1977**]. They use the same arrangement as that in Figure 8, except that geopotentials are carried at the interfaces of sigma layers. As the hydrostatic equation they have simply

$$\Delta \phi_k = -RT_k \Delta \ln \sigma_k \tag{4.21}$$

for all the sigma layers. Here

$$\Delta A_k \equiv A_{k+1/2} - A_{k-1/2}$$

denotes the vertical difference across the layer. The pressure gradient force retains the form (4.12), with the layer values of geopotential, needed by (4.12), prescribed as

$$\phi_k \equiv \tfrac{1}{2}\left(\phi_{k-1/2} + \phi_{k+1/2}\right). \tag{4.22}$$

To construct this scheme using the described three-step procedure, again equations (4.13) are needed as the second step. However, now the finite-difference hydrostatic equation implied by these equations is the same as the hydrostatic equation used to calculate geopotentials (4.21). Thus, in this sense, the scheme is hydrostatically consistent.

The staggered schemes of Janjić [**1977**], and Arakawa and Suarez [**1983**], in the same way as the Burridge and Haseler scheme achieve hydrostatic consistency through hydrostatic equations of the type in (4.21), in which a single temperature (or potential temperature) governs the change of geopotential across a layer. As already mentioned, they have additional features which shall be discussed later on to some extent.

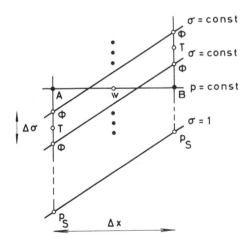

FIGURE 9. An example of hydrostatically inconsistent evaluation of the pressure gradient force by schemes such as the Burridge-Haseler scheme [**1977**], (4.21) and (4.12), (Janjić [**1977**]).

(Reprinted by permission of Gordon and Breach Science Publishers from F. Mesinger, *On the convergence and error problems of the calculation of the pressure gradient force in sigma coordinate models*, Geophys. Astrophys. Fluid Dynamics **19** (1982), 105–117.)

However, the second step equations of the type (4.13) can of course maintain hydrostatic consistency only as long as extrapolation is performed along linear segments of the profile of geopotential. Thus, as pointed out by Janjić [1977], hydrostatically consistent schemes of this type lose consistency in a situation as shown in Figure 9, when points at which the pressure surface p^* intersects the neighboring grid verticals are outside of the considered layer. These points are denoted by A and B in the figure.

As seen from the figure, for the hydrostatic consistency of this group of schemes one should require that

$$|\delta_x \phi|_\sigma \Delta x \leqslant |\delta_\sigma \phi| \Delta \sigma. \tag{4.23}$$

Thus, increasing the steepness of model mountains and increasing the vertical resolution may lead to a violation of the hydrostatic consistency of the scheme.

5. Convergence. The property of hydrostatic inconsistency appears to be under certain conditions related to the lack of convergence of the scheme (Mesinger, [1982]). An obvious example is the Arakawa scheme [1972], in which the geopotential of the lowest level, and consequently, also those of all levels above the lowest level, depend on the temperature of all model levels. Since, as pointed out in the introductory section, the

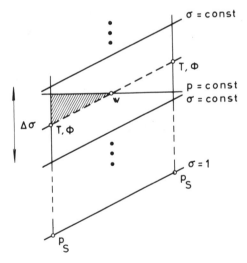

FIGURE 10. Illustration of the convergence problem of the Corby et al. scheme. The shaded area represents the section of the domain of dependence of the true value of the pressure gradient force not included in the domain of dependence of its finite-difference value.

(Reprinted by permission of Gordon and Breach Science Publishers from F. Mesinger, *On the convergence and error problems of the calculation of the pressure gradient force in sigma coordinate models*, Geophys. Astrophys. Fluid Dynamics **19** (1982), 105–117.)

true value of the pressure gradient force does not depend on temperatures above the considered pressure surface, in such a situation the difference between the true value of the pressure gradient force and its difference approximation can be arbitrarily great. This cannot be removed by the reduction of the grid intervals, and hence, the scheme is not convergent (Arakawa and Suarez [1983]).

A different kind of problem exists with the Corby et al. scheme. It is illustrated with the help of Figure 10. Note that the "domain of dependence" of the true value of the pressure gradient force at the v point in that figure is the region below the line $p = $ const shown in the figure. On the other hand, the finite-difference value calculated at the same point using the Corby et al. scheme depends on values at grid points on and below the dashed line connecting the two T, ϕ grid points. In this way, there exists a section of the domain of dependence of the true value of the pressure gradient force which is not included in the domain of dependence of its finite-difference value. If the temperature gradient is permitted to be a discontinuous function of space coordinates, the error of the scheme can again be arbitrarily great.

With schemes using the staggered calculation of geopotential, a similar situation occurs when the consistency condition (4.23) is violated. Then again, there exists a section of the domain of dependence of the true value of the pressure gradient force not included in the domain of dependence of its finite-difference value. In Figure 9, this is the section bounded by the vertical line at point A, the isobar and the uppermost constant sigma line shown in the figure.

6. The error problem: a numerical example. A numerical example calculated by Mesinger [1982] will be included here as an illustration of possible quantitative implications of some of the preceding considerations. Following a number of previous authors, an atmosphere in hydrostatic equilibrium is considered, so that the error is entirely due to the pressure gradient force approximation. Errors are calculated for the Corby et al. and for the Burridge-Haseler schemes, taken as possibly the simplest examples of hydrostatically inconsistent and consistent schemes, respectively.

Before describing the example, one more remark regarding the Corby et al. scheme should be made. Recall that the scheme was constructed to achieve exact cancellation of the two terms when temperature is a linear function of $\ln p$, (4.5). The procedure used included insertion of sigma level temperatures, (4.9). An objection can, however, be raised regarding this step. Namely, schemes used to initialize finite-difference calculations are typically not based on an analysis of temperature; rather, they are based on the analysis of geopotential. Note that values of geopotential represent information on the vertically integrated temperature. Thus, observed values of geopotential, defining the true value of the pressure gradient force, except for observational errors, cannot be recovered from a limited number of the local observations of temperature. Therefore, by analyzing temperature to initialize a primitive equation model, needlessly pressure gradient force errors are introduced in addition to those due to the observational errors and to the finite-difference scheme.

Accordingly, instead of using (4.9), it would have been more appropriate to use (4.7) to define the sigma level geopotentials, and then calculate temperature from the difference equations (4.1) and (4.2). Temperatures obtained in this way would differ from those defined by (4.9), and, therefore, the two terms of (4.12) would not cancel. Similar consideration is applicable to error estimates of a number of other authors (e.g., Nakamura [1978], Simmons and Burridge [1981]).

To have the numerical example include an estimate of the resulting error, Mesinger has calculated errors for the reference atmosphere of Corby et al., in which temperature is a linear function of ln p. It seemed of interest to have a look also at errors in a case of more irregular variations of temperature; for example, error calculations of Janjić [1977] showed maximum errors at tropopause levels. Therefore, a temperature variation including an inversion was also considered. Recalling a situation with cold air impinging at one side of a mountain barrier, an "inversion case" was defined, with an inversion below a presumed mountain height. In this inversion case, except for the inversion level, temperature was still assumed to be linear in ln p.

Following the error calculations of Phillips [1974] and Janjić, two neighboring surface pressure points were considered located along the direction of the x-axis at pressures of 1000 mb and 800 mb, respectively. The temperature at the 800 mb level was taken to be 0°C, and those at 1000 mb to be 10°C ("no inversion case") and –10°C ("inversion case"). Temperature above 800 mb in the "inversion case" was taken to be the

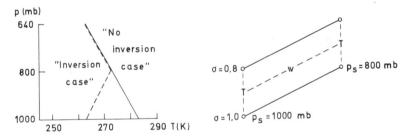

FIGURE 11. The temperature profiles used to calculate the errors of the Corby et al. and the Burridge-Haseler pressure gradient force schemes (left panel), and the location of the grid point at which the errors were calculated (right panel) (Mesinger [1982]).

same as in the "no inversion case". These two temperature profiles are shown in the left panel of Figure 11. Errors were calculated for the velocity point located in between the two surface pressure points, at (or, approximately at) the level $\sigma = 0.9$, as sketched in the right panel of the figure. Calculations were performed for $\Delta\sigma = 1/5$ (below the $\sigma = 0.8$ interface), $\Delta\sigma = 1/(3 \times 5)$, $\Delta\sigma = 1/(5 \times 5)$, etc., and in the limit as $\Delta\sigma \to 0$. These values are chosen to keep the velocity point at the same altitude, at $\sigma \cong 0.9$.

The errors obtained for the two schemes and the two temperature profiles are displayed in Table 1. The values shown are in units of geopotential. To obtain the values of the pressure gradient force they should be divided by the horizontal grid interval.

Quite an appreciable error is found to be associated with the Corby et al. scheme in the "no inversion case", when supposedly, the scheme should have had no error. The error, of course, results from the procedure used here to prescribe geopotentials, rather than temperatures. If the horizontal grid interval is taken to be 150 km, the obtained error of about 150 m^2s^{-2}, expressed in terms of geostrophic wind, corresponds to an error of the order of 10 ms^{-1}. Since with the temperature profile of the "no inversion case" the error shown for the Corby et al. scheme is entirely due to the discretization of the hydrostatic equation, the error vanishes in the limit as the thicknesses of sigma layers tend to zero.

With modest vertical resolution ($\Delta\sigma = 1/5$), the Corby et al. scheme is seen to give an error of about the same magnitude in the "inversion

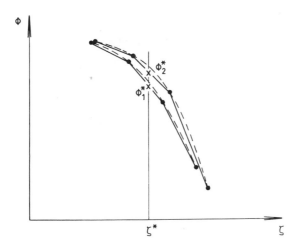

FIGURE 12. A schematic representation of the geopotential profiles at two adjacent grid points in the sigma system (Janjić [1979]).

TABLE 1.

Errors of the pressure gradient force analogs obtained using the Corby et al. and the Burridge-Haseler schemes, for the "no inversion case" and the "inversion case"; see text for details. Values are given in increments of geopotential ($m^2 s^{-2}$), between two neighboring grid points, along the direction of the increasing terrain elevations. (Note that some of the numbers in the last two lines are slightly different from those published in the referred paper; this is a result of the removal of an error that Mesinger has found in his program for calculation of the Burridge-Haseler scheme values. The numbers published previously actually represented errors of a scheme which, within the geopotential gradient term, used geopotentials of the $\sigma = 0.9$ surface rather than values defined by (4.22).)

$\Delta\sigma =$	1/5	1/15	1/25	$\cdots$	$\lim_{\Delta\sigma \to 0}$
Corby et al. scheme "no inversion case"	151.2	−48.7	29.0	$\cdots$	0
Corby et al. scheme "inversion case"	−159.6	−159.6	−159.6	$\cdots$	−159.6
Burridge-Haseler scheme "no inversion case"	0	0	0	$\cdots$	0
Burridge-Haseler scheme "inversion case"	0	−142.1	−153.3	$\cdots$	−159.6

case". However, with the temperature not being a single linear function of ln p, the error now does not vanish as the thicknesses of sigma layers tend to zero. In fact, it can be demonstrated that in this particular example the vertical resolution has no effect on the error. Thus, rather large errors are shown to be possible with the Corby et al. scheme irrespective of the vertical discretization problem.

With the hydrostatically consistent Burridge-Haseler scheme there is no error in the "no inversion case". Having no error in the limit as $\Delta\sigma \to 0$ should come as no surprise; recall that the two schemes differ only in the discretization of the hydrostatic equation and this difference vanishes as the thicknesses of sigma layers tend to zero. The absence of an error in case $\Delta\sigma = 1/5$ can be understood in view of the hydrostatic consistency of the scheme, and the special configuration of the sigma layer in this case (the right-hand panel of Figure 11), such that the pressure surface p^* passes through geopotential grid points at interfaces of the layer. With the geopotential interpolated in a consistent way, along its piecewise linear profile defined by the finite-difference hydrostatic equation, exact interface values of geopotential are recovered. The reason for the absence of errors at higher (but still not infinite) vertical resolution, when the consistency condition (4.23) is violated, is at this point not obvious. We shall return to an analysis of this case in the following section.

In the "inversion case", there is, of course, again no error for $\Delta\sigma = 1/5$; note that the hydrostatic consistency argument just given does not involve the prescribed temperature profile. However, as the vertical resolution is increased, the consistency condition is violated, and an appreciable error is once more obtained. The asymptotic error is entirely due to hydrostatic inconsistency: correct values of geopotential of the sigma surface, obtained by a vertical integration of the hydrostatic equation, are used to calculate the increment of geopotential within the first term of the pressure gradient force; while a local (incorrect) calculation of the sigma-to-pressure increments of geopotential is performed within the second term.

One should note that this rather large asymptotic error is a result of only the upper half of the inversion shown in Figure 11. In the considered limit the two schemes are not affected by the actual temperature profile within the lower half of the assumed inversion layer; a different profile in this lower half of that layer, provided the geopotential of the 900 mb remained the same, would have given the same asymptotic errors.

The asymptotic error of this example illustrates the nature of the pressure gradient force problem: it was not possible to remove this error by an increase in the accuracy of vertical differencing. In fact, it was demonstrated that an increase in the accuracy of vertical differencing can be associated with an *increase* in the error. The errors considered are related to the chosen horizontal grid length, but only through the difference in surface pressure between neighboring grid points. Therefore, these errors cannot be removed by an increase in formal accuracy of the horizontal differencing. Thus, it is seen that the difficulty is *not* due to the *truncation error* of the approximations to the space derivatives involved, as it is frequently believed. Considered errors result from the two-point temperature average used within the second term of the considered schemes, and also cannot be removed by an increase in formal accuracy of the temperature analog used within that term.

7. Reduction of the error. On the basis of arguments presented so far we believe the use of hydrostatically consistent schemes certainly appears advisable. However, this generally still does not accomplish elimination of the error, even in the simple case of a resting atmosphere. Let us, therefore, proceed with the analysis of the error.

Consider the reason for the error in case of a hydrostatically consistent scheme and an atmosphere at rest. Recall the argument explaining the absence of the error of the Burridge-Haseler scheme in case $\Delta\sigma = 1/5$ of the preceding example: a consistent interpolation of geopotential was performed up to the ends of two linear segments of the defined geopotential profile, so that exact geopotential values were recovered. In the more general case, such as illustrated in Figure 5, even though it is hydrostatically consistent, interpolation will have an error, because the actual profile of geopotential is not the same as that defined by the finite-difference hydrostatic equation. Thus, the values ϕ^* will generally have errors. Furthermore, errors of ϕ_2^* and ϕ_1^* will generally be different, so that

$$-\frac{\phi_2^* - \phi_1^*}{\Delta x}$$

will have an error.

In most cases, the geopotential profile defined by the finite-difference hydrostatic equation should be considered to be a piecewise linear function such as e.g. shown in Figure 6. The errors of ϕ^* would then be due to the deviation of the actual geopotential profile from that piecewise linear one. This situation is visualized in Figure 12. The dashed lines represent the "true" profiles at two adjacent points of the horizontal grid. The heavy dots correspond to the values of geopotential at the interfaces of the sigma layers, while the two piecewise linear curves represent the geopotential profiles obtained by integration of the finite-difference hydrostatic equation. As before, the values ϕ_2^* and ϕ_1^*, denoted by crosses, are used to calculate the pressure gradient force at the pressure level defined by ζ^*.

So far we have discussed the pressure gradient force error in the case of hydrostatically consistent calculations. The situation corresponding to the hydrostatic inconsistency is schematically represented in Figure 13. The notation is the same as in Figure 12. This time the error results from the linear extrapolation of geopotential beyond the range of validity of the finite-difference approximation to $\partial\phi/\partial\zeta$. Apparently, in this case we should expect larger errors.

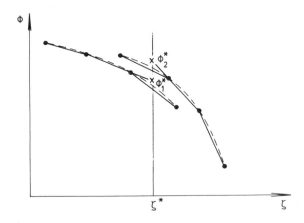

FIGURE 13. Same as in Figure 12, but for the case of hydrostatic inconsistency.

A similar analysis can be performed to show that analogous problems are encountered in the calculations using nonstaggered schemes for calculation of geopotential.

An obvious question at this point is: can one eliminate this error? It has been pointed out by Janjić [1979] that linear interpolation/extrapolation as illustrated by preceding figures will be error-free if ζ is chosen so as to have ϕ a linear function of ζ. In principle, this of course can be done for any given geopotential profile. However, note that in the "no inversion case" of the previous section for the Burridge-Haseler scheme, for which $\zeta = \ln p$, no errors were obtained even though geopotential was a *quadratic* and not a linear function of ζ. One may wonder whether some additional guidance can be obtained from an understanding of the reason for the absence of the error in this case.

Consider, therefore, a general quadratic function

$$\phi = \alpha + \beta\zeta + \gamma\zeta^2 \tag{7.1}$$

where α, β and γ are constant, and $\gamma \neq 0$. Note that, without loss of generality, this can be rewritten as

$$\phi = C + \frac{1}{2B}(A + B\zeta)^2, \tag{7.2}$$

with A, B, C being a new set of constants. We are interested in the errors of the scheme (3.8)

$$-\delta_x \overline{\phi}^{\,\eta} + \overline{\delta_\zeta \phi}^{\,x} \delta_x \overline{\zeta}^{\,\eta} \tag{7.3}$$

with geopotentials at interfaces being prescribed by (7.2).

When we insert (7.2) into the first term of (7.3) we readily obtain

$$-\delta_x \overline{\phi}^{\,\eta} = -\overline{\overline{(A + B\zeta)}^{\,x} \delta_x \zeta}^{\,\eta}. \tag{7.4}$$

Similarly, evaluation of the difference in geopotential across the layer yields

$$\Delta\phi_\eta = \overline{(A + B\zeta)}^{\,\eta} \Delta\zeta_\eta,$$

so that we at the same time have

$$\overline{\delta_\zeta \phi}^{\,x} \delta_x \overline{\zeta}^{\,\eta} = \overline{\overline{(A + B\zeta)}^{\,\eta} \delta_x \overline{\zeta}^{\,\eta}}. \tag{7.5}$$

Inspection of the right-hand sides of (7.4) and (7.5) shows that they will cancel when $\delta_x \zeta$ is not dependent on η; thus, in that case the scheme (7.3) will have no error. The Burridge-Haseler scheme is a special case of this situation: it is obtained from (7.3) inserting $\eta = p/p_S$ and $\zeta = \ln p$, which satisfies the requirement for $\delta_x \zeta = \delta_x \ln p_S$ to be independent of η.

This, therefore, explains the absence of the error of this scheme in the "no inversion case" of the preceding section. It is thus seen that with geopotential considered specified it is the Burridge-Haseler scheme that has no error when temperature is a linear function of ln p; while it was the Corby et al. scheme that in that case had no error with temperature considered specified.

Another special case for which the right-hand sides of (7.4) and (7.5) cancel is the case $B = 0$. However, note that, for $\gamma = 0$, ϕ is a linear function of ζ; thus, (7.3) will indeed have no error in this case as well.

It is not obvious that much guidance has been obtained from this consideration. The feature of a scheme to have no error for a general quadratic function of ζ, as compared to a linear function only, certainly seems attractive. However, there is no guarantee that this feature will be associated with acceptable errors in all, or most, cases of interest. Furthermore, various choices could substantially differ from the point of view of computational economy.

An attempt to deal with both of these difficulties has been made by Janjić [1977]. He has considered a family of schemes obtained from (7.3) by inserting

$$\eta = \frac{p - p_T}{p_S - p_T}, \qquad \zeta = (\ln p)^{1+m}, \qquad (7.6)$$

and has then sought to optimize the choice of the parameter m. To this end, for a given model resolution and vertical structure, he has calculated errors for a range of the values of m, in case of the terrain slope and an atmosphere at rest considered previously by Phillips [1974], and representing an approximation to the standard atmosphere. The obtained errors were generally rather small except at the uppermost layer, centered at the pressure of 300 mb. At that layer the error was quite large for $m = 0$ (when it amounted to the geopotential increment between neighboring grid points of more than 150 m^2s^{-2}); it then decreased with increasing values of m, and increased again when m was increased beyond the value of 1.2. Since the integer values of m were advantageous for reasons of economy, Janjić has chosen the value $m = 1$ as a compromise between the two requirements, that for minimization of the error and that for computational efficiency.

It is actually readily seen that with temperature having its typical tropospheric lapse rate, one should expect the errors of the scheme (7.6) for $m = 1$ to be smaller than those for $m = 0$. To this end, compare the hydrostatic equation (1.8) for $\zeta = \ln p$

$$\frac{\partial \phi}{\partial \ln p} = -RT, \qquad (7.7)$$

against that for $\zeta = \ln^2 p$

$$\frac{\partial \phi}{\partial \ln^2 p} = -\frac{RT}{2 \ln p}. \tag{7.8}$$

With tropospheric lapse rates, the slope of the geopotential profile defined by the discretization of (7.8) will be less variable than that defined by the discretization of (7.7), since on the right-hand side of (7.8) temperature is divided by another function which decreases with height at a rate of about the same order of magnitude. Thus, the pressure gradient force errors associated with (7.8) will be smaller, as calculations have indeed shown.

Finally, let us briefly consider the pressure gradient force error in case of *isentropic cooridinates*. Choosing $\eta = \theta$, (2.5) leads to

$$-\nabla_p \phi = -\nabla_\theta \left(\phi + c_p T \right). \tag{7.9}$$

Thus, in return for accepting the inconveniences of the lower boundary condition, in addition to other advantages, pressure gradient force problems due to mountain slope have been avoided. For example, there are no difficulties in simulating a resting atmosphere. Furthermore, with the pressure gradient force being a potential vector, its difference *formulation* can hardly be chosen in a hydrostatically inconsistent way. However, note that the pressure gradient force problems due to the slope of *coordinate surfaces* have *not* changed, in the sense that the condition (4.23) still has to be observed to maintain hydrostatic consistency. Otherwise, large errors are possible.

8. Vertical interpolation of initial pressure gradient force. Preceding considerations show that with sloping coordinate surfaces it appears not possible to construct a scheme able to simulate an arbitrary resting atmosphere without creating false pressure gradient forces. With a carefully chosen scheme the obtained pressure gradient force errors should be generally rather small; but in special situations (e.g. sharp inversions) large errors are possible.

The approach used so far assumed initial analysis of geopotential, or temperature. If this analysis were done on constant pressure or other surfaces not used as model coordinate surfaces, it would be followed by a vertical interpolation of the analyzed variable to obtain its values at model coordinate surfaces. Having the values of geopotential on coordinate surfaces, the initial temperature would be calculated using the hydrostatic equation of the model. If the temperature were interpolated, geopotential would be calculated in that way. Finally, the initial pressure gradient force would be evaluated.

It was pointed out by Sundqvist [1976] that a different sequence of initialization steps is possible. Following an analysis on pressure surfaces, pressure gradient force can be vertically interpolated to obtain its values on model coordinate surfaces. Assuming pressure gradient force to be known on coordinate surfaces, a set of elliptic equations can be derived, and the obtained numerical values used to solve this set of equations for temperature.

Apart from the problems of the analysis and the vertical interpolation error, the method proposed by Sundqvist appears to offer the possibility of simulating an arbitrary resting state without creating false motions due to the pressure gradient force errors. Rather than the initial pressure gradient force, it is the initial temperature that in this case would have an error. This error would be exactly such as to enable the correct initial pressure gradient force field to be recovered by the pressure gradient force scheme of the model; hopefully, at the same time, it would never be so large as to make the temperature field unreasonable.

Unfortunately, calculation of temperature using this approach is associated with difficulties. To derive his equation for temperature Sundqvist has applied the $\nabla_\sigma \cdot$ operator on the pressure gradient force. Had he applied the $\mathbf{k} \cdot \nabla_\sigma \times$ operator instead, he would have obtained a different equation, generally giving a different solution for temperature.

A simpler method of solving for temperature has been used by Mihailović [1981]. He has directly solved the finite-difference pressure gradient force expression for temperature, starting from known values of temperature at the boundaries. This, however, did not avoid the nonuniqueness problem: starting from two boundary points of a grid line, two generally different temperature values are obtained at each interior

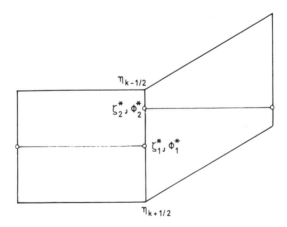

FIGURE 14. Schematic illustration of the nonuniqueness problem.

point. In the two-dimensional case, at each point two more values are obtained solving for temperature along the other set of grid lines. Furthermore, each solution obtained along a grid line starting from a boundary point at its one end violates the prescribed temperature at the other end of that line.

The following simple explanation of the nonuniqueness problem has been put forward by Janjić. Namely, let us for simplicity consider the one-dimensional problem. In this case the pressure gradient force contributions in between a grid point and two adjacent grid points are, in general, defined on two different ζ^* surfaces. Thus, at the central point, we shall have two values of ϕ^*. This situation is schematically represented in Figure 14. Since, however, geopotential may not be a linear function of ζ, a unique value of $\delta_\zeta \phi$ cannot yield both values of ϕ^*. Instead, we obtain two values of $\delta_\zeta \phi$ and, consequently, two temperatures. In the two-dimensional case two additional values will appear because the problem recurs in the direction of the other coórdinate axis.

In order to minimize the deviation of the initial sigma system pressure gradient force from the pressure gradient force obtained by interpolation from the pressure system, Janjić and Ničković used least squares fitting to define a unique linear geopotential profile within each sigma layer. This profile is then used to calculate unique temperature. However, a more elaborate variational approach is certainly also possible.

9. Blocking techniques for representation of mountains. In addition to the pressure gradient force problem, terrain-following coordinate systems have other difficulties. Dynamical processes may be distorted by the irregularities inherent to a sigma-type grid (Sadourny et al. [1981]). Several of the additional numerical problems (e.g., Simmons and Burridge [1981]; Simmons and Strüfing [1981]), while not fundamental, certainly are uncomfortable. Thus, recently it has been suggested a number of times that alternate approaches to the representation of mountains in numerical models may deserve more attention.

The pressure gradient force problem and, to a large extent also, the additional numerical problems result from the slope of coordinate surfaces. A number of definitions of the vertical coordinate have been introduced (e.g., Arakawa [1972] Simmons and Burridge [1981]) which have the vertical coordinate go to and eventually become equal to pressure as the height is increased. Unfortunately, such coordinates obviously can be only of limited help in alleviating the pressure gradient force problem, since the reduction of the slope of coordinate surfaces is limited.

A more drastic approach has been used by Egger [1972]. Egger's method consisted of vertical walls, in a sigma system model, placed so as to block the flow in a given sigma layer or layers and thus simulate the barrier effect of steep mountains. Even though remarkably successful, this method had obvious imperfections in not being able to properly accomodate the three-dimensional geometry of steep mountains, and in having steep mountains change their elevation as a function of time.

It has been pointed out by Mesinger [**1983**] that these two weaknesses can be removed by constructing model mountains so that they consist of three-dimensional grid boxes (Figure 15) and by defining the vertical coordinate in such a way as to have its surfaces remain at fixed elevations at places where they touch (and define) the ground surface. He has pointed out that this can be achieved by the coordinate

$$\eta \equiv \frac{p - p_T}{p_S - p_T} \frac{p_{rf}(z_S) - p_T}{p_{rf}(0) - p_T}. \tag{9.1}$$

Here $p_{rf}(z)$ is a suitably defined reference pressure as a function of z, the geometric height. To have, as stated, mountains formed of the grid boxes, the values of z_S are permitted to take only values given by

$$\eta_{k+1/2} = \frac{p_{rf}(z_S) - p_T}{p_{rf}(0) - p_T}, \qquad k = 1, 2, \ldots, N, \tag{9.2}$$

that is, by the values of η chosen for the interfaces of the η layers of the model.

Note that (9.1) implies that

$$\eta = 0 \quad \text{when } p = p_T \qquad \text{and} \qquad \eta = 1 \quad \text{when } z = 0.$$

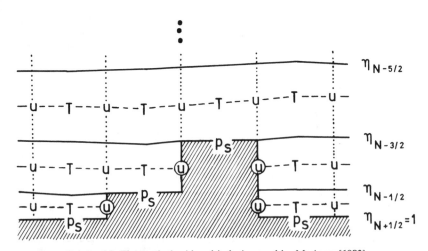

FIGURE 15. The vertical grid and indexing used by Mesinger [**1983**].

Furthermore,

$$\eta = \frac{p - p_T}{p_{rf}(0) - p_T} \quad \text{when } p_S = p_{rf}(z_S).$$

Thus, when pressure is a function of z only, and up to a level which includes the highest model mountains, this function is equal to the chosen reference pressure, η coordinate surfaces will all be horizontal. Then, for any reasonable choice of the finite-difference scheme, there will be no error in the pressure gradient force. This is a feature which with sigma type coordinates we saw as possible to achieve only under much more restrictive conditions. But, of course, also in other cases the slope of coordinate surfaces defined by (9.1) will be small, and errors of the pressure gradient force should be very much reduced.

The proposed system makes an effort to preserve, to the maximum extent possible, simplicity in the specification of the boundary condition. As a result, velocity components normal to the ground surface are readily set to zero. Further details of the wall boundary condition may, of course, be a matter of some complexity if particular features of the horizontal differencing are to be maintained. This obviously is a task to be considered for each scheme separately.

The coordinate of the type (9.1) is because of its dependence on z and because the top η surface is permiteed to be also at a level $p = \text{const} > 0$ more general than the η coordinate considered by Simmons and Burridge [1981]. As can be verified following Kasahara [1974], these two differences in themselves do not affect the continuous equations. There are, however, differences introduced by the steplike ground surface. One is a simple point of noting that in the pressure tendency equation the vertical integration is to be performed from top to the lowermost value of η, η_S, now not necessarily 1 (and, furthermore, *to the lowermost of the values of* η_S, along the sides of mountains).

Another difference is the need to consider horizontal discontinuities in $(p_{rf}(z_S) - p_T)/(p_S - p_T)$, resulting from discontinuities in ground elevation. Thus, the η surfaces as defined by (9.1) will, in fact, also be discontinuous.

This difficulty, as also pointed out by Mesinger, can be overcome by assuming that the values of $(p_{rf}(z_S) - p_T)/(p_S - p_T)$ in (9.1) are not the actual values, but rather values interpolated in such a way as to achieve the continuity of η surfaces. An additional property to be required of the interpolation algorithm is that these interpolated values tend to the actual values as the distance from the horizontal sides of the ground surface approaches zero; this is necessary in order to maintain the property of η surfaces to stay fixed to (and define) the horizontal sides of model mountains. It is not difficult to construct an interpolation algorithm having these two properties; however, there is no need to actually do it here, since the details of this algorithm will not be needed by the differencing scheme. The finite-difference scheme, in fact, can be considered to imply an interpolation algorithm with the very same properties.

Consequently, following Kasahara [1974] and Simmons and Burridge [1981], the governing equations for frictionless and adiabatic motion can be written as follows:

$$\frac{d\mathbf{v}}{dt} + f\mathbf{k} \times \mathbf{v} + \nabla\phi + \frac{RT}{p}\nabla p = 0, \tag{9.3}$$

$$\frac{dT}{dt} - \frac{\kappa T\omega}{p} = 0, \tag{9.4}$$

$$\frac{\partial}{\partial\eta}\left(\frac{\partial p}{\partial t}\right) + \nabla\cdot\left(\mathbf{v}\frac{\partial p}{\partial\eta}\right) + \frac{\partial}{\partial\eta}\left(\dot{\eta}\frac{\partial p}{\partial\eta}\right) = 0, \tag{9.5}$$

$$\frac{\partial\phi}{\partial\eta} = -\frac{RT}{p}\frac{\partial p}{\partial\eta}, \tag{9.6}$$

$$\omega \equiv \frac{dp}{dt} = -\int_0^\eta \nabla\cdot\left(\mathbf{v}\frac{\partial p}{\partial\eta}\right)d\eta + \mathbf{v}\cdot\nabla p, \tag{9.7}$$

$$\frac{\partial p_S}{\partial t} = -\int_0^{\eta_S} \nabla\cdot\left(\mathbf{v}\frac{\partial p}{\partial\eta}\right)d\eta, \tag{9.8}$$

$$\dot{\eta}\frac{\partial p}{\partial\eta} = -\frac{\partial p}{\partial t} - \int_0^\eta \nabla\cdot\left(\mathbf{v}\frac{\partial p}{\partial\eta}\right)d\eta. \tag{9.9}$$

The subscripts of the del operator have here been omitted; boundary conditions were assumed p = const at the top boundary $\eta = 0$, and $\dot{\eta} = 0$ at $\eta = 0$, and at horizontal parts of the ground surface $\eta = \eta_S$.

The blocking approach, as outlined here, has not yet been tested in actual integrations using comprehensive atmospheric models. However, two groups are presently involved in developing models of this type; thus, some experience in the performance of such models is likely to be available in a reasonably short time.

10. Conservation of angular momentum. For an analysis of the angular momentum and also the energy conservation feature we shall here use the notation of (9.3)–(9.9) and Figure 15. Note that these equations did not actually depend on the specific definition of eta apart from the stated boundary conditions, which are the same as those used with terrain-following coordinates. Thus, this system, as well as the figure, can be considered as rather general, covering both the terrain-following coordinates, which will have no step mountains, as well as the system including the presence of step mountains in the case eta is defined to be a coordinate of the type (9.1). We shall here present the analysis of Mesinger [**1983**], which, in turn, to a very large degree follows that of Simmons and Burridge [**1981**] (also Simmons and Strüfing [**1981**]). In addition to permitting a more general choice of vertical coordinate, it departs from their analysis (i) in not making a decision on the finite-difference analog of the hydrostatic equation, and (ii) in considering, from the start of the analysis, the effects of the horizontal discretization along with those of the vertical discretization—since in case of the step mountains a separation of these two discretization effects would not seem appropriate. To save space, Mesinger has restricted his consideration of the horizontal discretization to longitude only. Further generalization to two horizontal coordinates should in most cases present no difficulties.

In the continuous case, we want to evaluate the integral of the pressure gradient force with respect to pressure, and longitude, λ,

$$-\int_0^{2\pi}\int_0^{\eta_s}\left(\frac{\partial\phi}{\partial\lambda}+\frac{RT}{p}\frac{\partial p}{\partial\lambda}\right)\frac{\partial p}{\partial\eta}d\eta\,d\lambda, \qquad (10.1)$$

having in mind a possible existence of steplike mountains as shown in Figure 15. After use is made of

$$\frac{\partial\phi}{\partial\lambda}\frac{\partial p}{\partial\eta}=\frac{\partial}{\partial\lambda}\left(\phi\frac{\partial p}{\partial\eta}\right)-\frac{\partial}{\partial\eta}\left(\phi\frac{\partial p}{\partial\lambda}\right)+\frac{\partial\phi}{\partial\eta}\frac{\partial p}{\partial\lambda},$$

of the boundary condition $p=$ const along the surface $\eta=0$, and of (9.6), we obtain

$$\int_s\phi\frac{\partial p}{\partial s}ds. \qquad (10.2)$$

Here s is defined to follow the ground surface at the considered latitude, increasing eastward; and up the western and down the eastern sides of step mountains in case a choice of eta of the type (9.1) is being used. Thus, the contribution of the pressure gradient force to the change of angular momentum will be a "mountain torque" term proportional to (10.2).

To have no false production of angular momentum due to the finite-difference scheme, the analog to (10.1) has to be chosen so that it results in an analog to (10.2). Following Simmons and Burridge [**1981**], the analog to (10.1) is considered to be of the form

$$-\sum_u\left[\delta_\lambda\phi_k+\left(\frac{RT}{p}\frac{\partial p}{\partial\lambda}\right)_k\right]\overline{\Delta p_k}^\lambda\,\Delta\lambda. \qquad (10.3)$$

The summation here refers to all u points having u as a time-dependent variable; thus, it does not include the mountain side u points. Also as in Simmons and Burridge, geopotential is carried at interfaces, and ϕ_k is defined by

$$\phi_k \equiv \phi_{k+1/2} + \alpha_k R T_k. \tag{10.4}$$

This definition would appear to permit a rather general class of the pressure gradient force (and hydrostatic equation) schemes; for example, it includes the family of schemes proposed by Janjić [1977].

Transformation of (10.3), with the help of

$$\delta_\lambda(AB) = \overline{A}^\lambda \delta_\lambda B + \overline{B}^\lambda \delta_\lambda A,$$

results in a term with contributions from T points which all cancel except for those from T points next to the mountain sides. Further transformation of the second term obtained in this way, with the help of (10.4) and another general rule

$$A_{k+1/2}\delta_\lambda \Delta B_k = \Delta(A\delta_\lambda B)_k - \Delta A_k \delta_\lambda B_{k-1/2},$$

similarly gives terms which all cancel except for those from u points next to the ground surface. In this way, it is seen that (10.3) gives

$$\sum_{u,a} \overline{\phi_{k+1/2}}^\lambda \delta_\lambda p_{k+1/2} \Delta\lambda - \sum_{T,w} \phi_k \Delta p_k + \sum_{T,e} \phi_k \Delta p_k$$

$$- \sum_u \left[\overline{\Delta\phi_k}^\lambda \delta_\lambda p_{k-1/2} - \overline{\alpha_k R T_k}^\lambda \delta_\lambda \Delta p_k + \left(\frac{RT}{p} \frac{\partial p}{\partial \lambda} \right)_k \overline{\Delta p_k}^\lambda \right] \Delta\lambda. \tag{10.5}$$

Here the first three pairs of summation subscripts denote that summations are to be performed over all the u points immediately above, T points immediately west, and T points immediately east of the ground surface, respectively. With the grid as shown in Figure 15, the first three terms in (10.5) are in this way readily recognized to represent the simplest possible analog to the mountain torque integral (10.2). Thus, for a given definition of $\Delta\phi_k$ and α_k, the angular momentum conservation will be achieved if, within the analog to the pressure gradient force it is chosen

$$\left(\frac{RT}{p} \frac{\partial p}{\partial \lambda} \right)_k = -\frac{1}{\overline{\Delta p_k}^\lambda} \left(\overline{\Delta\phi_k}^\lambda \delta_\lambda p_{k-1/2} - \overline{\alpha_k R T_k}^\lambda \delta_\lambda \Delta p_k \right), \tag{10.6}$$

so as to make the expression in the square bracket of (10.5) equal to zero.

The requirement (10.6) reduces to that obtained by Simmons and Burridge [1981] (their equation (3.14)), for their special choice of the finite-difference analog to the hydrostatic equation. Thus, the more general definition of the vertical coordinate, permitting *inter alia* a steplike representation of mountains, has been seen to have had no effect on the angular momentum conservation requirement.

Using the sigma coordinate defined as

$$\sigma \equiv \frac{p - p_T}{\pi}, \qquad \pi \equiv p_S - p_T, \tag{10.7}$$

and considering the vertical differencing only, Arakawa and Suarez [**1983**] show that the vertically integrated circulation feature (1.6) will be maintained when the pressure gradient force is of the form

$$-\nabla\phi_{k_\ell} - \frac{1}{\pi}\left[\phi_k - \frac{\Delta(\sigma\phi)_k}{\Delta\sigma_k}\right]\nabla\pi. \tag{10.8}$$

It can readily be demonstrated that the one-dimensional version of (10.8) is one member of the family of schemes obtained by removing the horizontal discretization from (10.6). Namely, the vertical differencing scheme thus obtained reduces to the one-dimensional version of (10.8) for the special choice of the vertical coordinate, (10.7), used by Arakawa and Suarez.

11. Conservation of energy. The rate of kinetic energy generation by the pressure gradient force, due to eastward motion, in a vertical plane located at latitude φ and bounded by the surface $\eta = 0$ and by longitudes λ_1, λ_2, per unit distance in meridional direction is equal to

$$-\frac{1}{g}\int_{\lambda_1}^{\lambda_2}\int_0^{\eta_S} \frac{u}{a\cos\varphi}\left(\frac{\partial\phi}{\partial\lambda} + \frac{RT}{p}\frac{\partial p}{\partial\lambda}\right)\frac{\partial p}{\partial\eta}\,d\eta\,d\lambda. \tag{11.1}$$

Here u is the eastward velocity component and a the radius of the earth. As it was done in the preceding section, this consideration is also restricted to one horizontal coordinate only.

The geopotential gradient part of the integrand of (11.1) can with the help of the two-dimensional version of (9.5), and (9.6), be transformed into

$$\frac{1}{a\cos\varphi}\frac{\partial}{\partial\lambda}\left(u\phi\frac{\partial p}{\partial\eta}\right) + \frac{\partial}{\partial\eta}\left[\phi\left(\frac{\partial p}{\partial t} + \dot{\eta}\frac{\partial p}{\partial\eta}\right)\right] + \frac{RT}{p}\left(\frac{\partial p}{\partial t} + \dot{\eta}\frac{\partial p}{\partial\eta}\right)\frac{\partial p}{\partial\eta}. \tag{11.2}$$

The first term here integrates to zero for a closed region, and for the present boundary condition. With $p = \text{const}$ at the top boundary $\eta = 0$, and $\dot\eta = 0$ at the top as well as at the lower boundary, the second term integrates to give a contribution

$$-\frac{1}{g}\int_{\lambda_1}^{\lambda_2} \phi_S \frac{\partial p_S}{\partial t} d\lambda. \tag{11.3}$$

The third term of (11.2) and the pressure gradient part of (11.1) are compensated by the terms arising from the thermodynamic equation

$$\frac{1}{g}\int_{\lambda_1}^{\lambda_2}\int_0^{\eta_S} c_p \frac{\kappa T \omega}{p} \frac{\partial p}{\partial \eta} d\eta\, d\lambda \tag{11.4}$$

$$= \frac{1}{g}\int_{\lambda_1}^{\lambda_2}\int_0^{\eta_S} \frac{RT}{p}\left(\frac{\partial p}{\partial t} + \dot\eta \frac{\partial p}{\partial \eta} + \frac{u}{a\cos\varphi}\frac{\partial p}{\partial \lambda}\right)\frac{\partial p}{\partial \eta} d\eta\, d\lambda.$$

To have no false production of energy in transformations between the kinetic and total potential energy this compensation has to be achieved in the discrete case as well.

In parallel with (10.3), we consider the analog to (11.1) to be of the form

$$-\frac{1}{g}\sum_u \frac{u_k}{a\cos\varphi}\left[\delta_\lambda\phi_k + \left(\frac{RT}{p}\frac{\partial p}{\partial \lambda}\right)_k\right]\overline{\Delta p_k}^\lambda \Delta\lambda. \tag{11.5}$$

As it is to be used to determine an analog to $\kappa T\omega/p$, carried at T points, we transform each of its parts to a sum over T points. Considering the geopotential gradient part, note that

$$\sum_u U\delta_\lambda P = -\sum_T P\delta_\lambda U$$

for variables U and P defined at u and T points, respectively, of a staggered grid as shown in Figure 15, bounded by u points at which $U = 0$. Thus, the geopotential gradient part of (11.5) is transformed into

$$\frac{1}{g}\sum_T \frac{\phi_k}{a\cos\varphi}\delta_\lambda(u_k\Delta p_k)\Delta\lambda, \tag{11.6}$$

which accomplishes the equivalent of integrating to zero the first term of (11.2). Further transformation of (11.6) in the manner of (11.2) requires use of the finite-difference analog of the continuity equation. We choose

$$\Delta W_k + \frac{1}{a\cos\varphi}\delta_\lambda\left(u_k\overline{\Delta p_k}^\lambda\right) = 0, \tag{11.7}$$

where for brevity

$$W \equiv \frac{\partial p}{\partial t} + \dot\eta \frac{\partial p}{\partial \eta}.$$

In addition, making use of

$$A_k \Delta B_k = \Delta (AB)_k + (A_{k-1/2} - A_k) B_{k-1/2} + (A_k - A_{k+1/2}) B_{k+1/2}$$

and of the boundary conditions on p and $\dot{\eta}$, we see that (11.6) is transformed into

$$-\frac{1}{g} \sum_{PS} \phi_S \frac{\partial p_S}{\partial t} \Delta \lambda - \frac{1}{g} \sum_T \left[(\phi_{k-1/2} - \phi_k) W_{k-1/2} \right. \tag{11.8}$$

$$\left. + (\phi_k - \phi_{k+1/2}) W_{k+1/2} \right] \Delta \lambda,$$

containing, as its first term, an analog to (11.3).

Writing the analog to (11.4) as

$$\frac{1}{g} \sum_T c_p \left(\frac{\kappa T \omega}{p} \right)_k \Delta p_k \Delta \lambda, \tag{11.9}$$

we recognize that the expression in the square bracket of (11.8) is to define one (" vertical") part of the analog of $\kappa T \omega / p$. This part

$$\frac{1}{c_p \Delta p_k} \left[(\phi_{k-1/2} - \phi_k) W_{k-1/2} + (\phi_k - \phi_{k+1/2}) W_{k+1/2} \right], \tag{11.10}$$

making use of (10.4) and (11.7), we see can also be written as

$$\frac{1}{c_p \Delta p_k a \cos \varphi} \left[\Delta \phi_k \sum_{r=1}^{k-1} \delta_\lambda \left(u_r \overline{\Delta p_r}^\lambda \right) - d_k R T_k \delta_\lambda \left(u_k \overline{\Delta p_k}^\lambda \right) \right]. \tag{11.11}$$

Regarding the pressure gradient part of (11.5), note that, again with $U = 0$ at boundary points, we have

$$\sum_u U \overline{P}^\lambda = \sum_T P \overline{U}^\lambda.$$

Thus, it is found that the required cancellation will be achieved if the remaining ("horizontal") part of the analog to $\kappa T \omega / p$ is chosen to be of the form

$$\frac{1}{c_p a \cos \varphi} \overline{u_k \left(\frac{RT}{p} \frac{\partial p}{\partial \lambda} \right)_k}^\lambda, \tag{11.12}$$

where the analog to $(RT/p) \partial p / \partial \lambda$ is the same as that used in the momentum equation.

Alternatively, if this analog to $(RT/p) \partial p / \partial \lambda$ is restricted to the usual sigma system form

$$R \overline{T_k}^\lambda \left(\frac{1}{p} \frac{\partial p}{\partial \lambda} \right)_k,$$

as chosen by Simmons and Burridge [**1981**], the transformation to the sum over T points can be done so as to obtain, instead of (11.12),

$$\frac{\kappa T_k}{\Delta p_k a \cos \varphi} \; \overline{u_k \Delta p_k}^\lambda \left(\frac{1}{p}\frac{\partial p}{\partial \lambda}\right)^\lambda_k. \qquad (11.13)$$

Thus, it is seen that in this case some freedom remains in the choice of the analog to $\kappa T \omega / p$.

The result (11.10)/(11.11), and (11.12) or (11.13), has no differences compared to that given by Simmons and Burridge [**1981**] except for those reflecting the absence of a specific choice of the hydrostatic equation and the inclusion of the horizontal differencing in the analysis given here.

Restricting the pressure gradient force to their result (10.8), Arakawa and Suarez find that in order to conserve energy $\kappa T \omega / p$ in the thermodynamic equation must be of the form

$$\frac{1}{c_p \pi}\left[\left(\phi_k - \frac{\Delta(\sigma\phi)_k}{\Delta\sigma_k}\right)\left(\frac{\partial}{\partial t} + \mathbf{v}_k \cdot \nabla\right)\pi \qquad (11.14)$$

$$+ \frac{1}{\Delta\sigma_k}\left((\pi\dot\sigma)_{k+1/2}(\phi_k - \phi_{k+1/2}) + (\pi\dot\sigma)_{k-1/2}(\phi_{k-1/2} - \phi_k)\right)\right].$$

As in the previous section, it is readily demonstrated that the one-dimensional version of (11.14) is a special case of the scheme obtained by removing the horizontal discretization from the sum of (11.10), and (11.12) or (11.13). The scheme thus obtained reduces to the one-dimensional version of (11.14) after insertion of the vertical coordinate (10.7), used by Arakawa and Suarez.

Acknowledgments. The authors are grateful to Mr. Nikola Opačić, for his careful drafting of the figures. The work was supported by the Association for Science of the Republic of Serbia.

BIBLIOGRAPHY

A. Arakawa (1972). *Design of the UCLA general circulation model*, Tech. Rep. No. 7, Dept. Meteorol., Univ. of California, Los Angeles.

A. Arakawa and V. R. Lamb (1977). *Computational design of the basic dynamical processes of the UCLA general circulation model*, Methods in Computational Physics, Vol. 17, General Circulation Models of the Atmosphere (J. Chang, ed.), Academic Press, New York, pp. 173–265.

A. Arakawa and M. J. Suarez (1983). *Vertical differencing of the primitive equations in sigma coordinates*, Monthly Weather Rev. **111**, 34–45.

D. M. Burridge and J. Haseler (1977). *A model for medium range weather forecasting: adiabatic formulation*, Tech. Rep. No. 4, European Centre for Medium Range Weather Forecasts, Reading, U.K.

G. A. Corby, A. Gilchrist and R. L. Newson (1972). *A general circulation model of the atmosphere suitable for long period integrations*, Quart. J. Roy. Meteorol. Soc. **98**, 809–832.

G. A. Corby, A. Gilchrist and P. R. Rowntree (1977). *United Kingdom Meterological Office five-level general circulation model*, Methods in Computational Physics, Vol. 17, General Circulation Models of the Atmosphere (J. Chang, ed.), Academic Press, New York, pp. 67–110.

J. Egger (1972). *Incorporation of steep mountains into numerical forecasting models*, Tellus **24**, 324–335.

J. M. Gary (1973). *Estimate of truncation error in transformed coordinate, primitive equation atmospheric models*, J. Atmospheric Sci. **30**, 223–233.

A. Gilchrist (1975). *The Meteorological Office general circulation model*, Seminars on Scientific Foundations of Medium Range Weather Forecasts, European Centre for Medium Range Weather Forecasts, Reading, U.K., pp. 594–661.

Z. I. Janjić (1977). *Pressure gradient force and advection scheme used for forecasting with steep and small scale topography*, Contrib. Atmospheric Phys. **50**, 186–199.

——— (1979). *Numerical problems related to steep mountains in sigma coordinates*, Workshop on Mountains and Numerical Weather Prediction, European Centre for Medium Range Weather Forecasts, Reading, U.K., pp. 48–88.

D. R. Johnson (1980). *A generalized transport equation for use with meteorological coordinate systems*, Monthly Weather Rev. **108**, 733–745.

D. R. Johnson and L. W. Uccellini (1983). *A comparison of methods for computing the sigma-coordinate pressure gradient force for flow over sloped terrain in a hybrid theta-sigma model*, Monthly Weather Rev. **111**, 870–886.

A. Kasahara (1974). *Various vertical coordinate systems used for numerical weather prediction*, Monthly Weather Rev. **102**, 509–522.

Y. Kurihara (1968). *Note on finite difference expressions for the hydrostatic relation and pressure gradient force*, Monthly Weather Rev. **96**, 654–656.

F. Mesinger (1982). *On the convergence and error problems of the calculation of the pressure gradient force in sigma coordinate models*, Geophys. Astrophys. Fluid Dynamics **19**, 105–117.

——— (1983). *A blocking technique for representation of mountains in atmospheric models*, Riv. Meteorol. Aeronautica **43**.

D. T. Mihailović (1981). *Calculation of initial temperature from interpolated pressure gradient force in σ coordinate models*, Arch. Meteorol. Geophys. Bioclimat. Ser. A **30**, 239–251.

H. Nakamura (1978). *Dynamical effects of mountains on the general circulation of the atmosphere*. I: *Development of finite-difference schemes suitable for incorporating mountains*, J. Meteorol. Soc. Japan (2) **56**, 317–340.

N. A. Phillips (1957). *A coordinate system having some special advantages for numerical forecasting*, J. Meteorol. **14**, 184–185.

——— (1973). *Principles of large scale numerical weather prediction*, Dynamic Meteorology (P. Morel, ed.), Reidel, Dordrecht, pp. 1–96.

——— (1974). *Application of Arakawa's energy conserving layer model to operational numerical weather prediction*, Office Note No. 104, National Meteorol. Center, NOAA/NWS, U.S. Dept. of Commerce, Washington, D.C.

D. Rousseau and H. L. Pham (1971). *Premiers résultats d'un modèle de prévision numérique à courte échéance sur l'Europe*, La Météorologie **20**, 1–12.

R. Sadourny, O. P. Sharma, K. Laval and J. Canneti (1981). *Modelling of the vertical structure in sigma coordinate: a comparative test with FGGE data*, Proc. Internat. Conf. on Preliminary FGGE Data Analysis and Results (Bergen, 1980), World Meteorological Organization, Geneva, pp. 294–302.

A. J. Simmons and D. M. Burridge (1981). *An energy and angular-momentum conserving vertical finite-difference scheme and hybrid vertical coordinates*, Monthly Weather Rev. **109**, 758–766.

A. J. Simmons and R. Strüfing (1981). *An energy and angular momentum conserving finite-difference scheme, hybrid coordinates and medium-range weather prediction*, Tech. Rep. No. 28, European Centre for Medium Range Weather Forecasts, Reading, U.K.

J. Smagorinsky, R. F. Strickler, W. E. Sangster, S. Manabe, J. L. Holloway, Jr., and G. D. Hembree (1967). *Prediction experiments with a general circulation model*, Proc. Internat. Sympos. on Dynamics of Large Scale Atmospheric Processes (Moscow, 1965), Izdatel'stvo Nauka, Moscow, pp. 70–134.

J. D. Stackpole, L. W. Vanderman and J. G. Sela (1980). *U.S.A. National Meteorological Center (NMC) numerical prediction models*, Catalogue of Numerical Atmospheric Models for the First GARP Global Experiment, World Meteorological Organization, Geneva, pp. 216–274.

H. Sundqvist (1975). *On truncation errors in sigma-system models*, Atmosphere **13**, 81–95.

———— (1976). *On vertical interpolation and truncation in connexion with use of sigma system models*, Atmosphere **14**, 37–52.

T. Tokioka (1978). *Some considerations on vertical differencing*, J. Meteorol. Soc. Japan (2) **56**, 98–111.

K. Tomine and S. Abe (1982). *A trial to reduce truncation errors of the pressure gradient force in the sigma coordinate systems*, J. Meteorol. Soc. Japan (2) **60**, 709–716.

DEPARTMENT OF METEOROLOGY, UNIVERSITY OF BELGRADE, YUGOSLAVIA

FEDERAL HYDROMETEOROLOGICAL INSTITUTE, BELGRADE, YUGOSLAVIA

Current address: National Meteorological Center, W/NMC2, WWB Room 204, Washington, DC 20233

Lectures in Applied Mathematics
Volume **22**, 1985

Discrete Shocks for Difference Approximations to Systems of Conservation Laws

Daniel Michelson

Given a system of conservation laws in several space dimensions,

$$\sum_{i=1}^{m} \left(f_i(u) \right)_{x_i} = 0, \tag{1}$$

where $f_i(u)$, $u(x)$ are n-dimensional vector-functions, $x = (x_1, \ldots, x_m) \in R^m$. In the case of a time dependent problem, one should denote $x_m = t$ and assume $f_m(u) = u$. By planar or oblique shock solution to (1) we mean a solution

$$u(x) = \begin{cases} u_L & \text{for } x \cdot s \leqslant 0 \\ u_R & \text{for } x \cdot s > 0, \end{cases} \tag{2}$$

where $s = (s_1, \ldots, s_m) \in R^m$ is the slope of the shock and u_L, u_R satisfy the Rankine-Hugoniot condition

$$f(u_L) = f(u_R), \quad \text{where } f(u) = \sum_{i=1}^{m} s_i f_i(u). \tag{3}$$

Let

$$G\left(\left\{ E_x^j u(x) \right\} \right) = 0, \quad j = (j_1, \ldots, j_m) \in J \subset \mathbf{Z}^m, \tag{4}$$

1980 *Mathematics Subject Classification.* Primary 76L05; Secondary 65L99, 65M99.

Key words and phrases. Discrete shocks, systems of conservation laws, conservative difference approximations, Conley index, centered manifold theorem.

be a difference approximation of the system (1). Here $E_x^j u(x) = u(x + jh)$, $u(x)$ is a grid function on a uniform grid R_h^m with mesh size h and the multi-index j varies over a finite set J. We assume that the approximation is *conservative*, i.e.

$$G\big(\{E_x^j u(x)\}\big) = \sum_{i=1}^m (E_{x_i} - I) G_i\big(\{E_x^j u(x)\}\big),$$

and *consistent* with (1)

$$G_i\bigg(\underbrace{\{u(x), \ldots, u(x)\}}_{|J|\text{ times}} \bigg) = f_i(u(x)).$$

Let the slope $s = (s_1, \ldots, s_m)$ be rational—$s_i = p_i/q$, where $q > 0$ is the least common denominator.

DEFINITION. The grid function $v(\tau)$, $\tau \in R_{h/q}^1$, is called a discrete planar shock solution of (4) with the slope s if the grid function $u(x) = v(x \cdot s)$, $x \in R_h^m$, is a solution of (4) and

$$\lim_{\tau \to -\infty} v(\tau) = u_L, \qquad \lim_{\tau \to +\infty} v(\tau) = u_R \qquad (5)$$

(by the conservation and consistency, necessarily $f(u_L) = f(u_R)$).

REMARK. The step size h, as well as the denominator q, does not play any role in the Definition. One can assume $h = q = 1$ and $s_i = p_i$ are relatively prime numbers. We do not consider irrational s since then $v(\tau)$ should be defined on a set of points dense in R^1.

Question. Given a rational slope s and the states u_L, u_R satisfying the Rankine-Hugoniot condition, does the difference scheme have discrete shock solutions with the above slope and asymptotic states?

We consider only *weak* shocks, such that $|u_L - u_R|$ is small. More precisely: fix s, define $f(u) = \sum s_i f_i(u)$, $A(u) = \partial_u f(u)$. Let u_0 be such that $A(u_0)$ has a simple zero eigenvalue $\lambda_0(u_0) = 0$ with the corresponding right and left eigenvectors $r_0(u_0)$, $l_0(u_0)$. Assume *genuine nonlinearity* of λ_0, i.e.

$$r_0(u_0) \cdot \nabla \lambda_0(u_0) \neq 0.$$

Then

LEMMA. *For any u_L close to u_0, there exists a unique $u_R \neq u_L$ close to u_0 such that $f(u_L) = f(u_R)$. The planar shock which corresponds to the above u_L, u_R with small $|u_L - u_0|$ and $|u_R - u_0|$ is called weak.*

DEFINITION. The weak shock satisfies (violates) the entropy condition if $\lambda_0(u_L) > 0 > \lambda_0(u_R)$ $(\lambda_0(u_L) < 0 < \lambda_0(u_R))$.

THEOREM. *Let the scheme G, when linearized at the constant state u_0 and restricted to the direction of s, be (roughly speaking) kth order accurate and $k + 1$ order dissipative approximation of the differential operator $A(u_0) \, \partial/\partial\tau$, where $k = 2l + 1$ is an odd number. Then for any u_L close to u_0 with $\lambda_0(u_L) > 0$, there exists a discrete shock solution which satisfies the entropy condition. For $k = 1, 3$ (and probably for any odd k) there is actually a one-parameter family of such solutions forming a continuous trajectory connecting u_L and u_R. The above discrete shocks approximate (in a sense) a solution of the differential problem*

$$\frac{d^k y}{dt^k} = \frac{1}{2}(y^2 - 1), \qquad \lim_{\tau \to \mp \infty} y(\tau) = \pm(-1)^l. \qquad (6)$$

REMARK. For even k the above problem has no solutions, therefore even order schemes like Lax-Wendroff may have no discrete shock solutions.

The solution of (6) is not monotone. For $k = 3$ it appears as in Figure 1. The solution vanishes only once, has "overshoot" and "undershoot", and tends exponentially to the states $y = \pm 1$. The width of the "shock layer" in the physical space x is proportional to $h \cdot |b(u_0)/\lambda_0(u_L)|^{1/k}$, where $b(u_0)$ is the coefficient of the dissipation while $\lambda_0(u_L)$ is proportional to the strength of the shock.

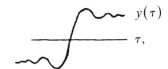

$y(\tau)$

$\tau,$

FIGURE 1.

Now we outline the accuracy and dissipativity assumptions for the scheme G. Let

$$G_\tau\big(\{ E^{j_0} v(\tau) \}\big) = G\big(\{ E_\tau^{j \cdot p} v(\tau) \}\big), \qquad j \in J,$$

where $E_\tau v(\tau) = v(\tau + h/q)$, $p = (p_1, \ldots, p_m) = s \cdot q$, and the index j_0 belongs to a set $J_0 \subset \mathbf{Z}$, $J_0 = \{ j \cdot p, j \in J \}$. Since G is conservative, so is G_τ, i.e.

$$G_\tau = (E_\tau - I) G_0.$$

The discrete shock solution $v(\tau)$ satisfies the equations

$$G_\tau\big(\{ E_\tau^{j_0} v(\tau) \}\big) = 0, \qquad \tau \in R_{h/q}^1; \, v(-\infty) = u_L, v(+\infty) = u_R, \qquad (7)$$

or, equivalently,

$$G_0\big(\{ E_\tau^{j_0} v(\tau) \}\big) - qf(u_L) = 0; \qquad v(-\infty) = u_L, v(+\infty) = u_R. \qquad (8)$$

Let dG_τ be the linearization of G_τ at the constant state u_0 and $d\hat{G}_\tau(\xi)$ its Fourier symbol.

ASSUMPTION 1. $d\hat{G}_\tau(\xi) \cdot r_0(u_0) = i\xi \cdot A(u_0)r_0(u_0) + O(\xi^{k+1})$ (this is k th order accuracy assumption in the direction of the eigenvector $r_0(u_0)$).

ASSUMPTION 2. $\det(d\hat{G}_\tau(\xi)) \neq 0$ for $0 < \xi < 2\pi$ (this is the dissipativity or regularity assumption).

ASSUMPTION 3. $l_0(u_0) \cdot d\hat{G}_\tau(\xi)r_0(u_0) = b(u_0)(iq\xi)^{k+1} + O(\xi^{k+2})$, where $(-1)^{(k+1)/2}b(u_0) > 0$ (i.e. positive dissipation). Here we suppose that the vectors $l_0(u_0)$ and $r_0(u_0)$ are normalized so that $l_0(u_0) \cdot r_0(u_0) = 1$ and $\nabla\lambda_0(u_0) \cdot r(u_0) = 1$.

EXAMPLE. Given a system of conservation laws, $(f_1(u))_{x_1} + (f_2(u))_{x_2} = 0$ in two space dimensions. Approximate $\partial/\partial x_1$, $\partial/\partial x_2$ by a central difference formula of order $2l + 2$ and add the dissipation of the form $h^{-1}(-\Delta)^{l+1}$, where $\Delta = (E_{x_1} + E_{x_1}^{-1} - 2) + (E_{x_2} + E_{x_2}^{-1} - 2)$. The resulting scheme is $2l + 1$ order accurate and $2l + 2$ dissipative, and satisfies the above assumptions. To arrive at the stationary shock one can solve also the time dependent problem $u_t + (f_1)_{x_1} + (f_2)_{x_2} = 0$. For a stable approximation one should use an implicit scheme like the Crank-Nicolson or backward Euler, while the spacial discretization is as above.

Let us outline the idea of the proof. (8) is rewritten in a two-step form

$$H(w(\tau), E_\tau w(\tau)) = 0, \tag{9}$$

where $w(\tau) = (v(\tau), E_\tau v(\tau), \ldots, E_\tau^{N-1}v(\tau), u_L(\tau))$ with $u_L(\tau) = u_L$ being a constant grid function. (Here we assume the set $J_0 = \{0, 1, 2, \ldots, N\}$.) The function of two vector variables $H(w_1, w_2)$ has a fixed (or zero) point

$$w_1 = w_2 = (u_0, \ldots, u_0). \tag{10}$$

In general, $\partial H/\partial w_1$ or $\partial H/\partial w_2$ may be singular so that one cannot solve the equation $H(w_1, w_2) = 0$ in terms of w_1, w_2. (This is the case of upwind difference schemes.) Nevertheless, at the fixed point (10) the characteristic polynomial $\det(\lambda \partial H/\partial w_1 + \partial H/\partial w_2)$ does not vanish identically. In such a situation we prove a Central Manifold Theorem for an Implicit Map. In our case this manifold has dimension $k + 1$ (or k after fixing u_L). On this manifold, after proper rescaling, (9) looks like

$$\frac{E_\tau \bar{y}(\tau) - \bar{y}(\tau)}{\varepsilon} = F(\bar{y}, \varepsilon), \tag{11}$$

where $\bar{y} = (y^{(1)}, \ldots, y^{(k)})$, $\varepsilon = q^{-1}|\lambda_0(u_L)/b(u_0)|^{1/k}$ and $F(\bar{y}, 0) = (y^{(2)}, \ldots, y^{(k)}, \frac{1}{2}(y^{(1)})^2 - \frac{1}{2})$. Clearly, (11) is a difference approximation with the step size ε of the differential system $d\bar{y}/d\tau = F(\bar{y}, 0)$. Thus the problem of weak discrete shocks is linked to the differential problem

$$d\tau/d\bar{y} = F(\bar{y},0), \quad \lim_{\tau \to -\infty} \bar{y}(\tau) = \bar{y}_L = \left((-1)',0,\ldots,0\right),$$
$$\lim_{\tau \to +\infty} \bar{y}(\tau) = \bar{y}_R = -\bar{y}_L. \tag{12}$$

(Recall that $k = 2l + 1$.) The last problem is equivalent to the scalar problem (6) for the component $y = y^{(1)}$. Kopell and Howard [1] proved the existence of an odd solution of (6) for $k = 3$, which was generalized and simplified by Mock [2]. Later, Conley [3] using his index proved the existence for any odd k. Recently, McCord [4] proved the uniqueness for $k = 3$. For $k = 3$ it is also known that the unstable manifold of $\bar{y}_L$ and the stable manifold of $\bar{y}_R$ intersect transversally along the solution of (12). Such a result is not proved for $k > 3$. The transversality implies easily the existence of a one-parameter family of solutions to the approximating difference problem (11). For $k > 3$, without transversality, we use Conley's index applied directly to the discrete problem (11). This way we prove the existence of at least one discrete shock solution, but not a one-parameter family of them.

The discrete shocks were studied first by Jennings [5], who proved their existence and stability for scalar monotone schemes. Majda and Ralston [6] obtained results similar to ours but only for first order schemes and not including upwind schemes. Osher [7] proved existence of stationary shocks in one space dimension for his scheme (3 grid points shock layer). There are no results about strong discrete shocks for systems. Mock [8] proved the existence of strong viscous shock profiles for some differential dissipative approximations of systems $u_t + f_x = 0$ which possess an entropy function.

Conclusion. For weak shocks the higher odd order schemes may be useful. The overshoots and undershoots are a matter of cosmetics since the asymptotic states u_L and u_R are correct. There is, however, no reason to expect that the global higher order accuracy is maintained in the cases of a nontrivial geometry, like the flow around a blunt body. For strong shocks the discrete profile might not exist. The shock layer is more narrow for stronger shocks and less dissipative schemes. In a generic situation, i.e. weak oblique shock, the first order upwind difference schemes are second order dissipative and smear the shock like any other 1st order scheme.

BIBLIOGRAPHY

1. Kopell, N. and L. N. Howard, *Bifurcations and trajectories joining critical points*, Adv. in Math. **18** (1975), 306–358.

2. Mock, M. S., *On fourth-order dissipation and single conservation laws*, Comm. Pure Appl. Math. **29** (1976), 383–388.

3. Conley, C., *Isolated invariant sets and the Morse index*, CBMS Regional Conf. Ser. in Math., no. 38, Amer. Math. Soc., Providence, R. I., 1978.

4. McCord, C. K., *Uniqueness of connecting orbits in the equation $y^{(3)} = y^2 - 1$*, preprint, Dept. of Math., Univ. of Wisconsin, Madison, 1983.

5. Jennings, G., *Discrete shocks*, Comm. Pure Appl. Math. **27** (1974), 25–37.

6. Majda, A. and J. Ralston, *Discrete shocks for systems of conservation laws*, Comm. Pure Appl. Math. **32** (1979), 445–482.

7. Osher, S. and F. Solomon, *Upwind difference schemes for hyperbolic systems of conservation laws*, Math. Comp. **38** (1982), 339–374.

8. Mock, M., *A topological degree for orbits connecting critical points of autonomous systems*, J. Differential Equations **38** (1980), 176–190.

DEPARTMENT OF MATHEMATICS, UNIVERSITY OF CALIFORNIA, LOS ANGELES, CALIFORNIA 90024

Lectures in Applied Mathematics
Volume **22**, 1985

Initial-Boundary Value Problems
for Incomplete Singular Perturbations
of Hyperbolic Systems

Daniel Michelson

Consider a flow of a slightly viscous perfect gas. This flow is governed by the compressible Navier-Stokes equations

$$\rho_t + v \cdot \nabla\rho + \rho \operatorname{div} v = 0,$$

$$v_t + v \cdot \nabla v + \frac{\nabla p}{\rho} + \left[-\nu\Delta v - \left(\mu + \frac{1}{3}\nu \right) \nabla \cdot \operatorname{div} v \right] = 0, \qquad (1)$$

$$T_t + v \cdot \nabla T + (\gamma - 1)T \operatorname{div} v + [-\chi\Delta T - \Phi] = 0.$$

Here ρ is the density, $p = R\rho T$, pressure, T, temperature, $v = (v_1, v_2, v_3)$ velocity,

$$\Phi = c_v^{-1} \left[\frac{1}{2}\nu \sum_{i,k} \left(\frac{\partial v_i}{\partial x_k} + \frac{\partial v_k}{\partial x_i} - \frac{2}{3}\delta_{i,k} \operatorname{div} v \right)^2 + \mu(\operatorname{div} v)^2 \right]$$

is the dissipation function, ν and μ are the normalized viscosity coefficients, χ is the thermal conductivity coefficient, $\gamma = c_p/c_v$ is the ratio of the specific heats and R is the gas constant. It is assumed that μ, ν and χ are proportional to a small parameter $\varepsilon > 0$. Thus the terms in the square brackets in (1) may be viewed as an incomplete elliptic perturbation of a

1980 *Mathematics Subject Classification*. Primary 35B25, 76N05, 35Q10; Secondary 35G15, 35G30, 35L50.

Key words and phrases. Singular perturbations, boundary layers, Navier-Stokes equations.

system of hyperbolic equations. In addition to equations (1), suitable initial and boundary conditions should be prescribed. The physical boundaries are usually formed by solid bodies with nonslip boundary conditions. However, in numerical computations it is necessary to limit the domain of computation by artificial boundaries. These boundaries are usually noncharacteristic and of inflow-outflow type. The number of boundary conditions which should be imposed for the incomplete parabolic system (1) is 5 for inflow boundary and 4 for outflow boundary. At the same time the unperturbed hyperbolic system requires 5 boundary conditions for supersonic inflow, 4 for subsonic inflow, none for supersonic outflow and one for subsonic outflow. The questions we are concerned with are the following:

(1) For which boundary conditions are the solutions of the perturbed problem (i.e. $\varepsilon > 0$) well defined in some time interval $0 \leqslant t \leqslant T_0$ and uniformly bounded there in an appropriate norm as $\varepsilon \to 0$?

(2) What is the limit of the above solutions and what are their asymptotics as $\varepsilon \to 0$?

It is well known that singular perturbation problems like the one considered here exhibit a boundary layer phenomenon. Let x_1 be the coordinate normal to the boundary, $x_- = (x_2, x_3)$—the tangential coordinates and denote the solution vector, (ρ, v, T) of (1) by $u(x, t, \varepsilon)$. Then one expects that

$$u(x, t, \varepsilon) = u^{(1)}(x, t, \varepsilon) + u^{(2)}(x_1/\varepsilon, x_-, t, \varepsilon), \qquad (2)$$

where $u^{(1)}$ and $u^{(2)}$ depend smoothly on their arguments. Here $u^{(1)}$ represents the smooth part of the solution while $u^{(2)}$ corresponds to the boundary layer and decays exponentially with respect to the x_1/ε variable. The aymptotic expansion of order N for $u^{(1)}$ and $u^{(2)}$ would be

$$u^{(1)}(x, t, \varepsilon) = \sum_{i=0}^{N} u_i^{(1)}(x, t)\varepsilon^i + O(\varepsilon^{N+1}),$$

$$u^{(2)}(x, t, \varepsilon) = \sum_{i=0}^{N} u_i^{(2)}\left(\frac{x_1}{\varepsilon}, x_-, t\right)\varepsilon^i + O(\varepsilon^{N+1}). \qquad (3)$$

We say that the boundary layer is of degree d if $u_i^{(2)} \equiv 0$ for $0 \leqslant i \leqslant d - 1$, and the boundary layer is weak if $u_0^{(2)} \equiv 0$. Suppose that the boundary conditions for the perturbed problem (1) are of the form

$$S(\{(\varepsilon D)^{\alpha} u\}_{\alpha \in \mathscr{A}}; x, t, \varepsilon) = 0, \qquad (4)$$

where $(\varepsilon D)^{\alpha} = \varepsilon^{|\alpha|} D_{x_1}^{\alpha_1} D_{x_2}^{\alpha_2} D_{x_3}^{\alpha_3} D_t^{\alpha_4}$, and $\alpha = (\alpha_1, \alpha_2, \alpha_3, \alpha_4)$ is a multi-index in some finite set $\mathscr{A} \subset \mathbf{Z}^4$. The weakness of the boundary layer means that there exists a smooth function $u(x, t)$ which is actually $u_0^{(1)}(x, t)$ in (3), and which satisfies the equations (1) and (4) to the zero order in ε. Take, for example, the boundary conditions suggested by Gustafsson and Sundström in [1]. In the case of subsonic inflow they are

$$v_1 + \frac{2\sqrt{\gamma RT}}{\gamma - 1} = g_1, \qquad v_2 = g_2, \qquad v_3 = g_3,$$

$$\left(\frac{4}{3}\nu + \mu\right)D_{x_1}v_1 - \chi\frac{R}{\gamma T}D_{x_1}T = 0,$$

$$(5)$$

where g_1, g_2, g_3 are prescribed functions. The first three boundary conditions in (5) together with the unperturbed system in (1) for $\varepsilon = 0$ form a well-posed hyperbolic problem. The solution $u_0^{(1)}(x, t)$ of this problem satisfies equations (1) and (5) to the zero order since ν, μ and χ are $\sim \varepsilon$, so that the boundary layer is weak. Instead of the last equation in (5), one may suggest a condition $T = g_4$. If this condition is consistent with the above solution $u_0^{(1)}(x, t)$, the boundary layer is still weak. In the case of a weak boundary layer, in order to obtain the functions $u_i^{(1)}$ and $u_i^{(2)}$ one should substitute the ansatz (2), (3) into equations (1) and (4), expand the resulting expressions in power series with respect to ε and compare equal powers of ε and equal scales x_1 or x_1/ε. Then $u_0^1(x, t)$ is a solution of the quasilinear hyperbolic system in (1) ($\varepsilon = 0$) with the boundary conditions $S(\{(0 \cdot D)^\alpha u\}; x, t, 0) = 0$ (it is assumed that such a solution does exist). The subsequent functions $u_i^{(1)}(x, t)$ are solutions of some nonhomogeneous linear hyperbolic initial-boundary value problems. The relevant equations are obtained by taking $\varepsilon = 0$ in (1) and (4) and linearizing around $u = u_0^{(1)}(x, t)$. On the other hand, the functions $u_i^{(2)}(x_1/\varepsilon, x_-, t)$ solve some boundary value problems in x_1 direction for a system of linear O.D.E.'s. The last ones are obtained by taking in (1) the terms with the x_1 derivative only and linearizing around $u_0^{(1)}(x, t)$ at $x_1 = 0$. Once the functions $u_i^{(1)}, u_i^{(2)}, 0 \leqslant i \leqslant N$, are defined, there is a task of justifying the expansion in (3). Namely, one should prove that the truncation error

$$u = u - \sum_{i=0}^{N}\left(u_i^{(1)} + u_i^{(2)}\right)\varepsilon^i,$$

$$(6)$$

is of the same order of magnitude in an appropriate norm as the next term $(u_{N+1}^{(1)} + u_{N+1}^{(2)})\varepsilon^{N+1}$ in the asymptotic expansion.

The main result which we announce here is the following. In the case of a *weak* boundary layer and under certain restrictions imposed on the boundary operator S, the solution u of the singularly perturbed problem (1), (4) is uniformly bounded as $\varepsilon \to 0$ and the asymptotic expansion in (3) is justified. The restrictions on S are of two types. First, S should satisfy a structural condition which, roughly speaking, assures that for a fixed ε (say $\varepsilon = 1$) the problem (1), (4) decouples, modulo lower order terms, into two boundary value problems: one for the "inviscid" component ρ and another for the "viscous" components v_1, v_2, v_3, T. Second, the linearized problem (1), (4), when frozen at each boundary point and solved in a half-plane, should satisfy so-called uniform Lopatinsky or Kreiss condition. This is essentially an algebraic condition, although not easily verifiable. The last condition is "almost" necessary for the uniform well-posedness of the problem (1), (4). Namely, we exclude hyperbolic boundary value problems which are ill posed in the sense of Kreiss and well posed in the sense of Hersh (see [2, p. 280]). As we have already mentioned, Gustafsson and Sundström in [1] suggested some dissipative boundary conditions for the equations in (1). These boundary conditions satisfy our two assumptions and imply a weak boundary layer (of degree 1). Thus, there is at least one set of inflow-outflow subsonic or supersonic boundary conditions for which the solution is uniformly bounded and the asymptotic expansion (3) holds.

In the case of a *strong* boundary layer the problem is completely open. However, for a linearized problem we have a positive result. Namely, let the coefficients which multiply the derivatives in (1) be smooth functions of x and t. Then, under the same restrictions on the (linear) boundary operator S, the solution of the corresponding initial-boundary value problem is uniformly bounded as $\varepsilon \to 0$ and the asymptotic expansion (2), (3) holds with $u_0^{(2)} \neq 0$.

The equations of slightly viscous gas provide, indeed, an important example of an incomplete singular perturbation of a hyperbolic system. The general problem is formulated as follows. There is a system of equations

$$\left[D_t + \sum_{j=1}^{k} A_j D_{x_j} + \varepsilon \left(0 \oplus \sum_{i,j=1}^{k} B_{ij} D_{x_i} D_{x_j} \right) \right] u = F, \qquad (7)$$

where A_j and B_{ij} depend on the arguments $u, D_x u_2, x, t, \varepsilon$, $F = F^{(1)}(u, x, t) + \varepsilon F^{(2)}(u, D_x u_2, x, t, \varepsilon)$, u_2 is the "viscous" component of u, and everywhere the dependence on the arguments is sufficiently smooth. The operator $P_1 = D_t + \sum_{j=1}^{k} A_j D_{x_j}$ is *strictly hyperbolic* and so is its inviscid left upper block $P_1^{(11)}$, while the operator $\Sigma B_{ij} D_{x_i} D_{x_j}$ is positive elliptic. The strict hyperbolicity of P_1 could be replaced by a weaker assumption that the symbol

$$Q(i\omega) = i \sum A_j \omega_j - \sum (0 \oplus B_{ij}) \omega_i \omega_j \qquad (8)$$

for small $|\omega|$ is *diagonalizable*. The diagonalization transformation T should depend (locally) smoothly on $u, \varepsilon D_x u_2, x, t, \varepsilon$ and locally analytically on $|\omega|$ and $\omega/|\omega|$. Although the perturbation is incomplete, it is assumed that the elliptic part has a *dissipative* effect on all the eigenvalues of $Q(i\omega)$. Precisely, the eigenvalues $\lambda(\omega)$ of $Q(i\omega)$ for real ω should satisfy an estimate

$$\operatorname{Re} \lambda(\omega) \geqslant \delta \frac{|\omega|^2}{\left(1 + |\omega|^2\right)} \tag{9}$$

with some positive constant δ. Note that the system in (1) satisfies all the above assumptions, with the only exception when $\gamma \cdot \operatorname{Pr} = 1$, where $\operatorname{Pr} = \nu/\chi$ is the Prandtl number. In this case the symbol $Q(i\omega)$ is not diagonalizable! (Curiously enough, for the air $\gamma = 1.4$, $\operatorname{Pr} \approx 0.7$ and $\gamma \cdot \operatorname{Pr} \approx 1$.) Equation (7) is solved in a domain $x \in \Omega$, $t \geqslant 0$, where Ω is compact and has a smooth boundary $\partial\Omega$. The boundary $\partial\Omega$ should be *noncharacteristic* for the hyperbolic operator P_1 and its block $P_1^{(11)}$. The boundary condition is of the same form as in (4)

$$S\left(\left\{(\varepsilon D)^\alpha u\right\}_{\alpha \in \mathscr{A}}, x, t, \varepsilon\right) = 0, \qquad x \in \partial\Omega, t > 0, \tag{10}$$

where $\mathscr{A}$ is some finite set of multi-indices $\alpha \in \mathbf{Z}^{k+1}$. At $t = 0$ an initial condition

$$u(x, 0) = f(x, \varepsilon), \qquad x \in \Omega, \tag{11}$$

is given. At the time-space corner $t = 0$, $x \in \partial\Omega$ this condition should be *compatible* with (10) to a sufficiently high order. Finally, S should satisfy the structural condition and the Kreiss condition we have already mentioned. Assume that the boundary layer is *weak*, i.e. there exists a smooth function $u_0^{(1)}(x, t)$ which satisfies equations (7), (10) and (11) to the zero order in ε. Then the result is: there exists a time interval $0 \leqslant t \leqslant T_0$ such that the solution $u(x, t)$ of (7), (10), (11) is uniformly bounded in a certain norm as $\varepsilon \to 0$ and the asymptotic expansion (2), (3) is justified. In the *linear* case the above result holds also for a *strong* boundary layer. The norm we are talking about is quite complicated. In particular, it includes the terms

$$\eta \|u\|_{m_1, m_2, m_3, \eta}^2 + \varepsilon \|D_x u_2\|_{m_1, m_2, m_3, \eta}^2. \tag{12}$$

Here

$$\|u\|_{m_1, m_2, m_3, \eta}^2 = \sum \left\|(\varepsilon D, \varepsilon\eta)^\alpha (\chi D_{x_1}, D_{x_-}, D_t, \eta)^\beta D_{x_1}^\gamma e^{-\eta t} u\right\|_{L_2(\Omega \times [0, T_0])}^2, \tag{13}$$

$\gamma \leqslant m_3$, $|\beta| + \gamma \leqslant m_2$, $|\alpha| + |\beta| + \gamma \leqslant m_3$, $\eta > 0$ is a parameter and $\chi = \chi(x_1)$ is a fixed nondecreasing function of x_1 such that $\chi(x_1) = x_1$ for $x_1 < \frac{1}{2}$ and $\chi(x_1) = 1$ for $x_1 > 1$. The norm in (13) is written for the domain $\Omega = \{ x \in R^k, x_1 \geqslant 0 \}$, and could be easily generalized for any compact domain with smooth boundary. In the case of a boundary layer of degree $d \geqslant 1$, the norm in (12) for $m_3 = d$ and sufficiently large m_1 and m_2 is uniformly bounded as $\varepsilon \to 0$.

BIBLIOGRAPHY

1. B. Gustafsson and A. Sundström, *Incompletely parabolic problems in fluid dynamics*, SIAM J. Appl. Math. **35** (1978) 343–357.

2. H.-O. Kreiss, *Initial boundary value problems for hyperbolic systems*, Comm. Pure Appl. Math. **23** (1970), 277–298.

DEPARTMENT OF MATHEMATICS, UNIVERSITY OF CALIFORNIA, LOS ANGELES, CALIFORNIA 90024

Lectures in Applied Mathematics
Volume **22**, 1985

A Mixed Finite Element Method
for 3D Navier-Stokes Equations

J. C. Nedelec

1. Introduction. We are concerned here with the Navier-Stokes system of equations which is

Find $\vec{u} = (u_1, u_2, u_3)$ and p such that in a domain Ω bounded regular,

$$\frac{\partial u_i}{\partial t} - \nu \Delta u_i + \sum_{j=1}^{3} u_j D_j u_i - \frac{\partial p}{\partial x_i} = f_i, \qquad 1 \leqslant i \leqslant 3,$$

$$\text{div } \vec{u} = 0 \quad \text{in } \Omega, \tag{1}$$

$$\vec{u}\big|_{\Gamma} = 0.$$

For 2D *problems of Stokes*, two types of finite element approximations are commonly used.

The more classical one is associated to the following variational formulations:

Find $(\vec{u}, p)$ in $(H_0^1(\Omega))^{2 \text{ or } 3} \times L^2(\Omega)/R$ such that

$$\int_{\Omega} \left(\frac{\partial \vec{u}}{\partial t} \cdot \vec{v} \right) dx + \nu \sum_{j=1}^{3} \int_{\Omega} \left(\frac{\partial \vec{u}}{\partial x_j} \cdot \frac{\partial \vec{v}}{\partial x_j} \right) dx + \int_{\Omega} p \, \text{div } \vec{v} \, dx = \int_{\Omega} (\vec{f} \cdot \vec{v}) \, dx$$

$$\forall \vec{v} \in \left(H_0^1(\Omega) \right)^{2 \text{ or } 3},$$

$$\int_{\Omega} \text{div } \vec{u} \cdot q \, dx = 0 \qquad \forall q \in L^2(\Omega)/R.$$

$$\tag{2}$$

1980 *Mathematics Subject Classification.* Primary 65N15, 65N30.

We then introduce two subspaces of $(H_0^1(\Omega))^{2 \text{ or } 3}$ and $L^2(\Omega)$ associated to finite element. In 2D, for instance, we can use a triangulation of Ω and two different finite elements. These two spaces, call them V_h and Q_h, are not independent. We need the *stability of the approximate system* that is *one* B. *condition* such that

$$\underset{\vec{v}_h \in V_h}{\text{Sup}} \frac{\int_\Omega \text{div} \, \vec{v}_h q_h \, dx}{\|\vec{v}_h\|_{H_0^1}} \geq \beta |q_h|_{L^2(\Omega)/R}. \tag{3}$$

With such a condition, the approximation is convergent and we can obtain error estimates. There is wide literature on this subject.

In *most practical examples of such finite element*, the divergence of $\vec{u}_h$ is different from zero:

$$\text{div} \, \vec{u}_h \neq 0 \quad \text{although it is small.} \tag{4}$$

The *approximate problem is nonconservative*. This can be seen by the Stokes theorem.

If the space V_h is such that $\text{div} \, \vec{u}_h \equiv 0$, then

$$\vec{u}_h = \begin{bmatrix} -\dfrac{\partial \varphi_h}{\partial x_2} \\[2mm] \dfrac{\partial \varphi_h}{\partial x_1} \end{bmatrix}$$

and φ_h is still a polynomial on each triangle so that φ_h is in a *space built with C^1-finite elements*. Such finite elements are of high degree: 3 at least (see, for instance, the works of R. Scott on this subject).

In 2D, there exists an easy way of building an approximate problem exactly conservative. It consists of using a (Ψ, ω) formulation. We get

$$\vec{u} = \vec{\text{curl}} \, \Psi = -\partial \Psi / \partial x_2 \quad \text{or} \quad \partial \Psi / \partial x_1, \quad \Psi \text{ is the stream function,}$$

$$\omega = \text{curl} \, \vec{u} = \partial u_2 / \partial x_1 - \partial u_1 / \partial x_2.$$

$$\tag{5}$$

Then a variational formulation for the Stokes problem is:
Find $\Psi \in H_0^1(\Omega)$ and $\omega \in H^1(\Omega)$ such that

$$\int_\Omega \left(\frac{\partial}{\partial t} \vec{\text{curl}} \, \Psi \cdot \vec{\text{curl}} \, \phi \right) dx + \nu \int_\Omega \left(\vec{\text{curl}} \, \omega \cdot \vec{\text{curl}} \, \phi \right) dx = \int_\Omega \left(\vec{f} \cdot \vec{\text{curl}} \, \phi \right) dx$$

$$\forall \phi \in H_0^1(\Omega);$$

$$\int_\Omega \omega \Pi \, dx - \int_\Omega \left(\vec{\text{curl}} \, \Psi \cdot \vec{\text{curl}} \, \Pi \right) dx = 0 \quad \forall \Pi \in H^1(\Omega). \tag{6}$$

We have a new system. The approximate system is obtained by building two subspaces Φ_h and Π_h of $H_0^1(\Omega)$ and $H^1(\Omega)$, respectively. For instance, on triangles we can use classical P_k finite elements for both spaces.

The equation (6) being a system, there is again a B. condition to satisfy. It is not true with the right norm, so that we lose one power of h in the error estimate. In conclusion, for this approximative problem, the speed $\vec{u}_h = \vec{\text{curl}}\,\Psi_h$ is exactly conservative. We must pay for that in the fact that the tangential components of u_h across an edge are no longer continuous.

2. A $\vec{\phi}, \vec{\omega}$ formulation in 3D. Now let Ω be a bounded domain in $\mathbf{R}^3$ of regular boundary Γ, *simply connected* and connected. We have

THEOREM 1. *Let* $\vec{u} \in (H_0^1(\Omega))^3$ *satisfying*

$$\text{div}\,\vec{u} = 0. \tag{7}$$

There exists a unique $\vec{\phi} \in (H^2(\Omega))^3$ *such that*

$$\vec{u} = \vec{\text{curl}}\,\vec{\phi}, \qquad \text{div}\,\vec{\phi} = 0, \qquad \vec{\phi} \wedge \vec{n}\big|_\Gamma = 0. \tag{8}$$

See, for instance, A. Bendali, J. A. Dominguez and S. Gallic [**10**]. We introduce now the vorticity

$$\vec{\omega} = \text{curl}\,\vec{u}. \tag{9}$$

Using the key formula,

$$\Delta\vec{u} = -\,\text{curl}\,\,\text{curl}\,\vec{u} + \text{grad}\,\text{div}\,\vec{u}, \tag{10}$$

the Navier-Stokes system can be written as

$$\frac{\partial}{\partial t}\,\vec{\text{curl}}\,\vec{\phi} + \nu\,\vec{\text{curl}}\,\vec{\omega} + \text{N.L.} - \text{grad}\,p = \vec{f},$$
$$\vec{\omega} - \vec{\text{curl}}\,\vec{\text{curl}}\,\vec{\phi} = 0, \qquad \text{div}\,\vec{\phi} = 0. \tag{11}$$

We must add boundary conditions

$$\vec{\phi} \wedge \vec{n}\big|_\Gamma = 0, \qquad \vec{\text{curl}}\,\vec{\phi} \wedge \vec{n}\big|_\Gamma = 0. \tag{12}$$

This second boundary condition is

$$\vec{u} \wedge \vec{n}\big|_\Gamma = 0. \tag{13}$$

The first boundary condition implies

$$\vec{u} \cdot \vec{n}|_{\Gamma} = 0 \qquad (14)$$

by the following simple geometric lemma.

LEMMA 1.

$$\vec{\phi} \wedge \vec{n}|_{\Gamma} \text{ implies } \overrightarrow{\text{curl}}\,\vec{\phi} \cdot \vec{n}|_{\Gamma} = 0. \qquad (15)$$

We can now give a variational formulation of the system (11). We introduce the following Hilbert spaces:

$$H(\text{curl}) = \left\{ \vec{\Pi} \in \left(L^2(\Omega)\right)^3, \ \overrightarrow{\text{curl}}\,\vec{\Pi} \in \left(L^2(\Omega)\right)^3 \right\},$$

$$\mathscr{X} = \left\{ \vec{\Psi} \in H\left(\overrightarrow{\text{curl}}\right), \text{div}(\vec{\Psi}) = 0, \vec{\Psi} \wedge \vec{n}|_{\Gamma} = 0 \right\}.$$

From Friedrichs it results that

LEMMA 2. $\mathscr{X} \subset (H^1(\Omega))^3$ and the norm $|\overrightarrow{\text{curl}}\,\vec{\Psi}|_{(L^2(\Omega))^3}$ is an equivalent norm to the H^1 one in $\mathscr{X}$.

But our approximation will not be in H^1. A variational formulation for (11) is

$$\int_{\Omega} \left(\frac{\partial}{\partial t} \overrightarrow{\text{curl}}\,\vec{\phi} \cdot \overrightarrow{\text{curl}}\,\vec{\Psi} \right) dx + \nu \int_{\Omega} \left(\overrightarrow{\text{curl}}\,\vec{\omega} \cdot \overrightarrow{\text{curl}}\,\vec{\Psi} \right) dx$$

$$+ \int_{\Omega} \left(\vec{u} \cdot \nabla u \cdot \overrightarrow{\text{curl}}\,\vec{\Psi} \right) dx = \int_{\Omega} \left(\vec{f} \cdot \overrightarrow{\text{curl}}\,\vec{\Psi} \right) dx \qquad \forall \vec{\Psi} \in \mathscr{X}, \quad (16)$$

$$\int_{\Omega} (\vec{\omega} \cdot \vec{\Pi}) - \int_{\Omega} \left(\overrightarrow{\text{curl}}\,\vec{\phi} \cdot \overrightarrow{\text{curl}}\,\vec{\Pi} \right) dx = 0 \qquad \forall \vec{\Pi} \in H(\overrightarrow{\text{curl}}).$$

We can remark that, in this formulation, the boundary conditions are split into the normal one, which is "in the space", and the tangential ones which are implicit.

3. Finite elements in $H(\overrightarrow{\text{curl}})$ and $H(\text{div})$.

3.1. It is "natural", for the variational problem (16), to use finite element subspaces of $H(\overrightarrow{\text{curl}})$.

We have introduced such finite elements especially for solving Maxwell equations in 3D. We shall describe such *elements on tetrahedrons*.

The space of polynomials. Let k be an integer. We denote by P_k the space of polynomials of degree k. We denote by $\tilde{P}_k$ the space of homogeneous polynomials of degree k. We introduce S^k:

$$S^k = \left\{ \vec{u} \in (\tilde{P}_k)^3, (\vec{r} \cdot \vec{u}) \equiv 0 \right\}, \qquad \vec{r} = [x_1 x_2 x_3].$$

We denote by R^k the spaces

$$R^k = (P_{k-1})^3 \oplus S^k,$$

$$k = 1. \quad S^1 \equiv \{\vec{u} = \vec{\beta} \wedge \vec{r}, \vec{\beta} \subset (P_0)^3\},$$

$$R^1 \equiv \{\vec{\alpha} + \vec{\beta} \wedge \vec{r}\}, \qquad \dim R^1 = 6.$$

$k = 2$. A base for S^2 is

$$\begin{bmatrix} x_3^2 \\ -x_1 x_2 \\ 0 \end{bmatrix}, \begin{bmatrix} x \\ 0 \\ -x_2 x_3 \end{bmatrix}, \begin{bmatrix} -x_1 x_2 \\ x_1^2 \\ 0 \end{bmatrix},$$

$$\begin{bmatrix} -x_1 x_2 \\ 0 \\ x_1^2 \end{bmatrix}, \begin{bmatrix} x_3^2 \\ 0 \\ -x_1 x_3 \end{bmatrix}, \begin{bmatrix} 0 \\ x_3^2 \\ -x_2 x_3 \end{bmatrix}, \begin{bmatrix} x_2 x_3 \\ -x_1 x_3 \\ 0 \end{bmatrix}, \begin{bmatrix} 0 \\ x_1 x_3 \\ -x_1 x_2 \end{bmatrix},$$

$\dim R^2 = 20$.

The degrees of freedom. There are 3 types of degrees of freedom:

(1) $$\int_a \vec{u} \cdot \vec{\tau} q \, ds \qquad \forall q \in P_{k-1},$$

a an edge of the tetrahedron K with tangent $\vec{\tau}$.

(2) $k \geqslant 2$. $$\int_f (\vec{u} \wedge \vec{n} \cdot \vec{q}) \, d\gamma \qquad \forall \vec{q} \in (P_{k-2})^2,$$

f one face of K of normal vector $\vec{n}$.

If $k \geqslant 3$,

(3) $$\int_K (\vec{u} \cdot \vec{q}) \, dx, \qquad \forall \vec{q} \in (P_{k-3})^3.$$

$$k = 1. \quad R = \{\vec{\alpha} + \vec{\beta} \wedge \vec{r}\},$$

$$\mathscr{A} = \left\{ \int_a \vec{u} \cdot \vec{\tau} \, ds \right\} \quad \text{or} \quad \vec{u}(M_i) \vec{\tau}, \qquad M_i = \text{middle of } a_i.$$

See Figure 1.

LEMMA 3. *These finite elements are unisolvant and conforming in* $H(\vec{\text{curl}})$. *The associated interpolation operator is such that*

$$|\vec{u} - \vec{\Pi}u|_{H(\text{curl})} \leqslant Ch^k |\vec{u}|_{(H^{k+1}(K))^3} \tag{17}$$

when the tetrahedron K satisfies

$h/\rho \leqslant C$, h maximum diameter, ρ radius of the inscribed sphere.

$$\tag{18}$$

3.2. *Finite elements in H(div).*

$$H(\text{div}) = \left\{ \vec{u} \in \left(L^2(\Omega) \right)^3, \, \text{div } \vec{u} \in L^2(\Omega) \right\}.$$

Space of polynomials.

$$\mathscr{D}^k = \left(\mathbf{P}_{k-1} \right)^3 \oplus \tilde{\mathbf{P}}_{k-1} \cdot \vec{r}.$$

$$k = 1. \quad \mathscr{D}^1 = \left\{ \vec{\alpha} + \beta \vec{r} \right\}, \qquad \dim \mathscr{D}^1 = 4.$$

$$k = 2, \quad \dim \mathscr{D}^2 = 15.$$

Degrees of freedom.

(1)
$$\int_f \vec{u} \cdot \vec{n} q \, d\gamma \qquad \forall q \in P_{k-1}.$$

If $k \geqslant 2$,

(2)
$$\int_K \vec{u} \cdot \vec{q} \, dx \qquad \forall \vec{q} \in \left(\mathbf{P}_{k-2} \right)^3.$$

$k = 1.$
See Figure 2.

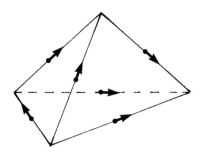

FIGURE 1.

$(\vec{u} \cdot \vec{n})(G_i)$
G: center of gravity
of the face f_i.

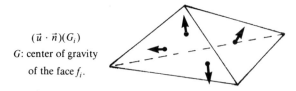

FIGURE 2.

LEMMA 4. *These finite elements are unisolvent and conforming in $H(\text{div})$.*
The corresponding interpolation operator satisfies

$$|\vec{u} - \Pi\vec{u}|_{H(\text{div})} \leqslant Ch^k|\vec{u}|_{(H^{k+1}(K))^3}, \tag{19}$$

if the tetrahedron satisfies

$$h/\rho \leqslant C \qquad (\text{div } u \equiv 0 \Rightarrow \text{div } \Pi u \equiv 0). \tag{20}$$

4. Spaces of approximation. Let $\mathcal{T}_h$ be a triangulation of the domain Ω by tetrahedrons.

Let us consider the following spaces of finite elements:

$$W_h = \left\{ \vec{\omega}_h \in H\left(\overrightarrow{\text{curl}}\right), \vec{\omega}_h|_K \in R^k \right\},$$

$$W_h^0 = \left\{ \vec{\omega}_h \in W_h, \vec{\omega}_h \wedge \vec{n}|_\Gamma = 0 \right\},$$

$$\Theta_h^0 = \left\{ \theta_h \in H_0^1(\Omega), \theta_h|_K \in \mathbf{P}_k \right\},$$

$$U_h = \left\{ \vec{u}_h \in H(\text{div}), \vec{u}_h|_K \in \mathcal{D}^k, \right.$$
$$\left. \text{div } \vec{u}_h = 0, \vec{u}_h \cdot \vec{n}|_\Gamma = 0 \right\}.$$

In fact, these $\vec{u}_h$ are of degrees P_{k-1}.

THEOREM 2. *The domain Ω being connected and simply connected, there exists for each $\vec{u}_h$ in U_h, a unique $\vec{\phi}_h \in W_h^0$ such that*

$$\overrightarrow{\text{curl}}\,\vec{\phi}_h = \vec{u}_h, \tag{21}$$

$$\int_\Omega \left(\vec{\phi}_h \cdot \text{grad } \theta_h \right) dx = 0 \qquad \forall \theta_h \in \Theta_h^0, \tag{22}$$

and we have

$$\left|\vec{\phi}_h\right|_{H(\overrightarrow{\text{curl}})} \leqslant C|\vec{u}_h|_{(L^2(\Omega))^3}. \tag{23}$$

This theorem is in some sense a discrete version of the following.
For every $\vec{u} \in H(\text{div})$ such that

$$\text{div } \vec{u} = 0, \qquad \vec{u} \cdot \vec{n} = 0,$$

there exists a unique $\vec{\phi} \in (H^1(\Omega))^3$ such that

$$\text{div } \vec{\phi} = 0, \quad \vec{\phi} \wedge n = 0 \quad \text{and} \quad \vec{u} = \overrightarrow{\text{curl}}\,\vec{\phi}.$$

We just replace the div $\vec{\phi} = 0$ by the orthogonality to the grad θ.

The proof of Theorem 2 is based on the Euler-Poincaré relation. Theorem 2 suggests the introduction of the space

$$\mathcal{X}_h = \left\{ \vec{\Psi}_h \in W_h^0, \int_\Omega \vec{\Psi}_h \cdot \text{grad } \theta_h\, dx = 0\, \forall \theta_h \in \Theta_h^0 \right\},$$

and now the approximation for the stationary Stokes equation is
Find $\vec{\phi}_h \in \mathscr{X}_n$, $\vec{\omega}_h \in W_h$ such that

$$\nu \int_\Omega \left(\overrightarrow{\mathrm{curl}}\, \vec{\omega}_h \cdot \overrightarrow{\mathrm{curl}}\, \vec{\Psi}_h \right) dx = \int_\Omega \left(\vec{f} \cdot \overrightarrow{\mathrm{curl}}\, \vec{\Psi}_h \right) dx \qquad \forall \vec{\Psi}_h \in \mathscr{X}_h,$$

$$\int_\Omega \left(\vec{\omega}_h \cdot \vec{\Pi}_h \right) dx - \int_\Omega \left(\overrightarrow{\mathrm{curl}}\, \vec{\phi}_h \cdot \overrightarrow{\mathrm{curl}}\, \vec{\Pi}_h \right) dx = 0 \qquad \forall \vec{\Pi}_h \in W_h. \tag{24}$$

For the stationary case, without the nonlinear term, we can prove some error estimates.

THEOREM 3. Let $\vec{\Psi}_h$, $\vec{\omega}_h$ be the solution of (24), and $\vec{\Psi}$, $\vec{\omega}$ the solution of the continuous corresponding problem; then we have

$$\left| \vec{\Psi} - \vec{\Psi}_h \right|_{H(\mathrm{curl})} + \left| \vec{\omega} - \vec{\omega}_h \right|_{(L^2(\Omega))^3} \leqslant C h^{k-1} \left(\|\vec{\Psi}\|_{(H^{k+1}(\Omega))^3} + \|\vec{\omega}\|_{(H^{k+1}(\Omega))^3} \right) \tag{25}$$

so that

$$\left| \vec{u} - \vec{u}_h \right|_{(L^2(\Omega))^3} \leqslant C h^{k-1}. \tag{26}$$

The proof is based on Theorem 2. Exactly as for the 2D problem, we lose one power of h compared to the expected error.

5. Nonstationary Navier-Stokes equations. In order to approximate the complete incompressible Navier-Stokes equations, we will use a two-step procedure. The first step is a Stokes one, the second a pure transport done by characteristics.
Step 1 (*Stokes*).

$$\frac{1}{\Delta t} \int_\Omega \left(\left(\overrightarrow{\mathrm{curl}}\, \vec{\phi}_h^{n+1/2} - \vec{u}_h^n \right) \cdot \overrightarrow{\mathrm{curl}}\, \vec{\Psi}_h \right) dx$$

$$+ \nu \int_\Omega \left(\overrightarrow{\mathrm{curl}}\, \vec{\omega}_h^{n+1/2} \cdot \overrightarrow{\mathrm{curl}}\, \vec{\Psi}_h \, dx \right) = \int_\Omega \left(\vec{f} \cdot \overrightarrow{\mathrm{curl}}\, \vec{\Psi}_h \right) dx \qquad \forall \vec{\Psi}_h \in \mathscr{X}_h,$$

$$\int_\Omega \left(\vec{\omega}_h^{n+1/2} \cdot \vec{\Pi}_h \right) dx - \int_\Omega \left(\overrightarrow{\mathrm{curl}}\, \vec{\phi}_h^{n+1/2} \cdot \overrightarrow{\mathrm{curl}}\, \vec{\Pi}_h \right) dx = 0 \qquad \forall \vec{\Pi}_h \in W_h. \tag{27}$$

This step is conservative. The result is $\vec{u}_h^{n+1/2} = \overrightarrow{\mathrm{curl}}\, \vec{\phi}_h^{n+1/2}$.
Step 2. We solve a discrete version of

$$\frac{\partial \vec{u}}{\partial t} + (\vec{u} \cdot \nabla) \vec{u} = 0. \tag{28}$$

We can use characteristics for the equation

$$\frac{\partial \vec{u}}{\partial t} + \left(\vec{u}_h^{n+1/2} \cdot \nabla \right) \vec{u} = 0. \tag{29}$$

In the simplest case, $k = 1$, the degree of freedom for $\vec{u}$ can be chosen as $\vec{u} \cdot \vec{n}$ *at the gravity center of the face*. The characteristic associated to $u_h^{n+1/2}$ and starting from this point is a *union of segments and is continuous*. For value of $\vec{u} \cdot \vec{n}$ we take the value of $\vec{u}_h^{n+1/2} \cdot \vec{n}$ at the bottom of this characteristic.

This step is not conservative in general.

This step provides the value of $\vec{u}_h^{n+1}$ in the space U_h. In general we will have

$$\operatorname{div} \vec{u}_h^{n+1} \neq 0.$$

An alternative expression for equation (28) is

$$\tag{30} \frac{\partial \vec{u}_i}{\partial t} + \operatorname{div}(\vec{u}_i \cdot \vec{u}) = 0,$$

which is in conservative form and can be treated in a weak form.

6. Some remarks on practical implementation. The conditions $\int_\Omega \vec{\phi}_h \cdot \operatorname{grad} \theta_h \, dx = 0$ are not very convenient to handle.

They can be replaced by other ones that also determine a unique $\vec{\phi}_h$ (not the same but such that $\vec{u}_h$ is unchanged!).

Consider the case $k = 1$.

The number of such conditions is the number of interior vertices of the mesh that is also the number of edges in a *maximal tree* in the *graph of edges*. To describe $\vec{\phi}_h$ we simply build such a tree and suppress all the unknowns on these edges.

With this procedure, the matrix of the Stokes step can be solved via a *conjugate gradient method* using an adequate *preconditioning matrix*. It is symmetric nondefinite.

What are the main aspects of this kind of splitting?

The *first step* is concerned with the diffusion process. It is linear and the matrix is the same for all time steps. This allows us to design a good preconditioner that leads to reasonable costs of computation (and storage).

There is no artificial viscosity added at this step so that it works for every ν. The mass is conserved.

The *second step* is more local. It can be viewed as solving a conservative law problem. That means that there is some viscosity added and, in general, the div $\vec{u}$ will not remain zero.

It can be viewed also as a weak transport of the vortex $\vec{\omega}$.

When using different boundary conditions $\vec{u}\big|_\Gamma = \vec{u}_0 \neq 0$, we must compute the data for $\vec{\Psi}_h \wedge \vec{n}$ by solving an auxiliary problem on the boundary.

Comments and conclusion. Our finite element method is the exact generalization to the three-dimensional case of the (ϕ, ω) method introduced in the two-dimensional case by R. Glowinski [4] and P. G. Ciarlet and P. A. Raviart [1].

We have only presented here a particular case of a general family of finite elements using polynomials of degree k. In the general case, we can prove error estimates in h^{k-1}.

There exist also two families of finite elements conforming in $H(\text{rot})$ and $H(\text{div})$, respectively, which are associated to cubes.

REFERENCES

1. Ciarlet, P. G. and P. A. Raviart, *A mixed finite element method for the biharmonic equation*, Mathematical Aspects in Finite Element Equation (C. de Boor, ed.), Academic Press, New York, 1974, pp. 125–145.

2. Fortin, M., *Résolution numérique des équations de Navier-Stokes par des éléments finis de type mixte*, 2nd Internat. Sympos. on Finite Element Methods in Flow Problems (S. Margherita Ligure, Italy, 1976).

3. Girault, V. and P. A. Raviart, *Finite element approximation of the Navier-Stokes equations*, Lecture Notes in Math., vol 749, Springer-Verlag, Berlin, Heidelberg and New York, 1979.

4. Glowinski, R., *Approximations externes par éléments finis d'ordre un et deux du problème de Dirichlet pour Δ^2*, Topics in Numerical Analysis. I, (J. J. H. Miller, ed.), Academic Press, London, 1973, pp. 123–171.

5. Lions, J. L., *Quelques méthodes de résolution des problèmes aux limites non linéaires*, Dunod, Paris, 1969.

6. Nedelec, J. C., *Mixed finite elements in $\mathbf{R}^3$*, Rapport Interne no. 49, Centre de Mathématiques Appliquées, École Polytechnique, Palaiseau, 1979.

7. Raviart, P. A., *Méthodes d'éléments finis pour les équations de Navier-Stokes*, Cours de l'École d'eté EDF-CEA-IRIA, 1979.

8. Temam, R., *Navier-Stokes equations*, North-Holland, Amsterdam, 1977.

9. Scholz, R., *A mixed method for 4th order problems using linear finite elements*, RAIRO Anal. Numér. **12** (1978), 85–90.

10. Bendali, A., J. M. Dominguez and S. Gallic, *A variational approach for the vector potential formulation of the Stokes and Navier-Stokes problems in three dimensional domains*, Rapport interne, Centre de Mathématiques Appliquées de l'École Polytechnique, Palaiseau; J. Math. Anal. Appl. (to appear).

ÉCOLE POLYTECHNIQUE, CENTRE DE MATHEMATIQUE APPLIQUÉES, 91128 PALAISEAU CEDEX, FRANCE

Lectures in Applied Mathematics
Volume **22**, 1985

Stability of the Flow Around a Circular Cylinder to Forced Disturbances

V. A. Patel

ABSTRACT. The stability of the two-dimensional, viscous imcompressible flow past a circular cylinder is studied by rotating the cylinder in the steady flow at Reynolds numbers 20, 40 and 60. The propagation of this disturbance is studied by dividing the stream and vorticity functions into steady and perturbed stream and vorticity functions. The steady stream and vorticity functions are expanded in the finite Fourier sine series. Steady semianalytic solutions are obtained for the symmetrical flow as a limit of time-dependent equations. The perturbed stream and vorticity functions are expanded in the finite Fourier sine and cosine series and then, along with the steady stream and vorticity functions expansions, substituted in the Navier-Stokes equations. This leads to a system of coupled parabolic partial differential equations in the coefficients of Fourier series which are solved numerically. The disturbance amplifies initially but does not grow.

1. Introduction. The problem of viscous incompressible flow past a circular cylinder is a classical one with an extensive literature. Coutanceau and Bouard [3] and Gerrard [7] determined that at $\mathrm{Re}(= 2aU/\nu$, where a is the radius of the cylinder, U is the free stream velocity and ν is the kinematic viscosity) $= 34$, the standing vortices become asymmetric, while Taneda [14] found that the wake begins to oscillate at $\mathrm{Re} = 30$. The measurements of the Strouhal number by Gerrard [7] in the Reynolds number range up to 350 are lower than those of Rosko [12, 13] and Tritton [16], and so Gerrard suggested that this quantity may depend upon the disturbance level in the flow. Taneda [15] has shown that close behind a cylinder, the wake can be made to oscillate by oscillating the cylinder for Re barely above unity. This leads to study theoretically the

---

1980 *Mathematics Subject Classification*. Primary 76E30, 35Q10.

stability of the wake behind a circular cylinder at Re = 20, 40 and 60. The basic idea used to study hydrodynamic stability (see e.g., Drazin and Reid [6, p. 8]) is used here. First, steady solutions are obtained for Re = 20, 40 and 60 by using finite series truncation method (see e.g., Van Dyke [18]). These solutions are disturbed, and the propagation of the disturbance in space and time is observed in much the same way as in the laboratory experiment. The development of the disturbance is determined by the Navier-Stokes equations for two-dimensional, viscous and incompressible flows. This investigation is therefore restricted by the assumption of two dimensionality and, therefore, valid as long as two-dimensional effects play the dominant role in the physical flow.

Desai [5] obtained analytic solutions by expanding the stream and vorticity functions in finite Fourier series for steady flow of a viscous incompressible fluid past a circular cylinder for Re = 1 to 40. Using this technique, Patel [10, 11] obtained Kármán vortex street behind a circular cylinder for Re = 100, 200 and 500, and semianalytic solutions for the impulsively started elliptic cylinder at various angles of attack. Underwood [17], Dennis and Shimshoni [4] and Nieuwstadt and Keller [8] obtained the semianalytic solutions for steady flows. Collins and Dennis [2] and Patel [9] obtained the semianalytic solutions for impulsively started symmetric flows. Steady semianalytic solutions are obtained as a limit of time-dependent equations. The disturbed stream and vorticity functions are expanded in Fourier series, and then substituted in Navier-Stokes equations along with steady stream and vorticity expansions. This leads to a system of coupled parabolic equations in the coefficients functions. This system is solved numerically. The cylinder is rotated in the steady flow, and it has been observed that the disturbance caused by this rotation amplifies initially but does not grow.

2. Basic equations and analysis. Consider a steady laminar flow of a viscous incompressible fluid around a circular cylinder of radius a. At time $t = 0$, the flow is disturbed. The equations which are assumed to govern the subsequent motions are the Navier-Stokes equations and the equation of continuity, which are coupled and to be solved subject to the boundary conditions of no slip on the surface of the cylinder and irrotational flow at infinity. The governing equation of motion in nondimensional form can be written as

$$\frac{\partial \zeta}{\partial t} + \frac{1}{r}\left(\frac{\partial \Psi}{\partial \theta} \frac{\partial \zeta}{\partial r} - \frac{\partial \Psi}{\partial r} \frac{\partial \zeta}{\partial \theta} \right) = \frac{2}{\text{Re}} \nabla^2 \zeta, \qquad (2.1)$$

where

$$\nabla^2 = \frac{\partial^2}{\partial r^2} + \frac{1}{r} \frac{\partial}{\partial r} + \frac{1}{r^2} \frac{\partial^2}{\partial \theta^2}. \qquad (2.2)$$

The stream function Ψ, the vorticity ζ and the velocity components u and v in the r and θ directions are connected by the relations

$$\zeta = -\nabla^2\Psi, \tag{2.3}$$

$$u = \frac{1}{r}\frac{\partial\Psi}{\partial\theta} \quad\text{and}\quad v = -\frac{\partial\Psi}{\partial r}. \tag{2.4}$$

The boundary conditions that are imposed upon the body surface are the usual impermeability and no slip condition for all time. As the distance from the cylinder becomes very large, it is assumed that the flow will approach more and more that of an irrotational flow. These boundary conditions can be written as:

For $t \geq 0$,

$$\Psi = \partial\Psi/\partial r = 0 \quad\text{at } r = 1 \tag{2.5}$$

and

$$\Psi = -(r - 1/r)\sin\theta \quad\text{and}\quad \zeta = 0 \qquad\text{as } r \to \infty. \tag{2.6}$$

It is convenient to work with the deviation ψ from the irrotational flow instead of Ψ so let us write

$$\psi(r,\theta,t) = \Psi(r,\theta,t) + (r - 1/r)\sin\theta. \tag{2.7}$$

The logarithmic transformation $\xi = \ln r$ of the radial coordinate is desirable, since the cells of the log-polar grid are smaller near the cylinder where the largest gradients occur in the flow. Equations (2.1) and (2.2), by using (2.7) and $\xi = \ln r$, reduce to

$$\frac{\partial\zeta}{\partial t} + e^{-2\xi}\left[\frac{\partial\psi}{\partial\theta}\frac{\partial\zeta}{\partial\xi} - \frac{\partial\psi}{\partial\xi}\frac{\partial\zeta}{\partial\theta}\right.$$

$$\left. + (e^\xi + e^{-\xi})\sin\theta\,\frac{\partial\zeta}{\partial\theta} - (e^\xi - e^{-\xi})\cos\theta\,\frac{\partial\zeta}{\partial\xi}\right] = \frac{2}{\mathrm{Re}}\nabla^2\zeta, \tag{2.8}$$

where

$$\nabla^2 = \frac{\partial^2}{\partial\xi^2} + \frac{\partial^2}{\partial\theta^2} \quad\text{and}\quad \zeta = -e^{-2\xi}\nabla^2\psi. \tag{2.9}$$

Substitution of (2.7) and $\xi = \ln r$ reduces (2.5) and (2.6) to:

For $t \geq 0$,

$$\psi = 0 \quad\text{and}\quad \frac{\partial\psi}{\partial\xi} = 2\sin\theta \qquad\text{at } \xi = 0, \tag{2.10}$$

and

$$\psi = \zeta = 0 \quad\text{as } \xi \to \infty. \tag{2.11}$$

Consider a disturbance stream function $\hat{\psi}(\xi, \theta, t)$ to be superimposed on the steady laminar flow $\bar{\psi}(\xi, \theta)$. This leads to

$$\psi(\xi, \theta, t) = \bar{\psi}(\xi, \theta) + \hat{\psi}(\xi, \theta, t), \tag{2.12}$$

and

$$\zeta(\xi, \theta, t) = \bar{\zeta}(\xi, \theta) + \hat{\zeta}(\xi, \theta, t), \tag{2.13}$$

where $\bar{\psi}$ and $\bar{\zeta}$ satisfy (2.8)–(2.11). In other words,

$$\frac{\partial \bar{\psi}}{\partial \theta} \frac{\partial \bar{\psi}}{\partial \xi} - \frac{\partial \bar{\psi}}{\partial \xi} \frac{\partial \bar{\zeta}}{\partial \theta} + (e^{\xi} + e^{-\xi})\sin\theta \frac{\partial \bar{\zeta}}{\partial \theta} - (e^{\xi} - e^{-\xi})\cos\theta \frac{\partial \bar{\zeta}}{\partial \xi}$$

$$= \frac{2e^{2\xi}}{\mathrm{Re}} \nabla^2 \bar{\zeta}, \tag{2.14}$$

where

$$\bar{\zeta} = -e^{-2\xi} \nabla^2 \bar{\psi}. \tag{2.15}$$

$$\bar{\psi} = 0 \quad \text{and} \quad \frac{\partial \bar{\psi}}{\partial \xi} = 2\sin\theta \qquad \text{at } \xi = 0, \tag{2.16}$$

and

$$\bar{\psi} = \bar{\zeta} = 0 \quad \text{as } \xi \to \infty. \tag{2.17}$$

Substitution of (2.12) and (2.13) in (2.8)–(2.11) along with (2.14)–(2.17) leads to

$$\frac{\partial \hat{\zeta}}{\partial t} + e^{-2\xi} \left[\frac{\partial \bar{\psi}}{\partial \theta} \frac{\partial \hat{\zeta}}{\partial \xi} - \frac{\partial \bar{\psi}}{\partial \xi} \frac{\partial \hat{\zeta}}{\partial \theta} + \frac{\partial \hat{\psi}}{\partial \theta} \frac{\partial \bar{\zeta}}{\partial \xi} - \frac{\partial \hat{\psi}}{\partial \xi} \frac{\partial \bar{\zeta}}{\partial \theta} + \frac{\partial \hat{\psi}}{\partial \theta} \frac{\partial \hat{\zeta}}{\partial \xi} - \frac{\partial \hat{\psi}}{\partial \xi} \frac{\partial \hat{\zeta}}{\partial \theta} \right.$$

$$\left. + (e^{\xi} + e^{-\xi})\sin\theta \frac{\partial \hat{\zeta}}{\partial \theta} - (e^{\xi} - e^{-\xi})\cos\theta \frac{\partial \hat{\zeta}}{\partial \xi} \right] = \frac{2}{\mathrm{Re}} \nabla^2 \hat{\zeta}, \tag{2.18}$$

where

$$\hat{\zeta} = -e^{-2\xi} \nabla^2 \hat{\psi}. \tag{2.19}$$

For $t \geqslant 0$,

$$\hat{\psi} = \partial \hat{\psi}/\partial \xi = 0 \quad \text{at } \xi = 0, \tag{2.20}$$

and

$$\hat{\psi} = \hat{\zeta} = 0 \quad \text{as } \xi \to \infty. \tag{2.21}$$

Since the steady flow is assumed to be symmetrical, the stream function $\bar{\psi}(\xi, \theta)$ can be expanded in a Fourier sine series

$$\bar{\psi}(\xi, \theta) = \sum_{n=1}^{\infty} f_n(\xi)\sin n\theta. \tag{2.22}$$

Substitution of (2.22) in (2.15) leads to

$$\bar{\zeta}(\xi, \theta) = - \sum_{n=1}^{\infty} F_n(\xi)\sin n\theta, \qquad (2.23)$$

where

$$F_n(\xi) = e^{-2\xi}\left(\frac{d^2 f_n}{d\xi^2} - n^2 f_n\right). \qquad (2.24)$$

Substitution of (2.22) in (2.16) and (2.17) leads to

$$f_n(0) = 0 \quad \text{for } n = 1, 2, \ldots,$$

$$\frac{\partial f_1}{\partial \xi}(0) = 2 \quad \text{and} \quad \frac{\partial f_m}{\partial \xi}(0) = 0 \qquad \text{for } m = 2, 3, \ldots, \qquad (2.25)$$

and

$$f_n(\xi_\infty) = 0 \quad \text{for } n = 1, 2, \ldots. \qquad (2.26)$$

Let us assume that $\hat{\psi}(\xi, \theta, t)$ is a reasonably well-behaved function throughout the domain, so that it can be represented in a Fourier series

$$\hat{\psi}(\xi, \theta, t) = a_0(\xi, t) + \sum_{n=1}^{\infty} (a_n(\xi, t)\cos n\theta + b_n(\xi, t)\sin n\theta),$$

$$(2.27)$$

where a_0, a_n and b_n are the functions to be determined.
    The substitution of (2.27) in (2.19) yields

$$\hat{\zeta}(\xi, \theta, t) = -A_0(\xi, t) - \sum_{n=1}^{\infty} (A_n(\xi, t)\cos n\theta + B_n(\xi, t)\sin n\theta),$$

$$(2.28)$$

where

$$A_0(\xi, t) = e^{-2\xi}\frac{\partial^2 a_0}{\partial \xi^2}, \qquad A_n(\xi, t) = e^{-2\xi}\left(\frac{\partial^2 a_n}{\partial \xi^2} - n^2 a_n\right)$$

and

$$B_n(\xi, t) = e^{-2\varepsilon}\left(\frac{\partial^2 b_n}{\partial \xi^2} - n^2 b_n\right). \qquad (2.29)$$

Substitution of (2.22), (2.23), (2.27) and (2.28) in (2.18) leads to an equation in which the terms of $\sin n\theta$ and $\cos n\theta$ are linearly independent, so the equation is satisfied only if the coefficients of $\sin n\theta$ and

$\cos n\theta$ are identically equal to zero. This leads to

$$\frac{\partial A_0}{\partial t} - \frac{e^{-2\xi}}{2}\left\{\frac{4}{Re}\frac{\partial^2 A_0}{\partial \xi^2} + (e^\xi - e^{-\xi})\frac{\partial A_1}{\partial \xi} + (e^\xi + e^{-\xi})A_1\right.$$

$$- \sum_{m=1}^{\infty} m\left[A_m\frac{df_m}{d\xi} - a_m\frac{dF_m}{d\xi} - F_m\frac{\partial a_m}{\partial \xi} b_m\frac{\partial A_m}{\partial \xi}\right.$$

$$\left.\left. - a_m\frac{\partial B_m}{\partial \xi} - B_m\frac{\partial a_m}{\partial \xi} + A_m\frac{\partial b_m}{\partial \xi} + f_m\frac{\partial A_m}{\partial \xi}\right]\right\} = 0,$$

$$(2.30)$$

$$\frac{\partial A_1}{\partial t} - \frac{e^{-2\xi}}{2}\left\{\frac{4}{Re}\left(\frac{\partial^2 A_1}{\partial \xi^2} - A_1\right)\right.$$

$$+ (e^\xi - e^{-\xi})\left(2\frac{\partial A_0}{\partial \xi} + \frac{\partial A_2}{\partial \xi}\right) + 2(e^\xi + e^{-\xi})A_2$$

$$+ \sum_{m=1}^{\infty} m\left[a_m\left(\frac{dF_{m+1}}{d\xi} + \frac{dF_{m-1}}{d\xi} + \frac{\partial B_{m+1}}{\partial \xi} + \frac{\partial B_{m-1}}{\partial \xi}\right)\right.$$

$$- A_m\left(\frac{df_{m+1}}{d\xi} + \frac{df_{m-1}}{d\xi} + \frac{\partial b_{m+1}}{\partial \xi} + \frac{\partial b_{m-1}}{\partial \xi}\right)$$

$$- b_m\left(\frac{\partial A_{m+1}}{\partial \xi} + \frac{\partial A_{m-1}}{\partial \xi}\right) + B_m\left(\frac{\partial a_{m+1}}{\partial \xi} + \frac{\partial a_{m-1}}{\partial \xi}\right)$$

$$\left.\left. - f_m\left(\frac{\partial A_{m+1}}{\partial \xi} + \frac{\partial A_{m-1}}{\partial \xi}\right) + F_m\left(\frac{\partial a_{m+1}}{\partial \xi} + \frac{\partial a_{m-1}}{\partial \xi}\right)\right]\right\} = 0,$$

$$(2.31)$$

$$\frac{\partial A_n}{\partial t} - \frac{e^{-2\xi}}{2}\left\{\frac{4}{Re}\left(\frac{\partial^2 A_n}{\partial \xi^2} - n^2 A_n\right) + (e^\xi - e^{-\xi})\left(\frac{\partial A_{n-1}}{\partial \xi} + \frac{\partial A_{n+1}}{\partial \xi}\right)\right.$$

$$+ (e^\xi + e^{-\xi})[(n + 1)A_{n+1} - (n - 1)A_{n-1}]$$

$$+ \sum_{m=1}^{\infty} m\left[a_m\left(\frac{dF_{m+n}}{d\xi} + \frac{dF_{m-n}}{d\xi} + \frac{\partial B_{m+n}}{\partial \xi} + \frac{\partial B_{m-n}}{\partial \xi}\right)\right.$$

$$- A_m\left(\frac{df_{m+n}}{d\xi} + \frac{df_{m-n}}{d\xi} + \frac{\partial b_{m+n}}{\partial \xi} + \frac{\partial b_{m-n}}{\partial \xi}\right)$$

$$- b_m\left(\frac{\partial A_{m+n}}{\partial \xi} + \frac{\partial A_{m-n}}{\partial \xi}\right) + B_m\left(\frac{\partial a_{m+n}}{\partial \xi} + \frac{\partial a_{m-n}}{\partial \xi}\right)$$

$$\left.\left. - f_m\left(\frac{\partial A_{m+n}}{\partial \xi} + \frac{\partial A_{m-n}}{\partial \xi}\right) + F_m\left(\frac{\partial a_{m+n}}{\partial \xi} + \frac{\partial a_{m-n}}{\partial \xi}\right)\right]\right\} = 0$$

$$\text{for } n = 2, 3, \ldots, \qquad (2.32)$$

and

$$\frac{\partial B_n}{\partial t} - \frac{e^{-2\xi}}{2}\left\{\frac{4}{Re}\left(\frac{\partial^2 B_n}{\partial \xi^2} - n^2 B_n\right) + (e^\xi - e^{-\xi})\left(\frac{\partial B_{n-1}}{\partial \xi} + \frac{\partial B_{n+1}}{\partial \xi}\right)\right.$$

$$+ (e^\xi + e^{-\xi})[(n+1)B_{n+1} - (n-1)B_{n-1}]$$

$$+ \sum_{m=1}^{\infty} m\left[-a_m\left(\frac{\partial A_{m+n}}{\partial \xi} - \frac{\partial A_{m-n}}{\partial \xi}\right) + A_m\left(\frac{\partial a_{m+n}}{\partial \xi} - \frac{\partial a_{m-n}}{\partial \xi}\right)\right.$$

$$- b_m\left(\frac{dF_{m+n}}{d\xi} - \frac{dF_{m-n}}{d\xi} + \frac{\partial B_{m+n}}{\partial \xi} - \frac{\partial B_{m-n}}{\partial \xi}\right)$$

$$+ B_m\left(\frac{df_{m+n}}{d\xi} - \frac{df_{m-n}}{d\xi} + \frac{\partial b_{m+n}}{\partial \xi} - \frac{\partial b_{m-n}}{\partial \xi}\right)$$

$$\left.\left. - f_m\left(\frac{\partial B_{m+n}}{\partial \xi} - \frac{\partial B_{m-n}}{\partial \xi}\right) + F_m\left(\frac{\partial b_{m+n}}{\partial \xi} - \frac{\partial b_{m-n}}{\partial \xi}\right)\right]\right\} = 0$$

$$\text{for } n = 1, 2, \ldots, \qquad (2.33)$$

where

$$B_{-n} = -B_n, \quad A_{-n} = A_n, \quad F_{-n} = -F_n, \quad f_{-n} = -f_n,$$

$$a_{-n} = a_n, \quad b_{-n} = -b_n \quad \text{and} \quad B_0 = 0. \qquad (2.34)$$

Substitution of (2.27) into (2.20) and (2.21) yields:
For $t \geqslant 0$,

$$b_n(0, t) = \frac{\partial b_n}{\partial \xi}(0, t) = 0 \qquad \text{for } n = 1, 2, \ldots,$$

$$\tag{2.35}$$

$$a_m(0, t) = \frac{\partial a_m}{\partial \xi}(0, t) = 0 \qquad \text{for } m = 0, 1, \ldots,$$

and

$$b_n(\xi_\infty, t) = 0 \qquad \text{for } n = 1, 2, \ldots,$$

$$a_m(\xi_\infty, t) = 0 \qquad \text{for } m = 0, 1, \ldots,. \tag{2.36}$$

3. Numerical solutions technique. Equations (2.29)–(2.34) along with boundary conditions (2.35) and (2.36) form a coupled nonlinear infinite system of partial differential equations to be solved for a_0, a_n, b_n, A_0, A_n and B_n, for given f_n and F_n. Once this system is solved, the flow field is known, since the stream and vorticity functions can be reconstructed from the assumed series expansions

$$\psi(\xi, \theta, t) = a_0(\xi, t) + \sum_{n=1}^{\infty} [a_n(\xi, t)\cos n\theta$$

$$\tag{3.1}$$

$$+ (f_n(\xi) + b_n(\xi, t))\sin n\theta],$$

and

$$\zeta(\xi, \theta, t) = -A_0(\xi, t) - \sum_{n=1}^{\infty} \left[A_n(\xi, t)\cos n\theta \right.$$

$$\left. + (F_n(\xi) + B_n(\xi, t))\sin n\theta \right]. \tag{3.2}$$

The infinite system given by (2.29)–(2.34) is made finite by truncating the stream and vorticity functions series in (3.1) and (3.2) at N. That is, we set

$$f_n(\xi) = F_n(\xi) = a_n(\xi, t) = A_n(\xi, t)$$

$$= b_n(\xi, t) = B_n(\xi, t) = 0 \quad \text{for } n > N. \tag{3.3}$$

For computation, the mesh constant $h = \Delta\xi = \pi/40$ was selected and infinity boundary conditions were imposed at large distance $r_\infty = e^{59\pi/40} = 102.909$. The system of equations (2.29)–(2.34) was integrated numerically by the following steps:

(1) At time $t = 0$, we selected rotational disturbance which is modeled on a physically realizable system. In other words $a_0(\xi, 0)$, $a_n(\xi, 0)$, $b_n(\xi, 0)$, $A_0(\xi, 0)$, $A_n(\xi, 0)$ and $B_n(\xi, 0)$ are known initially.

(2) The system (2.29) along with boundary conditions

$$a_m(0, t) = a_m(\xi_\infty, t) = 0 \quad \text{for } m = 0, 1, \ldots, N,$$

$$b_n(0, t) = b_n(\xi_\infty, t) = 0 \quad \text{for } n = 1, 2, \ldots, N \tag{3.4}$$

form the system of two-point boundary value problems for a_m and b_n. For a given disturbance, because of finite boundary, far-away flow will not be necessarily symmetric. Therefore, smooth boundary conditions $\partial\hat{\psi}(\xi_\infty, t)/\partial\xi = \partial\hat{\zeta}(\xi_\infty, t)/\partial\xi = 0$ are used. This leads to

$$\frac{\partial a_m}{\partial\xi}(\xi_\infty, t) = \frac{\partial A_m}{\partial\xi}(\xi_\infty, t) = 0 \quad \text{for } m = 0, 1, \ldots, N,$$

$$\frac{\partial b_n}{\partial\xi}(\xi_\infty, t) = \frac{\partial B_n}{\partial\xi}(\xi_\infty, t) = 0 \quad \text{for } n = 1, 2, \ldots, N. \tag{3.5}$$

$\partial a_m(\xi_\infty, t)/\partial\xi$ is replaced by

$$\frac{\left(-3a_m(\xi_\infty, t) + 4a_m(\xi_{\infty-1}, t) - a_m(\xi_{\infty-2}, t)\right)}{2h}$$

in (3.5). This gives

$$a_m(\xi_\infty, t) = \frac{\left(4a_m(\xi_{\infty-1}, t) - a_m(\xi_{\infty-2}, t)\right)}{3} \quad \text{for } m = 0, 1, \ldots, N. \tag{3.6}$$

Similarly

$$A_m(\xi_\infty, t) = \frac{(4A_m(\xi_{\infty-1}, t) - A_m(\xi_{\infty-2}, t))}{3} \quad \text{for } m = 0, 1, \ldots, N.$$

(3.7)

The left-hand sides of (2.29) contain A_m and B_n, whose values at all points in the field of computation are known at each time step. This is solved by a Hermitian method as described by Collatz [1, p. 166]. Let h be the spatial step of discretization and y_i, y_i' and y_i'' be the values of y, its first and second derivatives at node i. Then

$$y_{i-1}'' + 10y_i'' + y_{i+1}'' = \frac{12}{h^2}(y_{i-1} - 2y_i + y_{i+1}) + O(h^4). \quad (3.8)$$

The equations $\partial^2 a_m/\partial\xi^2 - m^2 a_m = e^{2\xi}A_m$ for $m = 0, 1, \ldots, N$ can be written by using (3.8) as

$$\left(1 - \frac{m^2 h^2}{12}\right) a_m(\xi_{i-1}, t)$$

$$-2\left(1 + \frac{5m^2 h^2}{12}\right) a_m(\xi_i, t) + \left(1 - \frac{m^2 h^2}{12}\right) a_m(\xi_{i+1}, t)$$

$$= \frac{h^2}{12}\left[e^{2\xi_{i-1}}A_m(\xi_{i-1}, t) + 10e^{2\xi_i}A_m(\xi_i, t) + e^{2\xi_{i+1}}A_m(\xi_{i+1}, t)\right]$$

$$\text{for } m = 0, 1, \ldots, N. \quad (3.9)$$

Equation (3.9) with boundary conditions (3.4) and (3.6) reduce to a tridiagonal system which is solved by a decomposition method. Thus a_m is computed at all points in the field of computation. Similarly b_n can be computed.

(3) The values of A_m and B_n on the surface of the cylinder can be explictly obtained from (2.29) by using (2.35). These are given (Patel [10]) by

$$A_m(0, t) = \frac{2a_m(1, t)}{h^2} \quad \text{for } m = 0, 1, \ldots, N,$$

$$B_n(0, t) = \frac{2b_n(1, t)}{h^2} \quad \text{for } n = 1, 2, \ldots, N.$$

(3.10)

(4) Time was increased by Δt; i.e., $t_{\text{new}} = t_{\text{old}} + \Delta t$.

(5) Equations (2.30)–(2.34) are solved by using the Crank-Nicolson method. By taking the average value at t and $t + \Delta t$, (2.32) can be

written as

$$A_n^{t+\Delta t} =$$

$$A_n^t + \Delta t \left\{ \frac{e^{-2\xi}}{\text{Re}} \left[\left(\frac{\partial^2 A_n}{\partial \xi^2} - n^2 A_n \right)^{t+\Delta t} + \left(\frac{\partial^2 A_n}{\partial \xi^2} - n^2 A_n \right)^t \right] \right.$$

$$+ \frac{e^{-2\xi}}{4} \left[(e^\xi - e^{-\xi}) \left\{ \left(\frac{\partial A_{n-1}}{\partial \xi} + \frac{\partial A_{n+1}}{\partial \xi} \right)^{t+\Delta t} + \left(\frac{\partial A_{n-1}}{\partial \xi} + \frac{\partial A_{n+1}}{\partial \xi} \right)^t \right\} \right.$$

$$+ (e^\xi + e^{-\xi}) \left\{ \left((n+1) A_{n+1} - (n-1) A_{n-1} \right)^{t+\Delta t} \right.$$

$$\left. + \left((n+1) A_{n+1} - (n-1) A_{n-1} \right)^t \right\}$$

$$+ \sum_{m=1}^{\infty} m \left\{ \left[a_m \left(\frac{dF_{m+n}}{d\xi} + \frac{dF_{m-n}}{d\xi} + \frac{\partial B_{m+n}}{\partial \xi} + \frac{\partial B_{m-n}}{\partial \xi} \right) \right.\right.$$

$$- A_m \left(\frac{df_{m+n}}{d\xi} + \frac{df_{m-n}}{d\xi} + \frac{\partial b_{m+n}}{\partial \xi} + \frac{\partial b_{m-n}}{\partial \xi} \right)$$

$$- b_m \left(\frac{\partial A_{m+n}}{\partial \xi} + \frac{\partial A_{m-n}}{\partial \xi} \right)$$

$$+ B_m \left(\frac{\partial a_{m+n}}{\partial \xi} + \frac{\partial a_{m-n}}{\partial \xi} \right)$$

$$- f_m \left(\frac{\partial A_{m+n}}{\partial \xi} + \frac{\partial A_{m-n}}{\partial \xi} \right) \tag{3.11}$$

$$\left. + F_m \left(\frac{\partial a_{m+n}}{\partial \xi} + \frac{\partial a_{m-n}}{\partial \xi} \right) \right]^{t+\Delta t}$$

$$+ \left[a_m \left(\frac{dF_{m+n}}{d\xi} + \frac{dF_{m-n}}{d\xi} + \frac{\partial B_{m+n}}{\partial \xi} + \frac{\partial B_{m-n}}{\partial \xi} \right) \right.$$

$$- A_m \left(\frac{df_{m+n}}{d\xi} + \frac{df_{m-n}}{d\xi} + \frac{\partial b_{m+n}}{\partial \xi} + \frac{\partial b_{m-n}}{\partial \xi} \right)$$

$$- b_m \left(\frac{\partial A_{m+n}}{\partial \xi} + \frac{\partial A_{m-n}}{\partial \xi} \right)$$

$$+ B_m \left(\frac{\partial a_{m+n}}{\partial \xi} + \frac{\partial a_{m-n}}{\partial \xi} \right)$$

$$- f_m \left(\frac{\partial A_{m+n}}{\partial \xi} + \frac{\partial A_{m-n}}{\partial \xi} \right)$$

$$\left. \left. \left. + F_m \left(\frac{\partial a_{m+n}}{\partial \xi} + \frac{\partial a_{m-n}}{\partial \xi} \right) \right]^t \right\} \right] \right\}$$

$$\text{for } n = 2, 3, \ldots, N.$$

The space derivatives are replaced by central differences and the system along with boundary conditions (3.7) and (3.10) reduces to a tridiagonal system which is solved by a decomposition method.

(6) Step (2) was repeated.

(7) Step (3) was repeated.

All of these steps were repeated at further times.

4. Results, discussion, and conclusion. Since the flow becomes asymmetrical according to Coutanceau and Bouard [3] and Gerrard [7] at $Re = 34$, so $Re = 20$, 40 and 60 were selected to study the stability. The steady solutions are obtained as a limit of the solutions of the time-dependent equations by finite series method as basically described by Patel [9]. Some of the results for steady solutions are given in Table 1 where C_D is the drag coefficient, L is the distance from the center of the cylinder to the tip of the eddy, N is the number of terms, and ϕ is the angular distance of the point at which the streamlines leave the surface of the cylinder from the front stagnation point.

TABLE I

Re	N	C_D	L	ϕ
20	20	2.0138	2.8153	136.22°
40	40	1.5041	5.3394	126.37°
60	40	1.3079	7.8496	120.29°

In order to study the stability, we have to disturb the steady flow. The disturbance must satisfy the boundary conditions (2.20) and (2.21), and must be compatible with the present scheme. We consider here the rotational disturbance. This can be achieved in the following way:

$$u = 0, \quad v = \omega, \qquad \text{at } \xi = 0. \tag{4.1}$$

Substitution of (4.1) along with $r = \ln \xi$ in (2.4) leads with the use of (2.7), (2.12), (2.22), (2.25) and (2.27) to

$$\partial a_0 / \partial \xi = -\omega \quad \text{at } \xi = 0. \tag{4.2}$$

Replacing the first derivative with forward difference in (4.2) along with $a_0(0, t) = 0$, one gets

$$a_0(1, 0) = -\omega h. \tag{4.3}$$

For numerical reasons, the following initial values were used:

$$A_0(J, 0) = 0.0000001, \quad A_1(J, 0) = .0001 \quad \text{for } J = 2, 3, \ldots, 32,$$

$$\tag{4.4}$$

and the rest values of A_m, B_n, b_n and a_m were zeros.

154   V. A PATEL

Because of the present numerical procedure any value of ω in (4.3) will give the same result provided (4.4) is used; however, for computation $\omega = .05$ was used.

Figure 1 represents the development of streamlines with $N = 20$ for Re = 20 at various times. The amplification of the disturbance is clearly visible and one vortex is distorted at $t = 2$ in Figure 1(a). Asymmetrical

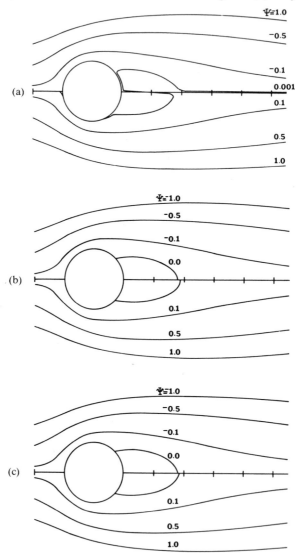

FIGURE 1. The development of streamlines with time for Re = 20 at (a) $t = 2$, (b) $t = 10$, (c) $t = 20$, (d) $t = 30$, (e) $t = 40$ and (f) $t = 50$.

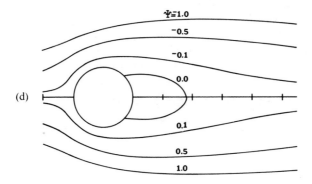

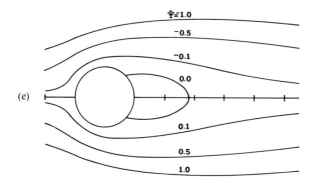

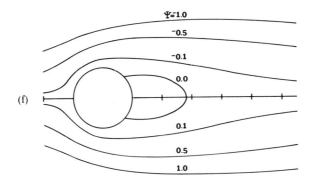

FIGURE 1. (continued)

vortices are seen at $t = 10$ and 20 in Figures 1(b) and 1(c), while in Figures 1(d), 1(e) and 1(f) symmetrical streamlines can be seen at $t = 30$, 40 and 50. In numerical calculations, the values of stream functions are not symmetrical even at $t = 50$. This can also be seen in Figure 2 which shows the development of the values of lift coefficient C_L with time for $\mathrm{Re} = 20$, 40 and 60. The values of the drag coefficient C_D and the lift coefficient C_L are computed following Patel [11] from

$$
\begin{aligned}
C_D &= \frac{4\pi}{\mathrm{Re}} \left(\frac{\partial F_1}{\partial \xi} + \frac{\partial B_1}{\partial \xi} - F_1 - B_1 \right)\Big|_{\xi=0}, \\
C_L &= \frac{4\pi}{\mathrm{Re}} \left(-2\frac{\partial A_0}{\partial \xi} + \frac{\partial A_1}{\partial \xi} - A_1 \right)\Big|_{\xi=0}.
\end{aligned}
\tag{4.5}
$$

The values of C_D remain the same up to the six digits for all considered Reynolds numbers, therefore, are not shown in the figure.

For $\mathrm{Re} = 40$, Figure 3 shows the development of the streamlines with $N = 40$ at various times. At $t = 1$, in Figure 3(a) the disturbance is amplified and one vortex is perturbed. At $t = 10$, 20, 30 and 40 vortices are not symmetrical in Figures 3(b), 3(c), 3(d) and 3(e). The disturbance is decaying. The values of lift coefficient C_L are decreasing in Figure 2 at $\mathrm{Re} = 40$. At $t = 50$, the symmetrical streamlines are seen in Figure 3(e) but numerical values are not symmetrical.

The development of the streamlines with $N = 40$ at various times are shown in Figure 4 for $\mathrm{Re} = 60$. In Figure 4(a), the amplified disturbance

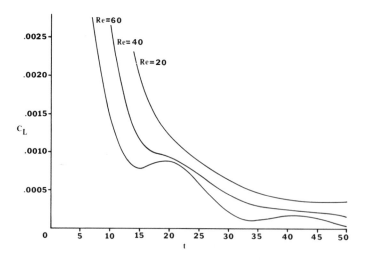

FIGURE 2. The development of C_L with time for $\mathrm{Re} = 20$, 40 and 60.

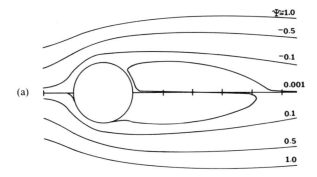

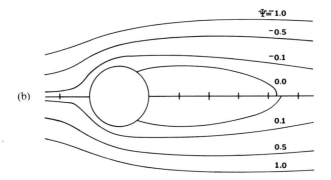

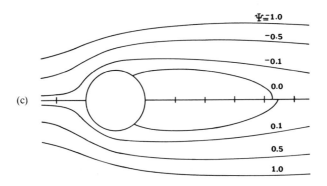

FIGURE 3. The development of streamlines with time for Re = 40 at (a) $t = 1$, (b) $t = 10$, (c) $t = 20$, (d) $t = 30$, (e) $t = 40$ and (f) $t = 50$.

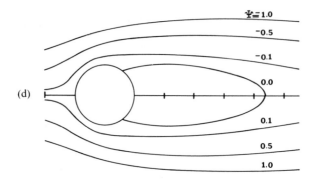

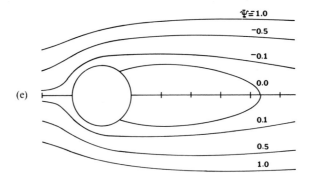

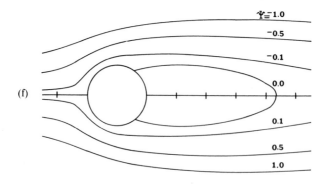

FIGURE 3. (continued)

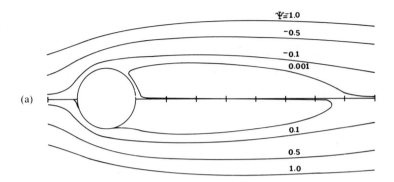

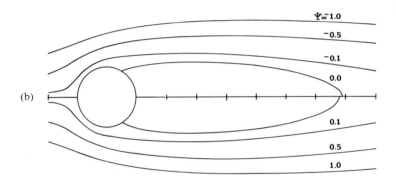

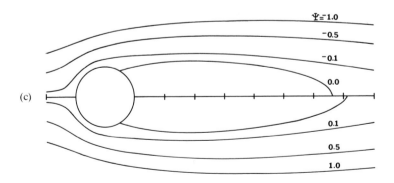

FIGURE 4. The development of streamlines with time for Re = 60 at (a) $t = 1$, (b) $t = 10$, (c) $t = 20$, (d) $t = 30$, (e) $t = 40$ and (f) $t = 50$.

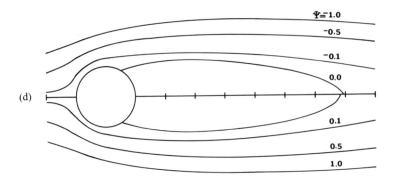

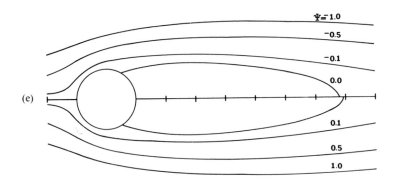

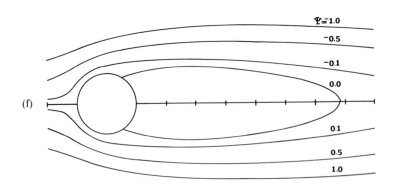

FIGURE 4. (continued)

is seen at $t = 1$. In Figures 4(b), 4(c), 4(d) and 4(e), vorticies oscillate at $t = 10, 20, 30$ and 40. The values of lift coefficient C_L also oscillate as seen in Figure 2 at Re = 60. At $t = 50$, Figure 4(f) shows the asymmetric vortices.

Initial disturbances do not grow at Re = 20, 40 and 60 but they do amplify initially. Time-dependent disturbances are under investigation. The author is thankful to Mr. R. J. Wilson of the Computer Center for his kind help with our Cyber 170-720 system.

REFERENCES

1. L. Collatz, *The numerical treatment of differential equations*, 3rd ed., Springer-Verlag, New York, 1966.

2. W. M. Collins and S. C. R. Dennis, *Flow past an impulsively started circular cylinder*, J. Fluid Mech. **60** (1973), 105–127.

3. M. Coutanceau annd R. Bouard, *Experimental determination of the main features of the viscous flow in the wake of a circular cylinder in uniform translation. Part 1. Steady flow*, J. Fluid Mech. **79** (1977), 231–256.

4. S. C. R. Dennis and M. Shimshoni, *The steady flow of a viscous fluid past a circular cylinder*, Aero. Res. Counc. Current Papers, No. 797, 1965.

5. S. M. Desai, *On a linear substructure underlying the Navier-Stokes equations as applied to the problem of flow of a viscous incompressible fluid past a circular cylinder*, Ph.D. Thesis, University of California, Berkeley, 1965.

6. P. G. Drazin and W. H. Reid, *Hydrodynamic stability*, Cambridge Univ. Press, Cambridge, 1981.

7. J. H. Gerrard, *The wakes of cylindrical bluff bodies at low Reynolds number*, Philos. Trans. Roy. Soc. London Ser. A **288** (1978), 351–381.

8. F. Nieuwstadt and H. B. Keller, *Viscous flow past circular cylinders*, Comput. & Fluids **1** (1973), 59–71.

9. V. A. Patel, *Time-dependent solutions of the viscous incompressible flow past a circular cylinder by the method of series truncation*, Comput. & Fluids **4** (1976), 13–27.

10. _____, *Kármán vortex street behind a circular cylinder by the series truncation method*, J. Comput. Phys. **28** (1978), 14–42.

11. _____, *Flow around the impulsively started elliptic cylinder at various angles of attack*, Comput. & Fluids **9** (1981), 435–462.

12. A. Rosko, *On the development of turbulent wakes from vortex streets*, N.A.C.A. Report 1191, 1954.

13. _____, *On the drag and shedding frequency of two-dimensional bluff bodies*, N.A.C.A. Technical note 3169, 1954.

14. S. Taneda, *Experimental investigation of the wakes behind cylinders and plates at low Reynolds numbers*, J. Phys. Soc. Japan **11** (1956), 302–307.

15. _____, *The stability of two-dimensional laminar wakes at low Reynolds number*, J. Phys. Soc. Japan **18** (1963), 288–296.

16. D. J. Tritton, *Experiments on the flow past a circular cylinder at low Reynolds numbers*, J. Fluid Mech. **6** (1959), 547–567.

17. R. L. Underwood, *Calculation of incompressible flow past a circular cylinder at moderate Reynolds numbers*, J. Fluid Mech. **37** (1969), 95–114.

18. M. Van Dyke, *Notes on computer-extended series in mechanics*, Prepared for the 20th Summer Res. Inst. of the Austral. Math. Soc., Canberra, 14 Jan.–8 Feb., 1980.

DEPARTMENT OF MATHEMATICS, HUMBOLDT STATE UNIVERSITY, ARCATA, CALIFORNIA 95521

Lectures in Applied Mathematics
Volume **22**, 1985

Some Contributions to the Modelling of Discontinuous Flows

Philip L. Roe

1. Introduction, including the first-order scheme. In 1980 I presented to the Seventh International Conference on Numerical Methods in Fluid Dynamics a paper [1] in which I advanced the rather large claim that "almost any" good numerical technique for solving the linear scalar wave equation ($u_t + au_x = 0$) could be converted, by a fairly simple mechanism, into an equally good numerical technique for solving quasi-linear systems of conservation laws ($\mathbf{u}_t + \mathbf{F(u)}_x = 0$). The particular attraction of that mechanism lay in its ability to generalise the class of asymmetric, or "upwinded" finite-difference schemes, in a simple and natural way. The present meeting seems a good opportunity to review and revise that claim, in the light of $2\frac{1}{2}$ years of additional experience. Some of that experience is my own, and some has been provided by close colleagues. Some has come from contributions made by workers who were able to develop the ideas of [1] in new ways, and some again from people who have quite independently arrived at similar viewpoints and developed related techniques. It must be said at once, however, that all I feel able to do is to place upon these developments a personal and provisional interpretation. The best combinations of these ideas are not yet clear, nor is their final competitive position relative to other classes of methods. Perhaps Bram van Leer's "Ultimate Conservative Differencing Scheme" remains something of a mirage, a dream city hovering in the distant haze. We cannot yet determine how far away it is, even though many of us feel

---

1980 *Mathematics Subject Classification.* Primary 35A40, 35L65, 76L05.

that its general outline and its directions are becoming steadily more apparent.

I will begin by discussing a fairly general form of the flux difference splitting concept, applicable to one-dimensional conservation laws

$$\frac{\partial}{\partial t}\mathbf{u} + \frac{\partial}{\partial x}\mathbf{F}(\mathbf{u}), \tag{1}$$

where a discrete solution is sought consisting of states $\mathbf{u}_i^n$ equal to the mean value of $\mathbf{u}$, at time level n, inside the ith member of a set of cells. The distinctive feature of flux difference splitting, in my view, is that before advancing to time level $(n + 1)$, we attempt to explain the events taking place at time level n by analysing the "fluctuations" [2], defined as

$$\boldsymbol{\phi}_{i+1/2}^n = \mathbf{F}_{i+1}^n - \mathbf{F}_i^n. \tag{2}$$

For hyperbolic systems, it is natural to "explain" these fluctuations as due to the passage of various waves from one cell to its neighbour (Figure 1). We can view such an explanation as a kind of subgrid model. Evidently the limited data does not make it sensible to seek very elaborate or detailed explanations, nor does it seem very meaningful to ask if one explanation is more accurate than another. What does seem to make sense is to ask for plausible (or, perhaps more accurately, probable) hypotheses which allow us to choose appropriate numerical methods with the least expenditure of computational effort.

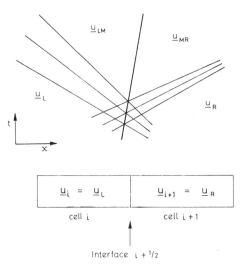

$$F_R - F_L = (F_R - F_{MR}) + (F_{MR} - F_{LM}) + (F_{LM} - F_L)$$

FIGURE 1. Schematic illustration of flux-difference splitting.

My own guess at the present time is that the amount of detail which is worth including in the explanation depends on the complexity of the equations being solved, but that for the unsteady Euler equations a useful explanation comprises the following elements. For each of the three waves $W^{(1)}$, $W^{(2)}$, $W^{(3)}$, we require:

(a) The change of state across $W^{(k)}$.
(b) The average speed of $W^{(k)}$.
(c) The width, or spreading rate, of $W^{(k)}$.

This information can be conveniently arranged into a table;

TABLE I

	$W^{(1)}$	$W^{(2)}$	$W^{(3)}$
Change in F_1 ($= \rho u$)	$\Delta \rho u^{(1)}$	$\Delta \rho u^{(2)}$	$\Delta \rho u^{(3)}$
Change in F_2 ($= p + \rho u^2$)	$(u - a)\Delta \rho u^{(1)}$	$u\Delta \rho u^{(2)}$	$(u + a)\Delta \rho u^{(3)}$
Change in F_3 ($= \rho u H$)	$(H - ua)\Delta \rho u^{(1)}$	$\frac{1}{2}u^2\Delta \rho u^{(2)}$	$(H + ua)\Delta \rho u^{(3)}$
Average wavespeed	$(u - a)$	u	$(u + a)$
Spreading rate	$\delta^{(1)}$	$\delta^{(2)}$	$\delta^{(3)}$

The table has been filled in here with the values which would be obtained analytically, in the case of an infinitesimal fluctuation. In that case we would have

$$\Delta \rho u^{(1)} = \frac{(u - a)}{2a^2}(\Delta p - \rho a \Delta u),$$

$$\Delta \rho u^{(2)} = u\left(\Delta \rho - \frac{1}{a^2}\Delta p\right), \tag{3}$$

$$\Delta \rho u^{(3)} = \frac{(u + a)}{2a^2}(\Delta p + \rho a \Delta u).$$

Here Δ is the difference between state i and $i + 1$, and all other symbols have their usual meaning, except for δ, which will be explained later.

I am going to describe some methods for solving time-marching problems; at each time-step the first action is to draw up a tabular explanation of each fluctuation, somewhat as in Table I. Such tables permit almost automatic extension of numerical methods from the scalar case to the system case. Think of the tables obtained by considering pairs of consecutive cells as being stacked in order like pages in a book. The information digested by a bookworm which selects a particular row (j) and column (k) is the information needed to find the effects of the kth wave on the jth conserved variable. We have split the problem up into nine nearly independent scalar subproblems. This gives us a great flexibility in designing numerical methods. I believe that in principle a different scalar algorithm could be used on each subproblem. Central differencing for one, upwind differencing for another, an implict method for a third, and so on. Such a combination could certainly be made conservative, and if it proved stable would work after a fashion.

However, it would seem rather pointless to indulge in these variations of technique without good motives. I cannot think of a sensible reason for applying different methods to the subproblems associated with a given wave; one feels that different effects of the same wave should be treated as similarly as possible. On the other hand, there do seem to be reasons for treating each wave independently of the others. The simplest is that algorithms which are stable applied to left-moving waves may not be stable for right-moving waves, and vice-versa. Those algorithms which are stable for both wave directions are suboptimal in many respects compared with those which reflect the appropriate bias. A subtler reason is that different waves may reflect different kinds of nonlinear behaviour. In the case of the Euler equations the sound waves with speed $u \pm a$ are what are called genuinely nonlinear waves, and the associated characteristics may spread to form expansion fans or converge into shock waves. Numerical methods which respond appropriately can be employed once the different types of behaviour have been recognised. The particle paths, whose speed is u, cannot of course collide, and so the associated contact discontinuity has an essentially linear behaviour, which may benefit from being treated by a different algorithm. These refinements were hinted at in [1], but have been developed more recently, and will be described below in more detail.

Now although I feel fairly sure that Table I contains about the right amount of information, I am not quite so sure exactly what form that information should take, or how it should be obtained. Taking the second point first, we can solve, exactly if we wish, the Riemann problem for $\mathbf{u}_L = \mathbf{u}_i$, $\mathbf{u}_R = \mathbf{u}_{i+1}$, and find the states LM, MR which appear between the waves in Figure 1. Then at least the first three rows of the table can be filled in, since the change in ρu due to the first wave is $(\rho u)_{LM} - (\rho u)_L$, and so on. There are by now several "approximate Riemann solvers" available. My own technique [3] is applicable to ideal compressible flow with an adiabatic equation of state, and its utility has been endorsed in a series of papers by Harten and coworkers [4–6]. A rather similar method has been proposed by Pandolfi [7]. Extensions to more general equations of state have been given by Lombard et al. [8, 9] and by Colella [10]. One could also adapt the method of Osher and Solomon [11] and of Osher [12] to the present purpose. The cited references describe the advantages claimed for each method.

There is also the other issue, of just what information the table should carry. In [13], it is argued that Table II is a better choice.

TABLE II

	$W^{(1)}$	$W^{(2)}$	$W^{(3)}$
Change in u_1 ($= \rho$)	$\Delta\rho^{(1)}$	$\Delta\rho^{(2)}$	$\Delta\rho^{(3)}$
Change in u_2 ($= \rho u$)	$(u-a)\Delta\rho^{(1)}$	$u\Delta\rho^{(2)}$	$(u+a)\Delta\rho^{(3)}$
Change in u_3 ($= e$)	$(H-ua)\Delta\rho^{(1)}$	$\frac{1}{2}u^2\Delta\rho^{(2)}$	$(H+ua)\Delta\rho^{(3)}$
Courant number	$(u-a)\Delta t/\Delta x$	$u\Delta t/\Delta x$	$(u+a)\Delta t/\Delta x$
Spreading rate	$\delta^{(1)}$	$\delta^{(2)}$	$\delta^{(3)}$

This table has also been filled in using the estimates based on small fluctuations; here

$$\Delta\rho^{(1)} = \frac{1}{2a^2}(\Delta p - \rho a \Delta u),$$

$$\Delta\rho^{(2)} = \Delta\rho - \frac{1}{a^2}\Delta p, \tag{4}$$

$$\Delta\rho^{(3)} = \frac{1}{2a^2}(\Delta p + \rho a \Delta u).$$

Table I is a rather literal interpretation of the phrase flux difference splitting; Table II will also describe a splitting of the flux difference if we choose to estimate the wave speeds in such a way that

$$\sum_k \nu^{(k)}\Delta\mathbf{u}^{(k)} = \frac{\Delta t}{\Delta x}\Delta\mathbf{F}, \tag{5}$$

where $\nu^{(k)}$ is the Courant number associated with the kth wave. As written, (5) could be taken as a set of three equations to be solved for $\nu^{(1)}$, $\nu^{(2)}$, $\nu^{(3)}$. Alternatively, we may be fortunate enough to find expressions for the wave speeds which make (5), together with (4) and Table II, into an identity. In [13], we find that this does indeed happen (uniquely) for the square root averaging procedure introduced in [3]. This observation makes the following algorithm a conservative differencing scheme.

(a) For every cell we store the conserved variables ρ, ρu, e.

(b) We also store, or possibly compute, u, H, p.

(c) For every pair of consecutive cells (L, R), we compute $\Delta\rho$, Δu, Δp, and the following mean values $\tilde{\rho}$, $\tilde{u}$, $\tilde{H}$, $\tilde{a}$, where

$$\tilde{\rho} = R\rho_L,$$
$$\tilde{u} = (Ru_R + U_L)/(R+1), \tag{6}$$
$$\tilde{H} = (RH_R + H_L)/(R+1),$$
$$\tilde{a}^2 = (\gamma - 1)(\tilde{H} - \tfrac{1}{2}\tilde{u}^2),$$

in which

$$R^2 = \rho_R/\rho_L. \tag{7}$$

(d) Using these differences and mean values, we fill in Table II, using equations (4).

(e) Let the change in the jth conserved variable due to the kth wave be $\Delta u_j^{(k)}$. Subtract $\nu^{(k)}\Delta u_j^{(k)}$ from the jth conserved variable in the right-hand cell if $\nu^{(k)} > 0$, or the left-hand cell if $\nu^{(k)} < 0$.

Algebraically, this algorithm is completely equivalent to the first-order algorithm described in [1, 3], but in the reformulation the actual flux differences do not explicitly appear. We work with a combination of conserved quantities and primitive variables. This produces slightly faster and neater coding, and gives rise to algebra which is somewhat closer to the more conventional theory of characteristics. In these developments, we have partly followed the ideas of Lombard et al. [8, 9].

Now although the above algorithm, in its original formulation, has been described as Roe's Method, we do not (and never have) recommended it for practical use. Its good features are

g(i) Under many circumstances it produces numerical results [3, 4, 14] almost identical to those obtained from the classical scheme of Godunov [15] but rather more cheaply and rather more systematically. It provides, therefore, an attractive alternative foundation on which to construct more elaborate schemes.

g(ii) It treats slowly-moving shock waves very well indeed. The limiting case of a stationary shock wave is always represented with at most intermediate point [2, 16].

g(iii) Although boundary conditions will not be discussed again in this paper, it is worth noting that if no special action is taken at boundaries, a very satisfactory radiation condition arises by default. Outgoing waves are found simply to disappear without reflection.

The bad features can be listed also.

b(i) First-order accuracy is not sufficient for realistic problems.

b(ii) Rapidly-moving shock waves are badly smeared.

b(iii) Contact discontinuities (unless stationary) diffuse without limit, as time increases, usually like $t^{1/2}$.

b(iv) Nonphysical solutions are obtained if the problem features an expansion fan which crosses a sonic point.

The remainder of this paper will be devoted to overcoming these defects, using only the information already generated in constructing Table II.

The first three faults are closely related. Until a few years ago it was thought to be difficult, if not impossible, to find any type of algorithm which achieved second-order accuracy without introducing unwanted oscillations near shock waves. We do not attempt to give here a history of this topic, which involves a complex interplay of practical and theoretical considerations. A majority opinion at the present time would probably be that stationary or slowly-moving shocks can be captured adequately by what may perhaps already be described as classical methods; either by adjusting the domain of dependence to suit subsonic or supersonic conditions [17, 18], or by devising nonlinear artificial viscosities, of which Jameson's is a recent and ingenious example [19, 20]. However, rapidly-moving shocks, across which no characteristic speed changes sign, completely defeat the first of these techniques, and seem so far to have largely eluded the second. More modern techniques all involve some kind of feedback mechanism which adapts the algorithm to the evolving solution. Best known and most widely used are those of van Leer [21–25], but see also Harten [4], Roe [26, 27], Roe and Baines [28]. It is this last work which will be followed, and enlarged upon, in §§2 and 3.

In those sections, we shall in fact be relying on the mechanism of [1] to extend to systems of equations results which are easily obtainable in the scalar case. This process of extension as tested by numerical experiments has been remarkably trouble-free and straightforward. In §4, we shall turn to the question of entropy-violating solutions. In this respect, the mechanism clearly cannot be complete, since it starts with the linear equations, where questions of physical correctness do not arise. Instead, the mechanism has to be started one stage nearer to its goal, with the nonlinear scalar equations. A modification of the first-order method will be demonstrated which gives physically correct solutions for a class of scalar conservation laws, and the incorporation of this modification, through the mechanism of [1], has given satisfactory results for the Euler equations.

2. The "monotone" second-order scheme. I put the word "monotone" in quotation marks to denote a certain deliberate vagueness about the properties I wish to claim. What we really want is a scheme which combines high accuracy with guaranteed exclusion of nonphysical oscillations or entropy-violating solutions. The scheme must apply to systems of equations in many dimensions. We are still very far away from being able to design such algorithms. Concepts such as monotonicity, control of total variation, and monotonicity preservation, all discussed in [4], appear to be only interesting preliminary attempts to obtain some mathematical grip on the problem. We follow a pragmatic approach, along the lines indicated by the theory, but often straying beyond the bounds of rigorous proof.

We begin by modifying the first-order algorithm given in §1 in a very simple way so as to achieve second-order accuracy in smooth regions of the flow. For each pair of cells, i, $(i + 1)$, define

$$\sigma^{(k)}_{1+1/2} = \operatorname{sgn} \nu^{(k)}_{i+1/2}. \tag{8}$$

Step (E) of the algorithm was to subtract, from the value of $\mathbf{u}$ in the cell $(i + \frac{1}{2} + \frac{1}{2}\sigma^{(k)}_{i+1/2})$, the quantity $\nu^{(k)}_{i+1/2}\Delta\mathbf{u}^{(k)}_{i+1/2}$. To develop a descriptive vocabulary, call this cell the target, and the subtracted vector quantity the signal, associated with the kth wave. Let us replace (E) with

(E′) Divide the signal into two unequal fractions, with weights $\frac{1}{2}(1 - \nu)$, $\frac{1}{2}(1 + \nu)$ (for typographical clarity, we have dropped the subscripts), to be subtracted, respectively, from cells i, $i + 1$.

Replacing (E) by (E′) produces an algorithm closely related to that of Lax and Wendroff, to which it reduces if $\mathbf{F} = A\mathbf{u}$ with A a constant matrix. Although the original, physically motivated, target continues to receive the larger share of the signal, the propagation of a "nonphysical" signal counter to the true direction creates typical Lax–Wendroff wiggles in regions of rapidly changing data. So far, therefore, we have failed to profit from the effort which went into explaining the fluctuations. We need to make the scheme a little less symmetrical. To begin with, we merely recast it. We delete (E′) and restore (E), but follow it by

(F) Define the smaller, nonphysical, fraction of the signal due to the kth wave to be

$$\Delta^*\mathbf{u}^{(k)} = \tfrac{1}{2}\big(1 - |\nu^{(k)}|\big)\nu^{(k)}\Delta\mathbf{u}^{(k)}. \tag{9}$$

(G) Subtract this from $\mathbf{u}_{i+1/2-\sigma/2}$; add it to $\mathbf{u}_{i+1/2+\sigma/2}$.

Evidently, (E)–(G) are, together, equivalent to (E′). The key to making the scheme "monotone" is to limit the values of the nonphysical transfers $\Delta^*\mathbf{u}^{(k)}$ in such a way as to rule out overshoots. One way to set about doing this [28] is to compare $\Delta^*\mathbf{u}$ in each pair of cells i, $i + 1$, with $\Delta^*_u\mathbf{u}$ in the "upwind" pair of cells i_u, $i_u + 1$, where $i_u = i - \sigma^{(k)}_{i+1/2} = i \pm 1$. Now in smooth regions of the flow, each of these vectors is of order Δx, and they differ by $O(\Delta x^2)$. In the transfer stage (G), $\Delta^*\mathbf{u}^{(k)}$ can be replaced by $\Delta^*_u\mathbf{u}^{(k)}$ and there is no loss of formal accuracy. Nor is there any loss of conservation (since *any* transfers are conservative). If we do make this replacement, we create an algorithm in which each wave system is handled in a manner which is second-order accurate and wholly upwinded, in the sense that if $\sigma^{(k)}$ is of one sign throughout some region, then the ith cell is influenced by the behaviour of the kth wave in the cells i, $i - \sigma^{(k)}$, $i - 2\sigma^{(k)}$. Properties of this scheme, in the constant-coefficient case, are given in, for example, [17, 29]. However, this upwinded scheme is not monotone either. A fairly good scheme, but still not monotone in any sense, results from transferring the arithmetic mean of Δ^* and Δ^*_u.

To explain the scheme which we have successfully used on systems of equations, it is probably easiest to revert to the linear scalar case

$$u_t + au_x = 0 \tag{10}$$

for which the Courant number ν is constant, and the nonphysical component of each signal is

$$b_{i+1/2} = \tfrac{1}{2}\nu(1 - |\nu|)(u_{i+1} - u_i). \tag{11}$$

Then at stage (G) we propose to transfer an amount

$$B_{i+1/2} = B(b_{i+1/2}, b_{i+1/2-\sigma}), \tag{12}$$

where $B(b_1, b_2)$ is some kind of averaging function which can be constructed to meet various desirable criteria. A simple example is the function which returns the value 0 if its arguments are of opposite sign, and otherwise returns the value of the argument with least modulus. Better examples than this can be devised and will form the substance of §3; meanwhile we note that if a B-function has been designed we can, in the spirit of [1], apply it to systems like this.

Let $b_{i+1/2}^{(k)}$, $b_{i+1/2-\sigma}^{(k)}$ be suitable norms for $\Delta_u^* \mathbf{u}^{(k)}$ in the cell $i + \tfrac{1}{2}$ and its upwind neighbour. Then in stage (G), replace the transferred quantity by

$$\Delta_u^* \mathbf{u}_{i+1/2}^{(k)} \frac{B\left(b_{i+1/2}^{(k)}, b_{i+1/2-\sigma}^{(k)}\right)}{b_{i+1/2}^{(k)}}. \tag{13}$$

A good practical choice for the norm appears to be the change in density due to the kth wave. The expression (13) is then easily evaluated using the information contained in Table II. Various B-functions can be supplied to the program as subroutines. It is to the design of these B-functions that we now turn.

3. Designing the B-functions. In this section some desirable properties of B-functions will be developed. We will analyse only the constant-coefficient case (10)–(12), relying on (13) to treat nonlinear systems. We will also make the simplifying assumption that B is a homogeneous function of b_1, b_2; i.e. for all μ,

$$B(\mu b_1, \mu b_2) = \mu B(b_1, b_2), \tag{14}$$

which expresses the desirable property that the algorithm remains invariant if the data is multiplied by a constant. Without real loss of generality we assume $a > 0$. The change of u in the ith cell is due to (i) the signal generated by the cell pair i, $i - 1$, (ii) the nonphysical transfer between $i + 1$ and i, and (iii) the nonphysical transfer between i and

$i - 1$. We write these three effects in terms of the fluctuations ϕ (equation (2)), and $k = \Delta t / \Delta x$:

$$u_i^{n+1} - u_i^n = -k\phi_{i-1/2} - \tfrac{1}{2}k(1 - \nu)B\big(\phi_{i+1/2}, \phi_{i-1/2}\big)$$
$$+ \tfrac{1}{2}k(1 - \nu)B\big(\phi_{i-1/2}, \phi_{i-3/2}\big). \tag{15}$$

It is apparent that any B-function will give rise to an algorithm which exactly convects constant or linear data. The case of quadratic data will give rise to fluctuations in arithmetic progression, say $\phi - \phi'$, ϕ, $\phi + \phi'$. It is easily shown that the LHS of (15) should then be

$$-k\nu\phi - \tfrac{1}{2}k(1 - \nu)\phi', \tag{16}$$

so that quadratic data will be convected exactly if the B-function is such that, for any ϕ, ϕ',

$$B(\phi + \phi', \phi) - B(\phi, \phi - \phi') = \phi'. \tag{17a}$$

This functional equation is satisfied by

$$B(b_1, b_2) = \lambda b_1 + (1 - \lambda)b_2, \tag{17b}$$

that is, by any linear averaging function.

For cubic data to be convected exactly, the value which the LHS of (15) should have can be found by cubic interpolation and put into the form

$$-k\phi_{i-1/2} + \tfrac{k}{6}(1 - \nu)\big[(1 + \nu)\phi_{i-3/2} + (2 - \nu)\phi_{i-1/2}$$
$$- (1 + \nu)\phi_{i-1/2} - (2 - \nu)\phi_{i+1/2}\big]. \tag{18}$$

This demonstrates that to achieve exact convection of cubics, B must be the particular linear averaging function

$$B(b_1, b_2) = \frac{(2 - \nu)}{3}b_1 + \frac{(1 + \nu)}{3}b_2. \tag{19}$$

To achieve third-order accuracy in the usual sense, it is sufficient to satisfy (19) within a tolerance of order $(b_1 - b_2)^2$, and to achieve second-order accuracy it is sufficient to satisfy (17b) within a tolerance of order $(b_1 - b_2)$.

However, if B is actually chosen to be an exactly linear average, then the whole scheme is linear, and it is well known (eg. [15, 22, 27, 4]) that linear schemes cannot be both monotone and accurate. The use of nonlinear schemes on a linear problem such as (10) does seem a little strange at first, but in fact many other problems are homogeneous rather than linear, for example, finding the maximum value or dominant frequency of a function. In the present context, it is not yet entirely clear what nonlinear behaviour we should be aiming for, but one condition we can impose is that the increment $(u_i^{n+1} - u_i^n)$ should be bounded between zero and $(u_{i-1}^n - u_i^n)$. This condition [22, 27] will certainly avoid overshoots, and if it is taken in conjunction with (15), leads to two constraints upon the B-functions. These are

$$\sup[\,B/b_2\,] - \inf[\,B/b_1\,] \leqslant 2/\nu,$$
$$\sup[\,B/b_1\,] - \inf[\,B/b_2\,] \leqslant 2/(1 - \nu). \tag{20}$$

These bounds give a clear indication that the ratio between the values of B and its smallest argument should not be permitted to become too large, but the possibility of more delicate constraints needs further investigation. We will outline below some new work which may well be applicable to other methods which make use of nonlinear limiting devices [4, 25, 30, 31].

We have already considered the case of fluctuations in arithmetical progression; let us consider the case of geometrical progression. In other words, we have exponentially varying data, with

$$\phi_{i-1/2}/\phi_{i-3/2} = \phi_{i+1/2}/\phi_{i-1/2} = r. \tag{21}$$

In this case, the total increment at the ith grid point can be written, using (14), (15) and (21), as

$$u_i^{n+1} - u_i^n = -k\phi_{i-1/2}\left[1 - \frac{1}{2}(1 - \nu)(1 - r)\frac{B(r, 1)}{r}\right]. \tag{22}$$

If the same exponential data is to be advected exactly by the algorithm, we can easily show that

$$u_i^{n+1} - u_i^n = \frac{r}{a(r - 1)}\phi_{i-1/2}(r^{-\nu} - 1). \tag{23}$$

Upon equating the expressions on the RHS of (22), (23) we obtain a rather striking result. If

$$B_\nu(r, 1) = \frac{2r}{\nu(1 - \nu)}\frac{\nu(1 - r) + r(1 - r^{-\nu})}{(1 - r)^2}, \tag{24}$$

then the resulting algorithm will yield the exact solution, after any number of time-steps, for any exponentially varying initial data. Note that (24) implies a more general result. Because of the homogeneous property (14), we have actually defined B as a function of two variables,

$$B(b_1, b_2) = b_2 B(b_1/b_2, 1), \tag{25}$$

but we can most easily illustrate these functions by their intersections with the plane $b_2 = 1$. Note also that (24) defines an algorithm in terms of two quantities of fundamental significance; the Courant number ν, and the parameter r which is often used to describe the smoothness of the data [22, 27].

It must be admitted that the algorithm defined by (24) would not be very practical. It would be an expensive computation because of the term $r^{-\nu}$. But (24), despite its rather shapeless appearance, has some intriguing mathematical properties. It is rather easier to analyse (24) after noting that it can be expressed as a hypergeometric function

$$B_\nu(r,1) = r\,{}_2F_1(1, 1+\nu, 3, 1-r), \tag{26a}$$

which converges for $0 < r < 2$. An alternative representation is

$$B_\nu(r,1) = {}_2F_1(1, 2-\nu, 3, (r-1)/r), \tag{26b}$$

which converges for $\frac{1}{2} < r < \infty$. Note that none of the expressions (24), (26a), (26b) succeed in defining B if r is negative. In fact, most work on limiters up to the present has made some assumption equivalent to setting $B = 0$ if $r < 0$. This causes some loss of accuracy near sharp extrema in the data, which will be discussed later. Meanwhile, we shall adopt this assumption, and note that the monotonicity constraints (20) are thereby simplified to

$$\sup_r B(r,1) \leqslant 2/\nu, \tag{27a}$$

$$\sup_r \tfrac{1}{r}B(r,1) \leqslant 2/(1-\nu). \tag{27b}$$

The function $B_\nu(r,1)$ is plotted against r in Figure 2, for various values of ν in the physically meaningful range $-1 \leqslant \nu \leqslant 2$. (For this range of ν the CFL rule does not prohibit stability.) In all cases B is a monotone increasing function of r, with monotone decreasing first derivative. The following properties of $B_\nu(r,1)$ can be readily established, either from first principles or by appeal to well-known properties of the hypergeometric function [32].

(i) The slope at the origin is

$$\frac{\partial}{\partial r}B_\nu(0,1) = \frac{2}{1-\nu},$$

which is the maximum value that it can have if (27b) is to be satisfied.

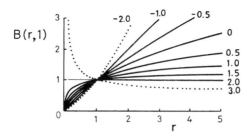

FIGURE 2. The nine special cases of Hyperbee.

(ii) There is a horizontal asymptote whose level is

$$\lim_{r \to \infty} B_\nu(r,1) = 2/\nu,$$

which is the maximum value that it can have if (27a) is to be satisfied.

(iii) $B(1,1) = 1$, and the slope there is

$$\frac{\partial}{\partial r} B_\nu(1,1) = \frac{2 - \nu}{3},$$

which is the slope required for third-order accuracy.

Clearly, the B-function defined by requiring the exact advection of exponential data has very distinguished properties. Anyone who is amused by such things may like to observe that the hypergeometric representation passes Kummer's test for the existence of a quadratic transformation [32, p. 560] if ν has one of the nine values $\{-2, -1, -\frac{1}{2}, 0, \frac{1}{2}, 1, 1\frac{1}{2}, 2, 3\}$, and then find the corresponding closed form expressions for B_ν. Probably the only useful outcome of that exercise, however, is to note the very simple result

$$B_{1/2}(r,1) = \frac{4r}{\left(1 + r^{1/2}\right)^2}. \tag{28}$$

This result is useful, because $\nu = \frac{1}{2}$ represents a worst case, in the sense that if $\nu = 0$ or 1 the first-order scheme is exact for any data, and the second-order terms vanish because of the factor $\nu(1 - \nu)$. The function defined by (28) is tangential, at $r = 1$, to the limiting function introduced by van Leer [22] which in our notation is

$$B(r,1) = \frac{r + |r|}{1 + r} \tag{29}$$

but (29) does not have any special properties with regard to exponential data. Subsequently van Leer has experimented with many different limiters [33, 34].

Analysis of the special function (24) (hereafter described as Hyperbee, for obvious reasons) began in an attempt to explain some remarkable experimental results, which will be described briefly below, and in more detail in [35]. Here we will give the explanation first, and the results later. Suppose we choose some B-function other than Hyperbee, and plot it on the same graph. At any point where the curves intersect, the value of r indicates a special case of exponential data, which would be advected exactly by the algorithm employing that B-function. All legitimate B-functions intersect Hyperbee at $r = 1$. This intersection corresponds to the rather trivial assertion that all our algorithms advect exactly all linear data, so call it the trivial intersection. Most algorithms in use make no nontrivial intersections, for example the basic choices (see Figure 3)

$$B(r,1) = r \quad \text{(Lax–Wendroff)},$$
$$B(r,1) = 1 \quad \text{(Warming–Beam)},$$

and also the simplest switching algorithm

$$B(r,1) = \min(r,1). \tag{30}$$

The unswitched scheme

$$B(r,1) = \tfrac{1}{2}(1 + r) \quad \text{(Fromm)}$$

generates one nontrivial intersection, unless $\nu = \tfrac{1}{2}$ when the intersections coincide. The unswitched third-order scheme corresponding to (19) always generates two coincident intersections. A B-function which always generates two widely separated intersections is the following:

$$B(r,1) = \max(r,1) \quad \text{if } \tfrac{1}{2} < r < 2,$$
$$B(r,1) = 2\min(r,1) \quad \text{for other positive values of } r. \tag{31}$$

This function, which we have nicknamed Superbee, was devised empirically whilst following up a suggestion by Woodward and Colella [31]

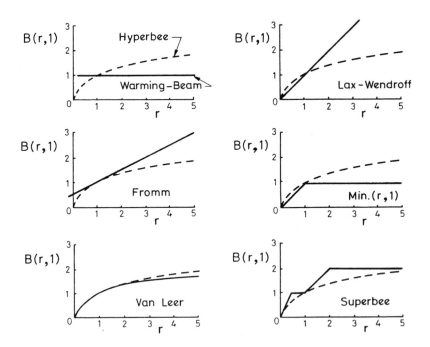

FIGURE 3. Hyperbee compared with six other limiters.

concerning a rather elaborate "discontinuity-sharpening" algorithm. I was trying to implant their idea into my own method, and to give it a simpler expression. In numerical experiments involving step data for the linear equation (10), it was found that the use of Superbee caused stable profiles to form which propagated essentially unchanged for many thousands of time steps (see Figure 4). When the numerical values were investigated, it was found that the head and tail of the profile corresponded rather closely to the values of r found at the two nontrivial intersections between Hyperbee and Superbee. The inflexion region which joins them is difficult to analyse theoretically but behaves very simply in practice. It just stays smooth and does not grow.

Some properties of Superbee have been investigated by Sweby [36], who also gives interesting results for test problems involving linear and nonlinear advection, and also for the Sod shock-tube problem. Along the lines of the discussion following Table I, he argues that the best combination of algorithms is to deploy Superbee on the second wave (or particle path) and (29) or (30) on the nonlinear waves. He also discusses the relationship between these limiters and the ones proposed by Harten [4] and by Chakravarthy and Osher [38].

Superbee has been applied to another one-dimensional test problem by the present author with the assistance of Paul Glaister (University of Reading). This is the blast-wave interaction problem proposed by Woodward and Colella [39]. A late stage in the evolution of this problem is shown in Figure 5, where density and velocity are both plotted against distance along a one-dimensional tube. Shock discontinuities are present in both plots, but contact discontinuities are only present in the density plots. Results using Superbee to compute all three wave fields are compared with results using (30) in the same way. A marked improvement of the contacts is noticeable when Superbee is used.

Lastly, we report that Superbee has been successfully incorporated into the two-dimensional aerofoil code [40] written at the Royal Aircraft

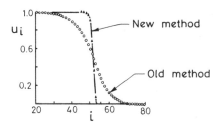

FIGURE 4. Linear advection of a step-function: 10,000 time steps at $\nu = 0.1$.

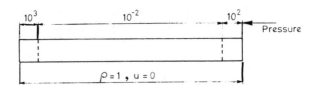

Initial conditions with discontinuities at $x/L = 0.1, 0.9.$

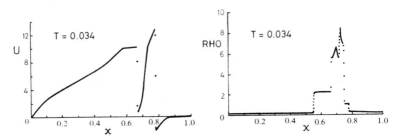

(a) Solution by present method with superbee: 1200 cells.

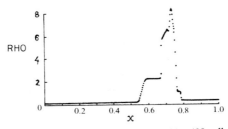

(b) Solution by present method with equation (30): 400 cells.

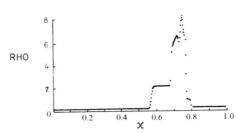

(c) Solution by present method with superbee: 400 cells.

FIGURE 5. The computed interaction between two blast waves.

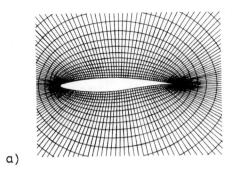

a)

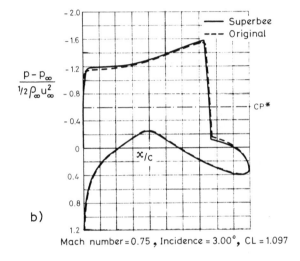

b)

Mach number = 0.75 , Incidence = 3.00°, CL = 1.097

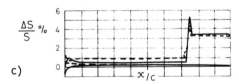

c)

FIGURE 6. Transonic flow past a two-dimensional aerofoil (RAE 2822). (a) Part of computing grid (b) Surface pressure (c) Surface entropy

Establishment by Dr. C. C. Lytton (formerly Dr. C. C. L. Sells). He finds
[41] that changing from a limiting procedure based on (30) to one based
on Superbee has significantly altered the results obtained from his code.
A typical experience is that gained on computations of the flow past on
RAE 2822 aerofoil at a freestream Mach number of 0.75 and an
incidence of 3°. This is a standard AGARD test case [42]. The aerofoil
and part of the 160 × 24 0-grid mesh are shown in Figure 6(a). The
following changes in the converged steady-state results were observed.

(i) The lift coefficient increased fractionally, from 1.0948 to 1.0974.

(ii) Originally there was a substantial discrepancy between two different ways of estimating the drag coefficients. An estimate based on
integrating surface pressures gave 0.0479, and an estimate based on
evaluating energy degredation through the shock gave 0.0385, a discrepancy of some 20%. The revised estimates are 0.0461 and 0.0452, with
a discrepancy of about 2%.

(iii) The pressure distribution over the aerofoil surface has changed
visibly (Figure 6(b)). Whether the overall distribution is an improvement
or not is of course nearly impossible to determine, but the shock wave is
slightly stronger and a little crisper.

(iv) The surface entropy distribution is definitely improved (Figure
6(c)). The quantity plotted is

$$\frac{p}{\rho^\gamma} \frac{\rho_\infty^\gamma}{p_\infty} - 1$$

which should of course be zero ahead of the shock wave. This particular
error has been reduced by a factor rather better than 2.0.

Putting all this evidence together, one is left with the definite impres-
sion that some slightly dissipative effects have been removed.

Another test has involved computing the supercritical flow past a
semicircular obstacle, a problem recently tackled by several authors [43].
The problem is attracting attention because of the observation that many
Euler codes spontaneously generate solutions containing regions of separated or recirculating flow. It is widely conjectured that these solutions
may be good approximations to the limiting solutions of the
Navier–Stokes equations at large Reynolds number. However, if closed
streamlines are present, this limit may not be unique [44]. Therefore, if
the results of different codes are compared, it is difficult to find grounds
for preferring one to another. Here, I merely record as a matter of
possible interest, that the code which incorporated (30) used to fail if
asked to compute the flow at any Mach number much greater than 0.40.
It is at about this Mach number that supersonic flow appears near the
equator, and separated flow almost simultaneously occurs near the rear
stagnation point. Using Superbee, supercritical flows with closed separa-
tions have been found at $M_\infty = 0.50$ and 0.60. These are the only

supercritical cases which have been attempted so far. Some results are shown in Figure 7. Note that the positions of the separation points have not been fixed in any way. The recirculating regions develop spontaneously in the course of a normal run.

4. On satisfying entropy. It is very well known that some finite-difference schemes admit stable steady solutions which do not satisfy any of the tests by which the physically correct members are selected from the

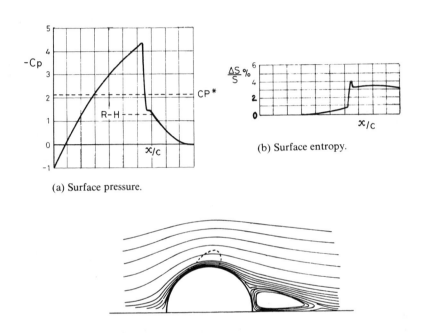

(a) Surface pressure.

(b) Surface entropy.

(c) Streamlines and sonic line.

7.1. Freestream Mach number = 0.50.

7.2. Freestream Mach number = 0.60.

FIGURE 7. Computations of transonic flow past a circular cylinder.

set of weak solutions. See, for example, [2, 4, 11, 16, 45, 46]. It is traditional to illustrate this point by considering the scalar case (the theory for
systems is still very undeveloped)

$$u_t + F_x = 0, \tag{32}$$

and then specialising to Burger's equation, $F = \frac{1}{2}u^2$, and initial data
$u(x, 0) = \text{sgn}(x)$. This problem admits the physically reasonable solution

$$u(x, t) = \begin{cases} \text{sgn}(x) & \text{if } |x|/t > 1.0, \\ x/t & \text{if } |x|/t < 1.0, \end{cases} \tag{33}$$

and also the solution

$$u(x, t) = \text{sgn}(x) \tag{34}$$

everywhere, which is physically unreasonable because characteristics
emerge out of the discontinuity along $x = 0$. Solutions like (34) are said
to violate an entropy condition. Now we consider the discrete problem
on some rectangular grid as in previous sections, and propose initial data
$u_i^0 = -1$ if $i < 0$, $u_i^0 = +1$ if $i \geq 0$. Many finite-difference schemes,
including all the schemes described so far, will accept this data as being
already in equilibrium (because all the fluctuations are zero) and will
therefore propagate it without change to all subsequent time levels.

My excuse for mentioning this very hackneyed example is to invite the
reader to consider a slight twist to the story. Suppose the differential
equation had actually involved the flux function $F = 1 - \frac{1}{2}u^2$? The
numerical data would remain unaltered, whether considered as a set of
u-values or as a set of F-values, but it would now comprise a valid,
realistic, steady solution, representing the exact continuum solution with
maximal resolution. So, what kind of finite-difference scheme will correctly distinguish between the two cases?

It seems to me that there are essentially two possibilities. One is that
the scheme contains some dissipative element which widens the transition
zone, and introduces values of u not present in the initial data. The
computer now has more information about the function $F(u)$. However,
there are still infinitely many functions $F(u)$ which coincide with the
data at the discrete samples, and so the computer still cannot be sure that
we have not chosen an $F(u)$ requiring an expansion fan. It has no option
but to widen the zone still further, introducing new values of u, and
acquiring more information about $F(u)$. A typical example of an algorithm which can be proved to converge, in this sense, to an entropy-
satisfying limit, as the grid is made finer is that of Lax and Friedrichs
[47]. However, no algorithm which operates like this can admit stationary
shocks with steady profiles of finite width, and the Lax–Friedrichs
scheme would today be considered far too dissipative for practical use.

The other possibility is that the computer is supplied directly with more complete information about the function $F(u)$. The prototype for such an algorithm is Godunov's scheme [15, 46] where the information derives from exactly solving the Riemann problem for all pairs of adjacent states. It is not possible to find this exact solution unless $F(u)$ is known for all values of u between the given values. It might be said that the computer is, in effect, supplied with a device which allows it to look inside the intervals without opening them up. Actually, for the scalar case, which is what we are discussing, this strategy is not as hard as it may seem. Osher [48] has shown that $F(u)$ can be sufficiently characterised by knowing all its extreme values; the only reason why an algorithm of the first kind must continually open up the intervals is the danger that a new extreme value may await discovery.

Osher's result is expressed in terms of interface flux functions $F_{i+1/2}$, such that

$$u_i^{n+1} - u_i^n = -\frac{\Delta t}{\Delta x}\left(F_{i+1/2}^n - F_{i-1/2}^n\right), \tag{35}$$

and states that convergence to an entropy solution is guaranteed by the following condition:

$$\begin{aligned}&\text{if } u_{i+1}^n > u_i^n, \text{ then } F_{i+1/2} \leqslant \inf F_{i+1/2},\\&\text{if } u_{i+1}^n < u_i^n, \text{ then } F_{i+1/2} \geqslant \sup F_{i+1/2},\end{aligned} \tag{36}$$

where $\sup F$, $\inf F$ are, respectively, the greatest and least values assumed by F in the range $\{u_i, u_{i+1}\}$. Osher calls such schemes E-schemes.

It is easy to translate this result into the present fluctuation-signal approach. Let each fluctuation $\phi_{i+1/2}$ be divided, with an obvious notation, into a left moving part and a right moving part; thus, dropping the n suffix

$$\phi_{i+1/2} = \overleftarrow{\phi}_{i+1/2} + \overrightarrow{\phi}_{i+1/2}, \tag{37}$$

where

$$\overleftarrow{\phi}_{i+1/2} = F_{i+1/2} - F_i, \tag{38a}$$

and

$$\overrightarrow{\phi}_{i+1/2} = F_{i+1} - F_{i+1/2}; \tag{38b}$$

then (36) implies that if $u_{i+1} > u_i$,

$$\overleftarrow{\phi}_{i+1/2} \leqslant \inf F_{i+1/2} - F_i, \qquad \overrightarrow{\phi}_{i+1/2} \geqslant F_{i+1} - \inf F_{i+1/2}, \tag{39a}$$

whereas if $u_{i+1} < u_i$,

$$\overleftarrow{\phi}_{i+1/2} \geqslant \sup F_{i+1/2} - F_i, \qquad \overrightarrow{\phi}_{i+1/2} \leqslant F_{i+1} - \sup F_{i+1/2}. \tag{39b}$$

It is easy to find geometric interpretations for these formulae, but there are several reasons why it may not be profitable to pursue these results too closely or too rigorously. An objection of principle is that a theory which applies only to scalar problems has at present to be extended by guesswork anyway. An objection of practice is that some of the schemes which can be demonstrated to converge to physically correct solutions actually do so in a rather disappointing way. Particularly it should be said that Godunov's method, sometimes thought of as the ideal 'physically based' scheme, actually gives expansion fans which have almost discontinuous behaviour at sonic points [5, 39, 50]. Also we cite the recent experience of Harten [4], who states that 'with reluctance' he has abandoned an earlier, rigorous approach to entropy enforcement [5, 46, 49] in favour of an empirical device which gives better numerical results for the Euler equations. We present below an account of a "spreading device" which has also proved successful in practice, but is motivated by physical arguments and contains no arbitrary parameters.

We begin by seeking estimates for the spreading rates associated with each wave in each fluctuation, i.e. the quantities $\delta^{(k)}$ in Tables I and II. For definiteness, consider $\delta^{(1)}$. Many estimates suggest themselves, and I suspect that it does not, in most cases, make much difference which we choose. For example,

$$\delta^{(1)}_{i+1/2} = \frac{1}{2}\left[(u-a)_{i+3/2} - (u-a)_{i-1/2}\right], \tag{40}$$

making use of wave speeds worked out for neighbouring interactions, or

$$\delta^{(1)}_{i+1/2} = (u-a)_{i+1} - (u-a)_i. \tag{41}$$

There are mild objections to both estimates. Equation (40) is less "local" than it could be, and (41) puts us to the trouble of computing wave speeds in each cell, which is expensive information for which we have no other use. A third estimate comes from considering the expansion fan, if it exists, to be a simple wave, across which the Riemann invariant does not change

$$a + \frac{\gamma - 1}{2}u = \text{constant.}$$

From this it follows at once that

$$\delta^{(1)} = \Delta(u-a)^{(1)} = \frac{\gamma + 1}{2}\Delta u^{(1)}, \tag{42a}$$

and similarly

$$\delta^{(3)} = \Delta(u+a)^{(3)} = \frac{\gamma + 1}{2}\Delta u^{(3)}, \tag{42b}$$

whilst, on physical grounds,

$$\delta^{(2)} = 0. \tag{42c}$$

These estimates use only local information which has already been generated. If (42a) or (42b) should turn out to be negative, then we know that we are dealing with a compression, and we set that spreading rate to zero. The part of the algorithm which deals with that wave will then be as described above, as of course will the part dealing with the particle path. A neat way to find the velocity jumps [13] is

$$\Delta u^{(1)} = \frac{1}{2}\left[\Delta u - \frac{\Delta p}{\tilde{\rho}\tilde{a}}\right], \tag{43a}$$

$$\Delta u^{(3)} = \frac{1}{2}\left[\Delta u + \frac{\Delta p}{\tilde{\rho}\tilde{a}}\right]. \tag{43b}$$

Now we make the (slightly arbitrary) decision that the standard form of the algorithm will also be applied to any expansion wave which runs wholly in one direction. (There is some justification for this; all the numerical results shown so far have completely neglected the distinction between compression and expansion waves.) Thus, no special action will be taken unless the spreading rate is positive and greater than twice the modulus of the mean wave speed. In that case only, we shall modify the *first-order* version of the algorithm, causing the fluctuation to send a signal to both targets, as in (37). If we assume that the characteristic field is centered at the interface between the cells (Figure 8(a)) it can be shown that we get an effect proportional to the square of the spreading rate. This is similar to the effect considered, and rejected, by Harten. However, if we assume that the characteristic field varies linearly across the cells (Figure 8(b)), then we find an effect directly proportional to the spreading rate. Not only is this a larger effect but also probably a more

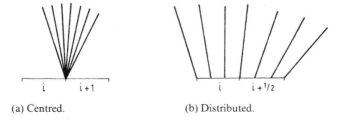

(a) Centred.   (b) Distributed.

FIGURE 8. The representations of a rarefaction.

realistic effect, it being in the nature of expansion fans to spread out and assume a fairly linear distribution. In fact, we obtain

$$\overleftarrow{\phi} = \frac{1}{2}\left(1 - \frac{1}{2}\frac{\delta}{\lambda}\right)\phi, \tag{44a}$$

$$\overrightarrow{\phi} = \frac{1}{2}\left(1 + \frac{1}{2}\frac{\delta}{\lambda}\right)\phi, \tag{44b}$$

where δ is the spreading rate, and λ is the mean velocity. The same concepts can be expressed in terms of an interface flux function,

$$\begin{aligned}
F_{i+1/2} &= \frac{1}{2}\left(1 + \frac{1}{2}\frac{\delta}{\lambda}\right)F_i + \frac{1}{2}\left(1 - \frac{1}{2}\frac{\delta}{\lambda}\right)F_{i+1} \\
&= \frac{1}{2}(F_i + F_{i+1}) - \frac{\delta}{4\lambda}(F_{i+1} - F_i) \\
&= \frac{1}{2}(F_i + F_{i+1}) - \frac{\delta}{4}(u_{i+1} - u_i) \\
&= \frac{1}{2}(F_i + F_{i+1}) - \frac{1}{4}(F'_{i+1} - F'_i)(u_{i+1} - u_i).
\end{aligned} \tag{45}$$

Equations (44a), (44b) and (45) apply very simply to Burger's equation. We get

$$\overleftarrow{\phi} = \tfrac{1}{2}u_i(u_{i+1} - u_i), \tag{46a}$$

$$\overrightarrow{\phi} = \tfrac{1}{2}u_{i+1}(u_{i+1} - u_i), \tag{46b}$$

$$F_{i+1/2} = \tfrac{1}{2}u_i u_{i+1}. \tag{47}$$

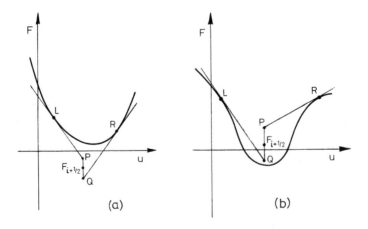

(a)   (b)

FIGURE 9. Examples of function for which equation (40) is (a) or is not (b) an E-scheme.

Equation (47) is to be inserted into the usual upwind scheme only if $u_i < 0 < u_{i+1}$, in which case $F_{i+1/2} < 0$, making the scheme an E-scheme, in Osher's sense [48]. Very satisfactory numerical tests were reported by van Leer in [14]. For more general equations, however, (45) may not be an E-scheme, as may be seen by a geometric representation (Figure 9(a)). LP, RQ are tangents to $F(u)$ at u_L, u_R; P, Q are at $\frac{1}{2}(u_L + u_R)$; $F_{i+1/2}$ is the midpoint of PQ, to be used if $F_i' < 0 < F_{i+1}'$. This makes (45) an

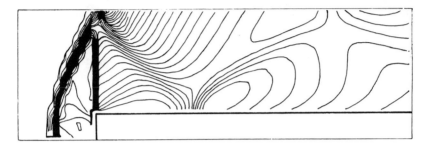

(a) Present method: first order, no spreading of expansion waves.

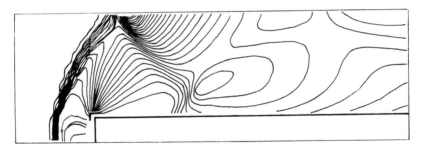

(b) Present method: first order, expansion waves spread.

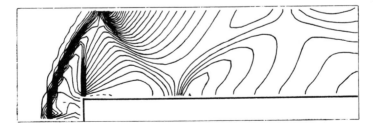

(c) Godunov's method [39].

FIGURE 10. Three versions of the unsteady flow in a stepped channel;   grid size 20×60.

E-scheme if $F(u)$ is any concave or convex function, but clearly it is possible to devise $F(u)$ such that (45) is not an E-scheme (Figure 9(b)).

Generalising the scheme to systems can be done very easily starting with (44a) and (44b) and applying the formulae to each column of Table II. That something along these lines may be necessary in practical cases can be seen by modifying the popular Sod test problem [37] by superposing on the initial conditions a uniform translation velocity of half a unit. Computations by Harten [4] and Sweby [50] show an unrealistic discontinuity in the middle of the expansion. For a two-dimensional example we exhibit Figure 10, due to Glaister [51], showing computations of another test problem proposed by Woodward and Colella [39]. This is the flow brought about by impulsive introduction of a step into a wind tunnel containing uniform flow at $M = 3.0$. The results are a snapshot, after the uniform flow has had time to travel four-step heights. Figure 10(a) is a simple first-order time-split computation, with no special action taken at sonic points, whereas Figure 10(b) incorporates a scheme very similar to the above and due to Sweby [50]. In Figure 10(a) the subsonic region is terminated by a one-dimensional expansion shock. Remarkably, when we look at Figure 10(c), taken from [39], the calculations made according to Godunov's method are nearly as bad as those in Figure 10(a).

Sweby has extended his scheme to second-order accuracy [50], but it seems quite possible that a first-order scheme may be good enough. The heuristic argument is that if we write the scalar equation as

$$u_t + a(u)u_x = 0, \tag{48}$$

then, in the cells adjacent to a smooth sonic point, $a(u)$ is of order Δx and so u_t is small anyway. If the sonic point is not in a region of smooth flow, then of course the formal accuracy of the scheme is irrelevant. In either case u_t has been estimated as accurately as we can justifiably demand. The tie-in with the programming logic of §2 is to set $\Delta^* \mathbf{u}^{(k)}$, as defined by (9), equal to zero, if the kth wave is an expansion containing a sonic point. Incidentally, the same heuristic argument suggests that we could afford to lose one order of accuracy near a smooth extremum (see the comment following (26b)).

Whilst on the general subject of guaranteeing the correct physical solution, I would like to draw attention to another feature of Figure 10. This is the reflection of the shock wave from the upper surface of the step, following its first reflection from the tunnel roof. In many of the computations reported by Woodward and Colella [39], this reflection takes place in a regular way. In Glaister's calculations, the reflection is regular when it first occurs, but develops with the passage of time into something more resembling the Mach reflection [52] which appears on the roof in both sets of calculations. This second Mach reflection is also seen in the preliminary results of Harten [4]. Which type of reflection is correct?

Let us note that something like Mach reflection is what is observed in experiments when the flow is viewed in sufficiently small detail [53], because there is always a thin boundary layer of low-energy gas. Now consider how the solid-wall boundary condition will be implemented in this particular problem. Since all boundaries are plane, it would seem sufficient to use fictitious image cells just outside the real flow, where pressure, density and parallel velocity are all set equal to the corresponding quantities just inside the real flow, but the normal velocity has its sign reversed. Indeed the only place where we might anticipate a problem is at the upper corner of the step, where the image cell has two different jobs to do.

I do not want to discuss the 'correct' way out of this local difficulty. Experience shows that it is easy to go wrong, but theory suggests that it will not be easy to tell if we have got it right. The exact inviscid solution around this corner may well be quite complicated, with a closed separation downstream of which the flow passes through an embedded shock wave (Figure 11). If this does happen, the final steady solution is probably not unique. Even if it does not happen, the body will be sheathed in an entropy layer, or entropy wake [54], of low density gas, which has passed through the strongest part of the bow shock wave. This layer might be expected to interact with the incident shock wave much as

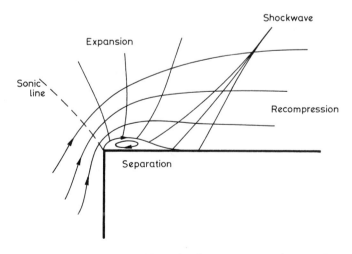

FIGURE 11. Possible high Reynolds number flow pattern around corner of step.

a boundary layer does. The point I wish to make is that even for relatively simple flows, it can be very hard to decide whether plausible looking results are actually reliable. These problems do not seem to be resolved by appeal to an entropy principle, or any other general principle that I know of.

5. Final remarks. At the time of writing [1], I hoped that a mathematical structure was revealing itself which would be sufficiently powerful to provide a complete link between studies of one-dimensional numerical advection, and completely arbitrary one-dimensional quasi-linear conservation laws. Today I am not quite so optimistic. The main theoretical obstacle to handling all conservation laws by one single method is the need to account for all possible modes of nonlinear behaviour. If these modes are complicated, then the scope for using approximate Riemann solutions is restricted. For if the approximate solution completely ignores the internal wave structure, as in [3], then some additional information must be added on. On the other hand, if the approximate solution aims also at describing the internal structure, as in [11], such a solution is unlikely to emerge in a computationally inexpensive form, except in the simplest cases. For some time to come, the compromise between economy and reliability probably has to be resolved individually for each set of equations we might wish to solve.

Where these equations have been well studied, however, and it is known that the modes of nonlinear behaviour are rather simple, then matters seem much clearer. The most valuable concept seems to be that of splitting the interaction between adjacent cells into a set of essentially independent physical effects, each of which can be modelled by an appropriate numerical method capable of modelling the structure of the corresponding wave. In this paper the emphasis has been on explicit schemes, but the value of the concept seems to be proving itself in implicit methods also [6]. A successful treatment of source terms and moving grids is described in [7], and of highly irregular grids in [2, 5, 50, 55]. Another possible application is to somewhat unorthodox explicit large time-step methods [56, 57], although I am not sure whether good numerical results for systems have yet been obtained by these means. The more modest claim that I would now propose is that systems of equations can be effectively solved, provided good numerical methods are available for scalar equations which exhibit the same modes of nonlinear behaviour.

Application of these ideas to multidimensional problems has been largely through the operator splitting concept [2, 3, 4, 6, 11, 40, 51]. The widespread feeling that it should be possible to improve on this is only slowly finding expression in algorithms. What has been done [2, 28, 35, 41] falls some distance short of the target, but the difficulty lies perhaps in the great variety of possibilities, and the need to find systematic methods of exploring them.

Perhaps the most valuable outcome of making a really clean job of the one-dimensional methods would be if it turned out to provide a good springboard into higher dimensions. Meanwhile the one-dimensional methods continue to evolve out of experience to a greater extent than they are deduced from axioms. I find the following quotation very apt [58]: "Science builds a house with bricks, which, once laid, are not touched again. Philosophy tidies a room, and so has to handle things many times. The essence of its procedure is that it starts with a mess, but we do not mind being hazy so long as the haze gradually clears".

REFERENCES

1. P. L. Roe, *The use of the Riemann problem in finite difference schemes*, Proc. Seventh Internat. Conf. on Numerical Methods in Fluid Dynamics, Stanford, 1980 (W. C. Reynolds and R. W. MacCormack, eds.), Lecture Notes in Phys., No. 141, Springer-Verlag, Berlin and New York, 1981, 354–359.

2. _____, *Fluctuations and signals, a framework for numerical evolution problems*, Numerical Methods for Fluid Dynamics (K. W. Morton and M. J. Baines, eds.), Academic Press, New York, 1982, 219–257.

3. _____, *Approximate Riemann solvers, parameter vectors, and difference schemes*, J. Comput. Phys. **43**, (1981), 357–372.

4. A. Harten, *High resolution schemes for hyperbolic conservation laws*, J. Comput. Phys. **49** (1983), 357–393.

5. A. Harten and J. M. Hyman, *A self-adjusting grid for the computation of weak solutions of hyperbolic conservation laws*, J. Comput. Phys. **50** (1983), 235.

6. H. C. Yee, R. F. Warming and A. Harten, *Implicit total variation diminishing (TVD) schemes for steady state calculations*, NASA TM 84342, March 1983.

7. M. Pandolfi, *A contribution to the numerical prediction of unsteady flows*, AIAA Paper 83-0121, January 1983.

8. C. K. Lombard, *Conservative supra-characteristics method for splitting hyperbolic systems of gasdynamics for real and perfect gases*, NASA CR 166307, January 1982.

9. C. K. Lombard, J. Oliger and J. Y. Yang, *A natural conservative flux-difference splitting for the hyperbolic systems of gasdynamics*, AIAA Paper 82-0976, 1982.

10. P. Colella, *Approximate solution of the Riemann problem for real gases*, preprint, Lawrence Berkeley Laboratory, University of California, May 1982.

11. S. Osher and F. Soloman, *Upwind schemes for hyperbolic systems of conservation laws*, Math. Comp. **38** (1982), 339–374.

12. S. Osher, *Shock modelling in aeronautics*, Numerical Methods for Fluid Dynamics (K. W. Morton and M. J. Baines, eds.), Academic Press, New York, 1982, 179–217.

13. P. L. Roe and J. Pike, *Efficient construction and utilisation of approximate Riemann solvers* (Comput. Methods Appl. Science Engrg., VI, R. Glowinski and J. L. Lions, eds.), North-Holland, 1984, 518.

14. B. van Leer, *On the relation between the upwind-differencing schemes of Godunov, Engquist-Osher and Roe*, SIAM J. Sci. Statist. Comput. **5** (1984), 1–20.

15. S. K. Godunov, *A difference method for the numerical calculation of discontinuous solutions of hydrodynamic equations*, Mat.-Sb. **47** (1959); English transl. U. S. Department of Commerce, JPRS 7225, 1960.

16. P. L. Roe, *On the internal structure of some captured shockwaves*, RAE report, in preparation.

17. R. F. Warming and R. W. Beam, *Upwind second-order difference schemes and applications in aerodynamic flows*, AIAA July 14, 1976.

18. T. H. Pulliam and J. L. Steger, *On implicit finite-difference simulations of three-dimensional flows*, AIAA Paper 78-10, 1978.

19. A. Jameson, W. Schmidt and E. Turkel, *Numerical solution of the Euler equations by finite-volume methods using Runge–Kutta time-stepping schemes*, AIAA Paper 81-1259, 1981.

20. A. Jameson, *Transonic aerofoil calculations using the Euler equations*, Numerical Methods in Aeronautical Fluid Dynamics (P. L. Roe, ed.), Academic Press, New York, 1982, 289-308.

21. B. van Leer, *Towards the ultimate conservative differencing scheme*. Part I, Proc. 3rd Intl. Conf. Num. Meth. in Fluid Dynamics.

22. _____ , *Towards the ultimate conservative differencing scheme*. Part II, J. Comput. Phys. **14** (1974), 361–370.

23. _____ , *Towards the ultimate conservative differencing scheme*. Part III, J. Comput. Phys. **23** (1977), 263–275.

24. _____ , *Towards the ultimate conservative differencing scheme*. Part IV, J. Comput. Phys. **23** (1977), 276–299.

25. _____ , *Towards the ultimate conservative differencing scheme*. Part V, Comput. Phys. **32** (1979), 101–136.

26. P. L. Roe, *An improved version of MacCormack's shock-capturing algorithm*, RAE TR 79041, 1979.

27. _____ , *Numerical algorithms for the linear wave equation*, RAE TR 81047, 1981.

28. P. L. Roe and M. J. Baines, *Algorithms for advection and shock problems*, Proc. 4th GAMM Conf. on Numerical Methods in Fluid Mechanics, Paris, 1981 (H. Viviand, ed.), Vieweg, Braunschweig, 1982, 281-290.

29. B. Gabutti, *On two upwind finite-difference schemes for hyperbolic equations in non-conservative form*, Comput. & Fluids **11** (1983), 207-230.

30. P. Colella, *A direct Eulerian MUSCL scheme for gasdynamics*, SIAM J. Sci. Statist. Comput. (to appear).

31. P. Colella and P. R. Woodward, *The piecewise parabolic method (PPM) for gas dynamical simulations*, J. Comput. Phys. **54** (1984), 174-201.

32. M. Abramowitz and Irene Stegun, *Handbook of mathematical functions*, U. S. Department of Commerce, Washington, D. C., 1964.

33. G. D. van Albada, B. van Leer and W. W. Roberts, *A comparative study of computational methods in cosmic gas dynamics*, ICASE Report 81-24, August 1981.

34. W. A. Mulder and B. van Leer, *Implicit upwind methods for the Euler equations*, AIAA paper 83-1930, 1983.

35. P. L. Roe and M. J. Baines, *Asymptotic behaviour of some non-linear schemes for linear advection*, Proc. 5th GAMM Conf. Num. Meth. in Fluid Mechanics, Rome, 1983 (M. Pandolfi and R. Piva, eds.), Vieweg, 1984, 283-290.

36. P. K. Sweby, *High resolution schemes using flux limiters for hyperbolic conservation laws*, preprint, University of California, Los Angeles, 1983.

37. G. A. Sod, *A survey of several finite difference methods for systems of non-linear hyperbolic conservation laws*, J. Comput. Phys. **27** (1978), 1.

38. S. Chakravarthy and S. Osher, *Computing with high-resolution upwind schemes for hyperbolic equations*, Large Scale Computations in Fluid Mechanics (Proc. 15th AMS-SIAM Summer Seminar on Applied Mathematics, La Jolla, 1983, B. J. Engquist, S. J. Osher and R. C. J. Somerville, eds.), AMS, Providence, 1985.

39. P. Woodward and P. Colella, *The numerical simulation of two-dimensional fluid flow with strong shocks*, J. Comput. Phys. **54** (1984), 115–173.

40. C. C. L. Sells, *Solution of the Euler equations for transonic flow past a lifting aerofoil*, RAE TR 80065, 1980.

41. C. C. Lytton, *Solution of the Euler equations for transonic flow past a tilting airfoil—the Bernouilli formulation*, RAE TR 84080, 1984.

42. R. C. Lock, *Test cases for numerical methods in two-dimensional transonic flows*, AGARD Report No. 575, 1970.

43. M. D. Salas, *Recent developments in transonic Euler flow over a circular cylinder*, NASA TM 83282, April 1982.

44. G. K. Batchelor, *On steady laminar flow with closed streamlines at large Reynolds number*, J. Fluid Mech. **1** (1956), 177–190.

45. A. Harten, J. M. Hyman and P. D. Lax, *On finite difference approximation and entropy conditions for shocks*, Comm. Pure Appl. Math. **29** (1976), 297–322.

46. A. Harten, P. D. Lax and B. van Leer, *On upstream differencing and Godunov-type schemes for hyperbolic conservation laws*, SIAM Rev. **25** (1983), 35–60.

47. P. D. Lax, *Weak solutions of non-linear hyperbolic equations and their numerical computation*, Comm. Pure Appl. Math. **7** (1954), 159–193.

48. S. Osher, *Riemann solvers, the entropy condition, and difference approximations*, SIAM J. Numer. Anal. **21** (1984), 217–235.

49. A. Harten and P. D. Lax, *A random-choice finite-difference scheme for hyperbolic conservation laws*, SIAM J. Numer. Anal. **18** (1981), 289–315.

50. P. K. Sweby, *Shock-capturing schemes*, Ph. D. Thesis, University of Reading, Reading, U. K., 1982.

51. P. Glaister, Private communication, University of Reading, 1983.

52. H. Polachek, *Shock wave interactions*, High Speed Aerodynamics and Jet-Propulsion. III. Fundamentals of Gas Dynamics, Section E (H. W. Emmons, ed.), Oxford Univ. Press, New York and London, 1958, 482–522.

53. L. F. East, *The application of laser anemometry to the investigation of shockwave-boundary layer interactions*, AGARD CPP 193, alos RAE TM 1666, 1976.

54. W. D. Hayes and R. F. Probstein, *Hypersonic flow theory*, Vol. I. Inviscid Flows, Academic Press, New York, 1966.

55. J. Pike, *Euler equation solutions on irregular grids*, RAE report, in preparation.

56. R. J. LeVeque, *A large time step generalisation of Godunov's method for systems of conservation laws*, preprint, University of California, Los Angeles, 1984.

57. A. Harten, *On a large time-step high resolution scheme*, ICASE Report 82–34, November 1982.

58. L. Wittgenstein, *Lectures, Cambridge* 1930–32 (D. Lee, ed.), Blackwell, Oxford, 1980.

ROYAL AIRCRAFT ESTABLISHMENT, BEDFORD, UNITED KINGDOM

Current address: College of Aeronautics, Cranfield Institute of Technology, Cranfield Bedford MK43 OAL, England

Lectures in Applied Mathematics
Volume **22**, 1985

Techniques for Numerical Simulation of Large-Scale Eddies in Geophysical Fluid Dynamics

Robert Sadourny

0. Introduction. The problem addressed here is the definition of horizontal diffusion operators designed to parameterize in some optimal way the dynamical effect of subgrid scales in numerical models of large-scale geophysical flows. By large-scale flow we mean the energetically dominant part of fluid motion whose horizontal scale L is significantly larger than the vertical scale D of the fluid—a few kilometers for the atmosphere or the ocean. Due to

(i) this small aspect ratio ($\varepsilon = D/L \ll 1$),

(ii) the hydrostatic assumption that vertical acceleration is negligible,

(iii) the two-dimensionalizing effect of the earth's rotation (*Poincaré* [**1910**], *Proudman* [**1916**], *Taylor* [**1917**]),

the large-scale dynamics of geophysical flows is essentially horizontal; the knowledge of its specific properties is a prerequisite for the design of adequate lateral diffusion operators in numerical models.

We proceed in the following way. In §1, the equations of large-scale geophysical motion—the so-called *primitive* equations—are described in a form which we appear later on as most convenient, using thermodynamic entropy as vertical coordinate. §2 deals with the classical *quasi-geostrophic* approximation, which simplifies the primitive equations to a point which makes possible the phenomenological investigation of large-scale dynamics. This phenomenology of the quasi-geostrophic *macroturbulence* is the subject of the §3. Finally, in §4, the phenomenological

1980 *Mathematics Subject Classification.* Primary 76C99, 86A05, 86A10.

theory is used as a guideline for deriving optimal *lateral diffusion schemes* for numerical simulation of large-scale geophysical flow.

1. The primitive equations in entropy coordinate. As soon as one assumes hydrostatic motion in a stably stratified ($\partial s/\partial z > 0$) atmosphere or ocean, thermodynamic entropy s can be used as the vertical coordinate instead of z. The advantage is that s is a true lagrangian variable ($Ds/Dt = 0$) for adiabatic motion. Denoting specific volume by α and pressure by p, we may write the equation of state in a very general form

$$\alpha = H_p(x, p), \tag{1}$$

where the index p refers to partial derivation with respect to p at constant s, of the thermodynamic function $H(s, p)$. Then the primitive equations read (Sadourny [**1984**])

$$D\mathbf{V}/Dt + f\mathbf{N} \times \mathbf{V} + \text{grad } S = 0, \tag{2}$$

$$\partial S/\partial s = H_s, \tag{3}$$

$$\partial r/\partial t + \text{div}(r\mathbf{V}) = 0, \tag{4}$$

respectively: the equation of motion, the hydrostatic equation, and the continuity equation; here $\mathbf{V}$ is horizontal velocity, f the Coriolis parameter, $\mathbf{N}$ the upward vertical unit vector, $S = H + \phi$ the *Montgomery potential*—ϕ is geopotential, and $r = -\partial p/\partial s$ a *pseudo-density* associated with the use of entropy as the vertical coordinate; grad and div are two-dimensional spherical operators; $\partial./\partial s$ refers to partial derivation with respect to s at constant x, y, t, while index s refers to partial derivation with respect to s at constant p for a thermodynamic function of s and p. Equations (2) and (3) yield the *potential vorticity* equation

$$D\eta/Dt = 0, \tag{5}$$

where η is the hydrostatic form of Ertel's potential absolute vorticity:

$$\eta = (f + \text{rot } \mathbf{V})/r. \tag{6}$$

The top boundary of the fluid is a moving surface $s = s_T(x, y, t)$; we take it as a *free* boundary with prescribed pressure:

$$Ds_T/Dt = 0, \tag{7a}$$

$$p(x, y, s_T, t) = p_T(x, y); \tag{7b}$$

the bottom boundary of the fluid is again a moving surface $s = s_B(x, y, t)$ which we take as a *rigid* boundary with prescribed height or geopotential:

$$Ds_B/Dt = 0, \tag{8b}$$

$$\phi(x, y, s_B, t) = \phi_B(x, y); \tag{8b}$$

note that the bottom boundary is actually moving in the s-coordinate frame of reference, although it remains fixed in physical space. This system of equations has the following invariants; total *energy*

$$E = \iint_D dx \, dy \left[p_B \phi_B + \int_{s_B}^{s_T} ds \left(H + rV^2/2 \right) \right],$$ (9)

and the integrals over isentropic surfaces of arbitrary functions of potential absolute vorticity

$$P_A(s) = \iint_{D'(s, t)} A(\eta, s) r \, dx \, dy,$$ (10)

among which we may select *potential enstrophy*

$$P(s) = \iint_{D'(s, t)} \frac{\eta^2}{2} r \, dx \, dy.$$ (11)

In (9), D is the horizontal domain of the flow; in (10) and (11), D' is defined as the part of the isentrope which, at time t, belongs to the fluid: $D'(s, t) = D \cap \{(x, y)|s_B(x, y, t) \ll s \ll s_T(x, y, t)\}$.

Note that solving (2), (3), (4), (7) and (8) involves solving a nonlinear elliptic problem (3), (7b), (8b) at every time step, in order to calculate the diagnostic variable S from the prognostic variable r and the boundary conditions. This is done easily by first integrating $\partial p/\partial s = -r$ downwards from the top boundary condition (7b); then pressure is determined everywhere and (3) can be integrated upwards from (8b).

2. The quasi-geostrophic approximation. We now give a brief outlook of the quasi-geostrophic approximation. We assume that the fluid departs only weakly from some resting state $\mathbf{V} = 0$, $p = \bar{p}(s)$, $r = \bar{r}(s) = -d\bar{p}/ds$, etc. When linearized around such a resting state, the primitive equations (2), (3), (4), (7), and (8) yield a denumerable set of eigenfrequencies ω_n associated with a complete set of eigenmodes. On the other hand, we have to consider the nonlinear advective frequencies μ_n, in addition to ω_n, if we want a full estimate of the time scales of the motion. Let us denote by U a root-mean-square measure of velocity, and by L_n the scale of eigenmode n; then we have the estimate $\mu_n = U/L_n$.

We first assume that essentially all the excitation of the flow is contained in the *slowly evolving modes*, or modes such that

$$|\omega_n|/\bar{f} = O(\varepsilon),$$ (12a)

$$\mu_n/\bar{f} = O(\varepsilon),$$ (12b)

with $\varepsilon \ll 1$; here $\bar{f}$ is a mean value of f over D. By imposing (12a) we exclude the fastest inertia-gravity modes. We infer from (12) that the acceleration is negligible compared to the Coriolis term in equation (2), which, to first order in ε, to reduces *geostrophic equilibrium*:

$$f\mathbf{N} \times \mathbf{V} + \text{grad } S = 0. \tag{13}$$

Another assumption we make is that the Coriolis parameter $f = \bar{f} + f'$ remains close to its mean value

$$f'/\bar{f} = O(\varepsilon). \tag{14}$$

Assumption (14) means that we exclude equatorial dynamics. Then, to first order in ε, f can be replaced by $\bar{f}$ in equation (13); it follows that the flow is, to first order, nondivergent and therefore describable in terms of a streamfunction ψ:

$$\mathbf{V} = \mathbf{N} \times \text{grad } \psi, \qquad \psi = S/\bar{f}. \tag{15}$$

The last assumption we have to make is that the fluctuations of the pseudo-density r are small with respect to its mean value:

$$r'/\bar{r} = O(\varepsilon). \tag{16}$$

Then the first order expansion of potential vorticity reads

$$\eta = (\Delta + \Lambda)\psi + f, \tag{17}$$

where Δ refers to the two-dimensional spherical laplacian and Λ to the vertical elliptic operator

$$\Lambda \cdot = \frac{\bar{f}^2}{\bar{r}} \frac{\partial}{\partial s} \left(\frac{1}{\bar{\alpha}_s} \frac{\partial \cdot}{\partial s} \right). \tag{18}$$

This *quasi-geostrophic potential vorticity* is just carried along with fluid particles in absence of dissipation or forcing. Equations (5) and (15) yield the quasi-geostrophic potential vorticity equation

$$\frac{\partial \eta}{\partial t} + \frac{\partial(\psi, \eta)}{\partial(x, y)} = 0, \tag{19}$$

with the advection operator in Jacobian form.

We have now to consider the boundary conditions. We redefine the upper and lower boundary surfaces by

$$s_T(x, y, t) = \bar{s}_T + s_T'(x, y, t),$$
$$s_B(x, y, t) = \bar{s}_B + s_B'(x, y, t). \tag{20}$$

Again, we assume that the fluctuations are small compared to appropriate vertical scales:

$$s_T' d(\text{Ln } \bar{p})/ds = O(\varepsilon), \qquad s_B' d(\text{Ln } \bar{\phi})/ds = O(\varepsilon). \tag{21}$$

Then boundary conditions (7) and (8) are approximated to first order in ε on the flat surfaces $s = \bar{s}_T$, $s = \bar{s}_B$; and the approximation reduces to advecting pressure and geopotential perturbations, respectively:

$$\frac{\partial p'}{\partial t} + \frac{\partial(\psi, p')}{\partial(x, y)} = 0 \quad \text{at } s = \bar{s}_T,$$

$$\frac{\partial \phi'}{\partial t} + \frac{\partial(\psi, \phi')}{\partial(x, y)} = 0 \quad \text{at } s = \bar{s}_B. \tag{22}$$

These pressure and geopotential perturbations have the following form:

$$p' = \frac{\bar{f} \, \partial\psi/\partial s}{\bar{\alpha}_s}, \qquad \phi' = \bar{f}\left(1 - \frac{\bar{\alpha}}{\bar{\alpha}_s}\frac{\partial}{\partial s}\right)\psi. \tag{23}$$

Equations (19) and (22) are the prognostic equations of the problem, which determine the advance in time of η (inside the fluid), p' (on the upper boundary) and ϕ' (on the lower boundary); when these are known, the elliptic problem (17), (23) is solved for the unknown ψ.

If taken in the appropriate form, all the invariants of the primitive equations (9), (10) and (11) are again invariants of the quasi-geostrophic problem. The *energy* invariant reads

$$E = \frac{1}{2}\iint_D dx\,dy\left[\left.\frac{\bar{f}^2}{\bar{\alpha}}\psi^2\right|_{\bar{s}_B} + \int_{\bar{s}_B}^{\bar{s}_T} ds\left(\bar{r}(\text{grad }\psi)^2 + \frac{\bar{f}^2}{\bar{\alpha}_s}\left(\frac{\partial\psi}{\partial s}\right)^2\right)\right]. \tag{24}$$

The generalized invariants related to potential vorticity are

$$P_A(s) = \iint_D dx\,dy\,\bar{r}A(\eta, s), \tag{25}$$

among which the *vertical structure* of *potential enstrophy*:

$$P(s) = \iint_D dx\,dy\,\bar{r}\frac{\eta^2}{2}. \tag{26}$$

In fact, the quasi-geostrophic system of equations has more invariants than the primitive equations. This is because of the simplified treatment of the upper and lower boundary conditions (22), which yields the two generalized boundary invariants

$$P_A(\bar{s}_T) = \iint_D dx\,dy\,\bar{r}A(p'), \qquad P_A(\bar{s}_B) = \iint_D dx\,dy\,\bar{r}A(\phi'), \tag{27}$$

among which we select the quadratic ones, which we call the *available potential energies* on top and bottom boundaries

$$P_T = \iint_D dx\, dy\, \frac{\eta_T^2}{2}, \qquad P_B = \iint_D dx\, dy\, \frac{\eta_B^2}{2}; \qquad (28)$$

from now on we use, instead of p' and ϕ', the normalized variables

$$\begin{aligned} \eta_T &= \bar{f} p' & \text{at } s &= \bar{s}_T, \\ \eta_B &= \bar{f}/\bar{\alpha}\phi' & \text{at } s &= \bar{s}_B. \end{aligned} \qquad (29)$$

With these variables an alternative form for (24) is

$$E = \frac{1}{2} \iint_D dx\, dy\left(\psi_T \eta_T + \psi_B \eta_B - \int_{\bar{s}_B}^{\bar{s}_T} ds\, \bar{r}\psi\eta\right). \qquad (30)$$

This form will be most convenient for the following analyses.

3. Phenomenology of quasi-geostrophic turbulence. Quasi-geostrophic turbulence appears as a rather straightforward extension of two-dimensional turbulence. In two-dimensional turbulence, vorticity (the horizontal laplacian of the horizontal streamfuntion) is advected by the horizontal, nondivergent flow. In quasi-geostrophic turbulence, potential vorticity (the three-dimensional laplacian of the horizontal streamfunction) is also advected by the horizontal, nondivergent flow, together with its natural Dirichlet–Neumann boundary conditions at top and bottom. What complicates the quasi-geostrophic theory, however, is the interplay between the dynamics on upper and lower boundaries, and the dynamics of the inside flow. We shall avoid this complication and limit our considerations to the following simple, but complementary cases.

Internal quasi-geostrophic turbulence is governed by potential vorticity advection within dynamically neutral upper and lower boundaries:

$$\begin{aligned} \eta_T &= 0 & \text{at } s &= \bar{s}_T, \\ \frac{\partial n}{\partial t} + \frac{\partial(\psi, n)}{\partial(x, y)} &= 0 & \text{for } \bar{s}_B &< s < \bar{s}_T, \\ \eta_B &= 0 & \text{at } s &= \bar{s}_B. \end{aligned} \qquad (31)$$

In the study of this case, we shall admit that the only invariants which play a significant dynamical role are *total energy* and *total potential enstrophy*:

$$E = -\frac{1}{2} \iint_D dx\, dy \int_{\bar{s}_B}^{\bar{s}_T} ds\, \bar{r}\psi\eta, \qquad P = -\frac{1}{2} \iint_D dx\, dy \int_{\bar{s}_B}^{\bar{s}_T} ds\, \bar{r}\eta^2. \qquad (32)$$

We thus neglect *a priori* all effects possibly related to the generalized invariants (25) or the detailed invariants (26).

External quasi-geostrophic turbulence, on the other hand, is governed by advection of pressure and geopotential perturbations on upper and lower boundaries, separated by a dynamically neutral fluid:

$$\frac{\partial \eta_T}{\partial t} + \frac{\partial(\psi, \eta_T)}{\partial(x, y)} = 0 \qquad \text{at } s = \bar{s}_T,$$

$$\eta = 0 \qquad \text{for } \bar{s}_B < s < \bar{s}_T, \qquad (33)$$

$$\frac{\partial \eta_B}{\partial t} + \frac{\partial(\psi, \eta_B)}{\partial(x, y)} = 0 \qquad \text{at } s = \bar{s}_B;$$

the invariants we shall consider for this case are just *energy* and *total available potential energy on boundaries*:

$$E = \frac{1}{2} \iint_D dx\, dy (\psi_T \eta_T + \psi_B \eta_B), \qquad P = \frac{1}{2} \iint_D dx\, dy (\eta_T^2 + \eta_B^2).$$

$$(34)$$

Here, like in the internal case, we suppose that the generalized invariants (27) play no particular dynamically important role; also, we admit that separate conservation of P_T and P_B (28) is unimportant.

With such simplifications we are now in position to investigate the phenomenological properties of quasi-geostrophic flow. Since these will be studied in spectral space, we first proceed to establish the spectral formulations. From now on the Coriolis parameter f will be considered constant: this means we shall not be interested in the interaction of Rossby waves (which arise from the Coriolis linear term in (17), (19)) with quasi-geostrophic macro-turbulence.

The problem is formulated in spectral form by expanding ψ and η, first horizontally, then vertically, on the orthonormal eigenfunctions of the elliptic problems

$$\mathscr{I}: \eta \to -\psi \qquad \text{(internal case)},$$
$$\mathscr{E}: (\eta_B, \eta_T) \to (\psi_B, \psi_T) \qquad \text{(external case)}.$$

For simplicity, we suppose from now on that D is the infinite plane. Then the (positive) eigenvalues of $\mathscr{I}$ and $\mathscr{E}$ will be denoted by $K_n^2(k)$, where k is the horizontal wave number and n is an index which identifies the vertical mode: $n = 0, 1, 2, \ldots$ in case $\mathscr{I}$, $n = 0, 1$ in case $\mathscr{E}$. The modes corresponding to $n = 0$ will be referred to as *barotropic modes*; the modes corresponding to $n = 1$ will be called the first *baroclinic modes*, and so on. These denominations correspond to the ordering

$$K_n(k) < K_{n'}(k) \quad \text{for } n < n', \qquad (35)$$

and $K_n(k)$ is referred to as an equivalent three-dimensional wave number.

The dependencies of $K_n(k)$ on the horizontal wave number k are displayed in Figure 1 for the two problems $\mathscr{I}$ and $\mathscr{E}$. At very large horizontal scales,

$$K_n(k) \to K_n^o > 0 \quad \text{for } k \to 0; \tag{36}$$

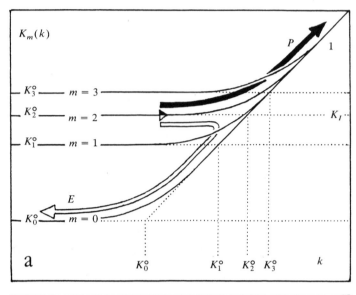

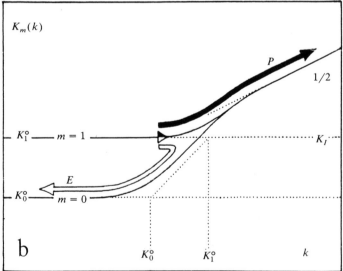

FIGURE 1. Distribution of eigenvalues $K_n(k)$ in the (a) internal and (b) external cases: logarithmic scales, asymptotic slopes indicated. P-flux diagrams are in black arrows; E-flux diagrams are in white arrows. Black and white triangles show injection scales.

K_n° are referred to as the successive deformation wave numbers. The simple case is case $\mathscr{S}$, where we have

$$K_n^2(k) = K_n^{\circ 2} + k^2; \tag{37}$$

case $\mathscr{E}$ is more involved, and we need a close look at the relation

$$(\Lambda - k^2)\Sigma_k = 0, \tag{38}$$

which, from (17), (18) and (33), defines the vertical structure of the eigenmodes, to verify that

$$\begin{aligned}K_n^2(k) &= K_n^{\circ 2} + O(k^2) &&\text{for } k \to 0, \\ K_n^2(k) &= O(k) &&\text{for } k \to \infty.\end{aligned} \tag{39}$$

Thus the two problems $\mathscr{S}$ and $\mathscr{E}$ look equivalent at very large scales (except for the limitation of $\mathscr{E}$ at two vertical modes), but different at very small scales.

Let us consider now a horizontally homogeneous, horizontally isotropic, internal or external quasi-geostrophic turbulence over an infinite plane. We assume a stationary thermal forcing at horizontal wave number k_I, in baroclinic mode n_I. We denote by ε the energy injection rate, which corresponds to a P-injection rate

$$\pi = K_I^2 \varepsilon, \tag{40}$$

with the definition $K_I = K_{n_I}(k_I)$.

We assume that the association of forcing with nonlinear dynamics induces a statistically stationary motion in an ever-increasing three-dimensional wave number band $\{K_-(t), K_+(t)\}$, with

$$K_0^\circ \leftarrow K_-(t) < K_I < K_+(t) \to - \quad \text{as } t \to \infty. \tag{41}$$

This is a reasonable assumption, sustained by numerical experimentation with statistical closure models. Let us now evaluate the E- and P-budgets within this stationary wave number band. For this purpose we use the spectral expansions of E and P, related by

$$\begin{aligned}P_n(k) &= K_n^2(k) E_n(k), \\ E &= \sum_n \int_0^\infty E_n(k)\, dk, \\ P &= \sum_n \int_0^\infty K_n^2(k) E_n(k)\, dk.\end{aligned} \tag{42}$$

By $\varepsilon_+(t)$, $\pi_+(t)$ we denote the E- and P-fluxes into $\{K_+(t), \infty\}$; and by $\varepsilon_-(t)$, $\pi_-(t)$ the E- and P-fluxes into $\{K_0^\circ, K_-(t)\}$. From the stationarity assumption we have

$$\varepsilon_+(t) + \varepsilon_-(t) = \varepsilon, \qquad \pi_+(t) + \pi_-(t) = \pi. \tag{43}$$

On the other hand, (42) yields

$$\pi_+(t) \geqslant K_+^2(t)\varepsilon_+(t), \qquad K_0^{\circ 2}\,\varepsilon_-(t) \leqslant \pi_-(t) \leqslant K_-^2(t)\varepsilon_-(t). \quad (44)$$

We naturally assume that all the fluxes we consider are positive, since the excitation wave number band is supposed to expand in time. Then it follows from (43) and (44) that, in the asymptotic limit ($t \to \infty$) of *fully developed* quasi-geostrophic turbulence,

$$\begin{aligned}
\pi_-(t) &\to K_0^{\circ 2}\,\varepsilon, & \varepsilon_-(t) &\to \varepsilon; \\
\pi_+(t) &\to \left(K_I^2 - K_0^{\circ 2}\right)\varepsilon, & \varepsilon_+(t) &\to 0.
\end{aligned} \quad (45)$$

Thus, in fully developed quasi-geostrophic turbulence, all the energy injected cascades back *to barotropic large scales*, while most of the *P*-injected cascades down to infinitesimal *three-dimensional* scales.

There is a weak point, however, in the above demonstration. The assumption

$$K_0^\circ \leftarrow K_-(t) \qquad (46)$$

in (41), prevents any reverse cascade of energy from taking place in baroclinic modes, since we assume stationarity for $K > K_-(t)$. Therefore, we still have to prove the impossibility of reverse baroclinic energy cascades. This can be done in the following way.

The driving mechanism of quasi-geostrophic turbulence is the straining of η-isolines by horizontal velocity gradients; and of course the straining is efficient only if the velocity gradient has a scale larger than the structure it tries to distort. Then, the *kinetic enstrophy* integral

$$\tau_n^{-1}(k) = \int_0^k \frac{p^4}{K_n^2(p)} E_n(p)\,dp, \qquad (47)$$

is a measure of the efficiency of the straining of a particular structure with wave number k, by the nth vertical mode; $\tau_n(k)$ is the associated time scale. It is then readily seen that, for very large horizontal scales, the most efficient straining agent will be the barotropic mode, because it contains the larger amount of kinetic energy. Then all baroclinic modes will behave like passive scalars: their energy will be drained towards smaller scales. The only mode which can exhibit a reverse energy cascade is the barotropic mode whose dynamics are self-contained.

We have now a schematic view of the cascade processes in quasi-geostrophic flow: they are illustrated on Figure 1 along with the eigenvalue diagrams. The corresponding spectral shapes are easily derived from (45), and (47). In the *barotropic reverse energy cascading range*, we have

$$E_0(k) = C_- \varepsilon^{2/3} k^{-5/3} \qquad \text{for } K_0^\circ < k < K_I,$$
$$E_0(k) = C_- 2^{1/3} \varepsilon^{2/3} k^{-7/3} \qquad \text{for } k < K_0^\circ . \tag{48}$$

In the *P-cascading range*, we have to distinguish between cases $\mathcal{I}$ and $\mathcal{E}$, since the $K_n(k)$ have different asymptotic forms (37), (39) for $k \to \infty$:

$$E(k) = C_+ \left\{ \left(K_I^2 - K_0^{\circ 2} \right) \varepsilon \right\}^{1/2} k^{-3} \left(\mathrm{Ln}\, \frac{k}{K_I} \right)^{-1/3} \qquad \text{for case } \mathcal{I},$$

$$E(k) = C_+ \left\{ \left(K_I^2 - K_0^{\circ 2} \right) \varepsilon \right\}^{2/3} k^{-8/3} \qquad \text{for case } \mathcal{E}. \tag{49}$$

All vertical modes are asymptotically equivalent for $k \to \infty$; therefore we give the *P*-inertial range spectra in terms of $E(k) = \Sigma_n E_n(k)$.

4. Formulation of lateral diffusion in numerical models. We can now summarize the phenomenology of quasi-geostrophic turbulence as follows. Potential enstrophy or available potential energy on top and bottom boundaries are drained towards infinitesimal scales; energy, on the contrary, is trapped in large scales and, therefore, strictly conserved, *P*-dissipation, associated with *E*-conservation, necessarily means that energy has to be converted from potential to kinetic form: this is the well-known process of *baroclinic instability*. The horizontal wave number band where this instability occurs and energy conversion takes place is the band $\{ K_1^\circ , K_I \}$.

A numerical model designed to handle such a flow must resolve or accurately model these conversion processes in order to get the large-scale barotropic eddies right. The easy way is, of course, to use a high resolution; if the cut-off wave number k_c is significantly larger than K_I, truncation effects will not interfere with the energy cycle, and the large-scale barotropic modes will be properly maintained.

The use of high-resolution models, however, is not always the best choice. Many problems of climate dynamics, for instance, necessitate very long-term integrations for which the use of heavy models would be prohibitive; another example is ocean models, which necessitate a mesh-size of a few kilometers if they have to resolve accurately the first deformation scale. On the other hand, using a low-resolution model is a delicate matter: truncation will interfere with the energy path as soon as k_c becomes of the order of K_I; the classical lateral diffusion schemes, being essentially dissipative, will inhibit the conversion process and weaken the reverse energy cascade (Sadourny and Hoyer [**1982**]). What we need therefore are *efficient parameterizations of baroclinic instability*.

On the basis of the above phenomenology, we must require that lateral diffusion operators be *E-conservative* and *P-dissipative*; then the proper amount of energy conversion will be automatically ensured, and energy will cascade towards larger barotropic scales at a rate exactly equal to the injection rate for any choice of the cut-off wave number. These two properties are easy to implement; they yield the model equation

$$\frac{\partial \eta}{\partial t} + \frac{\partial(\psi, .)}{\partial(x, y)}\left\{\eta - \theta \mathscr{F}\frac{\partial(\psi, \eta)}{\partial(x, y)}\right\} = 0. \tag{50}$$

The diffusion term is the second term in the bracket; this model is of course applicable to η, η_T or η_B indifferently. Equation (50) is strictly energy-conserving; it is P-dissipating as soon as $\mathscr{F}$ is positive definite:

$$\frac{dP(s)}{dt} + \iint_D dx\, dy\, \theta\left\{\frac{\partial(\psi, \eta)}{\partial(x, y)}\mathscr{F}\frac{\partial(\psi, \eta)}{\partial(x, y)}\right\} = 0. \tag{51}$$

When $\mathscr{F} = 1$, (50) reduces to an *upwind* technique for computing the advection of η. More generally, we shall impose that $\theta \mathscr{F}$ has a diagonal horizontal spectral transform $\theta(k)$. The interpretation in this case is again an upwind correction in the advection of η-structures, but now the time lag is wave number dependent; $\theta(k)$ can be identified with the characteristic time scales obtained from statistical closure modelling of the subgrid scales (Basdevant, Lesieur and Sadourny [1978]): it should vanish for $k \ll k_c$—large-scale structures are not directly affected by truncation, and increase sharply in the vicinity of k_x. An approximate form is

$$\theta(k) = \theta_0(k/k_c)^\beta, \tag{52}$$

where large values of β (for example, $\beta = 16$) ensure the expected cusplike behaviour in the vicinity of k_c. $\mathscr{F}$ is then proportional to a laplacian to the power $\beta/2$. Tests of this diffusion scheme for thermally forced two-layer baroclinic quasi-geostrophic flow are to be found in Sadourny and Basdevant [1984].

This lateral diffusion model is readily extended to the primitive equations, provided they are written in entropy-coordinate form. The energy-conserving diffusion term appears in the rotation term of the equation of motion:

$$DV/Dt + \{f - r\theta\mathscr{F}(\mathbf{V} \cdot \operatorname{grad} \eta)\}N \times \mathbf{V} + \operatorname{grad} S = 0. \tag{53}$$

Combined with the continuity equation (4), (53) yields the modified potential vorticity equation

$$D\eta/Dt - (1/r)\operatorname{div}\{\theta\mathscr{F}(\mathbf{V} \cdot \operatorname{grad} \eta)r\mathbf{V}\} = 0, \tag{54}$$

corresponding to (50). From (53), energy is exactly conserved; from (4) and (53), potential enstrophy at isentropic level s is dissipated according to

$$\frac{dP(s)}{dt} + \iint_D (\mathbf{V} \cdot \operatorname{grad} \eta)\theta\mathscr{F}(\mathbf{V} \cdot \operatorname{grad} \eta)r\, dx\, dy = 0, \tag{55}$$

i.e., proportionally to a positive quadratic form of the advective time change of potential vorticity.

REFERENCES

C. Basdevant, M. Lesieur and R. Sadourny, 1978. *Subgridscale modeling transfer in two dimensional turbulence*, J. Atmospheric Sci. **35**, 1028–1042.

W. Blumen, 1978. *Uniform potential vorticity flow. Part 1. Theory of wave interactions and two-dimensional turbulence*, J. Atmospheric Sci. **35**, 774–783.

J. G. Charney, 1971. *Quasi-geostrophic turbulence.*, J. Atmospheric Sci. **28**, 1087–1095.

H. Poincaré, 1910. *Leçons de mécanique céleste. Tome 3. Théorie des marées*, Gauthier-Villars, Paris, pp. 183–188.

J. Proudman, 1916. *On the motion of solids in a liquid possessing vorticity*,Proc. Roy. Soc. London Ser. A **92**, 408–424.

R. Sadourny, 1984. *Quasi-geostrophic turbulence: an introduction*, Turbulence and Predictability in Geophysical Fluid Dynamics and Climate Dynamics, North-Holland, Amsterdam, in press.

R. Sadourny and C. Basdevant, 1984. *Parameterization of sub-grid scale barotropic and baroclinic eddies in quasi-geostrophic models: the anticipated potential vorticity method*, submitted.

R. Sadourny and J. M. Hoyer, 1982. *Inhibition of baroclinic instability in low-resolution models*, J. Atmospheric Sci. **39**, 2138–2143.

R. Salmon, 1978. *Two-layer quasi-geostrophic turbulence in a simple special case*, Geophys. Astrophys. Fluid Dynamics **10**, 25–51.

_____ , 1980. *Baroclinic instability and geostrophic turbulence*, Geophys. Astrophys. Fluid Dynamics **15**, 167–212.

G. I. Taylor, 1917. *Motion of solids when the flow is not irrotational*, Proc. Roy. Soc. London Ser. A **93**, 99–113.

LABORATOIRE DE MÉTÉOROLOGIE DYNAMIQUE, ÉCOLE NORMALE SUPÉRIEURE, 75231 PARIS CEDEX 05, FRANCE

Lectures in Applied Mathematics
Volume **22**, 1985

Finite Difference Techniques for Nonlinear Hyperbolic Conservation Laws

Richard Sanders[1]

1. Introduction. We shall consider numerical approximations to the initial value problem for nonlinear systems of conservation laws

$$\frac{\partial u}{\partial t} + \sum_{i=1}^{d} \frac{\partial}{\partial x_i} f_i(x, t, u) = g(x, t, u), \qquad u(x, 0) = u_0(x). \quad (1.1)$$

Here $x = (x^1, \ldots, x^d) \in \mathbf{R}^d$, $u(x, t)$ is an m-vector of unknowns and each flux function $f_i(x, t, u)$ is vector-valued having m components. The system (1.1) is said to be hyperbolic when all eigenvalues of every real linear combination of the Jacobian matrices are real. It is well known that solutions of (1.1) may develop discontinuities in finite time, even when the initial data are smooth.

Conservation form finite difference schemes have proved to be the most successful numerical methods used to approximate discontinuous solutions of (1.1). The well-known theorem of Lax and Wendroff [9], shows that when a finite difference scheme is in conservation form, the limit approximations obtained will automatically satisfy the Rankine-Hugoniot jump condition. In addition, when the approximate solutions of (1.1) satisfy a discrete entropy inequality, nonphysical limit solutions can be shown not to appear. The notion of an entropy satisfying solution, that is, of a vanishing viscosity solution, has been completely treated by Kruzkov in [6] when (1.1) is a *single* equation. He shows that, under reasonable assumptions on f and g, entropy solutions of (1.1) exist and are unique.

1980 *Mathematics Subject Classification*. Primary 65M10, 65M15.

[1]Research supported by NSF Grant # MCS 82-00676.

This paper develops a class of conservation form finite difference schemes that are based on the finite volume method (i.e. the method of averages). These schemes do not fit into the classical framework of conservation form schemes treated in [9]. The finite volume schemes are specifically intended to approximate solutions of multidimensional problems in the absence of rectangular geometries. We further develop different schemes that utilize the finite volume approach for time discretization. Specifically, we discuss local time discretization and moving spatial grids.

No technical proofs are given in this paper. The interested reader is encouraged to see the cited references for additional details.

2. The finite volume method. We begin by motivating the finite volume approach. Let Ω be an open, bounded and connected subset of $\mathbf{R}^d$. Further, suppose that Ω has a Lipschitz boundary $\partial\Omega$. Assuming that (1.1) has a solution which has bounded variation (see §3), (1.1) and the divergence theorem combine to show that the "Ω average" of the exact solution evolves according to

$$\frac{d}{dt}\frac{1}{|\Omega|}\int_\Omega u(x,t)\,dx\Big|_{t=t_0} = -\frac{1}{|\Omega|}\int_{\partial\Omega}(F\cdot n)(x,t_0,\gamma^- u(x,t_0))\,d\sigma$$

$$+\frac{1}{|\Omega|}\int_\Omega g(x,t_0,u(x,t_0))\,dx. \qquad (2.1)$$

Throughout this paper, γ^- (γ^+) will denote the inner (outer) boundary trace operator and $n=(n_1,\ldots,n_d)$ will denote the unit *outward* normal of Ω. The key observation to make here is that $F\cdot n(x,t,u)$ is a one-space-dimensional flux function

$$F\cdot n = \sum_{i=1}^d n_i f_i.$$

We now consider a class of two-point, one-space-dimensional, *numerical* flux functions, which will be denoted by

$$h_{F\cdot n}(x,t,u_1,u_2). \qquad (2.2)$$

Two conditions are imposed on the numerical flux (2.2). First, a consistency condition is imposed:

$$h_{F\cdot n}(x,t,u,u) = F\cdot n(x,t,u). \qquad (2.3)$$

Second, a condition that guarantees conservation form is imposed:

$$h_{F\cdot(-n)}(x,t,u_1,u_2) = -h_{F\cdot n}(x,t,u_2,u_1). \qquad (2.4)$$

REMARK 2.1. The conservation form condition (2.4) is always satisfied when the numerical flux function (2.2) is derived from any one-space-dimensional, three-point, conservation form difference scheme. One must merely identify $h_{F\cdot(-n)}$ in the manner prescribed in (2.4).

To derive a semidiscrete finite volume approximation we do the following: Partition $\mathbf{R}^d$ into a union of nonoverlapping cells of the type above, $\mathbf{R}^d = \cup_j \Omega_j$. Define

$$u^\delta(x, t) = \sum_j u_j(t)\chi_{\Omega_j}(x), \tag{2.5}$$

where $\chi_{\Omega_j}(x)$ is the characteristic function of Ω_j. Formally insert (2.5) into (2.1) and approximate the boundary integral in (2.1) by

$$\int_{\partial\Omega_j} h_{F\cdot n}(x, t, \gamma^+ u^\delta, \gamma^- u^\delta)\, d\sigma.$$

The semidiscrete approximation is then obtained from the solution of the system of ordinary differential equations

$$\frac{d}{dt}u_j = -\frac{1}{|\Omega_j|}\int_{\partial\Omega_j} h_{F\cdot n}(x, t, \gamma^+ u^\delta, u_j)\, d\sigma$$

$$+ \frac{1}{|\Omega_j|}\int_{\Omega_j} g(x, t, u_j)\, dx. \tag{2.6}$$

REMARK 2.2. It might appear that the right-hand side of (2.6) is excessively difficult to evaluate. This is not the case for many physical applications. If the system (1.1) is invariant under rotation and if the flux $F \cdot n$ does not depend explicitly on x, a closed form expression for the boundary integral in (2.6) is easily computed. This is done in particular for the Osher upwind numerical flux applied to the Euler equations [12]. Also see [3].

The two most natural types of time discretization for the semidiscrete system (2.6) are forward and backward Euler. Denoting the right-hand side of (2.6) by $\mathscr{H}(x, t, u^\delta(t))$ and defining $u^\delta(t^n)$ in the usual fashion we arrive at an explicit method

$$u_j(t^{n+1}) = u_j(t^n) + \Delta t^n \mathscr{H}(x, t^n, u^\delta(t^n)), \tag{2.7}$$

and an implicit method

$$u_j(t^{n+1}) = u_j(t^n) + \Delta t^n \mathscr{H}(x, t^{n+1}, u^\delta(t^{n+1})). \tag{2.8}$$

Additional types of time discretization techniques are discussed in §4.

For the *scalar* equation it is not difficult to justify the constructions above when h_f corresponds to a monotone flux function. The numerical flux (2.2) is said to be monotone if it is nonincreasing in u_1 and nondecreasing in u_2.

3. Theoretical justification. In this section, we offer some theoretical justification for the scalar version of the finite volume algorithm developed in §2. We begin with a definition.

DEFINITION 3.1. The BV seminorm of a function u: $\mathbf{R}^d \to \mathbf{R}$ is given by

$$\|u\|_{BV} = \sup \int u \operatorname{div}(\phi) \, dx,$$

where the supremum is taken over all functions $\phi \in (C_0^\infty(\mathbf{R}^d))^d$ with Euclidean lengths less than or equal to one.

It is easily seen that for sufficiently smooth functions the definition above reduces to

$$\|u\|_{BV} = \int |\nabla u| \, dx.$$

This fact makes the following lemma apparent.

LEMMA 3.1. *The family* $\{ u \in L^1 : \|u\|_{BV} \leqslant K \}$ *is precompact in* $L^1_{\mathrm{loc}}(\mathbf{R}^d)$.

One further simple lemma is needed.

LEMMA 3.2. *Let* u^δ *be piecewise constant as defined by* (2.5). *Then*

$$\|u^\delta\|_{BV} = \frac{1}{2} \sum_{j,k} |u_j - u_k| \, |\partial\Omega_j \cap \partial\Omega_k|,$$

where $|\partial\Omega_j \cap \partial\Omega_k|$ *is the area of the common boundary shared by* Ω_j *and* Ω_k.

We now show that solutions of the implicit algorithm (2.8) satisfy a discrete entropy inequality when the numerical flux $h_{F \cdot n}$ is monotone. Recall that the numerical flux $h_{F \cdot n}(u_1, u_2)$ is said to be monotone if it is a nonincreasing function in its first argument and a nondecreasing function in its second argument. To simplify the exposition, we assume that (2.8) is autonomous and $g(x, t, u) \equiv 0$.

Explicit approximations from (2.8) also satisfy a similar discrete entropy inequality when monotone flux functions are employed and a suitable CFL restriction is obeyed; see [13].

The discrete entropy inequality. Recall the implicit finite volume algorithm

$$u_j^{n+1} = u_j^n - \frac{\Delta t^n}{|\Omega_j|} \int_{\partial\Omega_j} h_{F \cdot n}\left(\gamma^+ u^\delta(t^{n+1}), u_j^{n+1}\right) d\sigma. \tag{3.1}$$

Define the common side of Ω_j and Ω_k as $S_{j,k}$. Using the conservation requirement (2.4), (3.1) may be written as

$$u_j^{n+1} = u_j^n - \frac{1}{2}\frac{\Delta t^n}{|\Omega_j|}\sum_k \int_{S_{j,k}} \left(h_{F\cdot n_j^k}\left(u_k^{n+1}, u_j^{n+1}\right)\right.$$

$$\left. - h_{F\cdot n_k^j}\left(u_j^{n+1}, u_k^{n+1}\right)\right) d\sigma,$$

where n_j^k is the unit outward normal of Ω_j along $S_{j,k}$. Let c be any real constant. The divergence theorem shows that

$$\left(u_j^{n+1} - c\right) = \left(u_j^n - c\right)$$

$$- \frac{1}{2}\frac{\Delta t^n}{|\Omega_j|}\sum_k \int_{S_{j,k}} \left(h_{F\cdot n_j^k}\left(u_k^{n+1}, u_j^{n+1}\right) - h_{F\cdot n_j^k}(c, c)\right) d\sigma$$

$$+ \frac{1}{2}\frac{\Delta t^n}{|\Omega_j|}\sum_k \int_{S_{j,k}} \left(h_{F\cdot n_k^j}\left(u_j^{n+1}, u_k^{n+1}\right) - h_{F\cdot n_k^j}(c, c)\right) d\sigma.$$

$$(3.2)$$

The monotonicity of the numerical flux allows us to write

$$\int_{S_{j,k}} \left(h_{F\cdot n_j^k}\left(u_k^{n+1}, u_j^{n+1}\right) - h_{F\cdot n_j^k}(c, c)\right) d\sigma,$$

as

$$\text{sgn}\left(u_j^{n+1} - c\right)\int_{S_{j,k}} \left| h_{F\cdot n_j^k}\left(u_k^{n+1}, u_j^{n+1}\right) - h_{F\cdot n_j^k}\left(u_k^{n+1}, c\right)\right| d\sigma \qquad (3.3)$$

$$- \text{sgn}\left(u_k^{n+1} - c\right)\int_{S_{j,k}} \left| h_{F\cdot n_j^k}\left(u_k^{n+1}, c\right) - h_{F\cdot n_j^k}(c, c)\right| d\sigma \equiv A_{j,k} + B_{j,k},$$

which transforms the right-hand side of (3.2) into

$$\left(u_j^n - c\right) - \frac{1}{2}\frac{\Delta t^n}{|\Omega_j|}\sum_k \left(A_{j,k} + B_{j,k}\right) + \frac{1}{2}\frac{\Delta t^n}{|\Omega_j|}\sum_k \left(A_{k,j} + B_{k,j}\right). \quad (3.4)$$

Multiplying (3.4) by $\text{sgn}(u_j^{n+1} - c)$ and using the triangle inequality, we arrive at

$$\left|u_j^{n+1} - c\right| \leq \left|u_j^n - c\right| - \frac{1}{2}\frac{\Delta t^n}{|\Omega_j|}\sum_k \left(|A_{j,k}| - |B_{j,k}|\right)$$

$$+ \frac{1}{2}\frac{\Delta t^n}{|\Omega_j|}\sum_k \left(|A_{k,j}| - |B_{k,j}|\right),$$

$$(3.5)$$

which is what we shall call the *discrete entropy inequality*.

The entropy error follows from (3.5) in the usual way. For brevity, we merely outline its derivation.

PROPOSITION 3.1. *Suppose that approximations $u^\delta(x, t)$, obtained from the implicit scheme* (3.1), *have variation bounded uniformly for $\delta > 0$ on the strip $[0, T]$. Assume further that there exists a constant K independent of δ and points $x_j \in \Omega_j$ such that, for each j, $\Omega_j \subset B(x_j, K\delta)$. Then, for any $c \in \mathbf{R}$ and $\phi \geqslant 0$, $\phi \in C_0^\infty(\mathbf{R}^d \times (0, T))$, we have*

$$E_\phi(u^\delta) \equiv -\int_{\mathbf{R}^d \times \mathbf{R}^+} |u^\delta - c|\phi_t + \mathrm{sgn}(u^\delta - c)\big(F(u^\delta) - F(c)\big) \cdot \nabla\phi \, dx \, dt$$

$$\leqslant \mathrm{const}\big(\|\nabla\phi\|_\infty + \|\phi_t\|_\infty\big)(\delta + \Delta t).$$

The constant above depends in a linear way on $\max_{0 \leqslant t \leqslant T} \|u^\delta(t)\|_{\mathrm{BV}}$.

OUTLINE OF PROOF. Define $E_j^k(u_k, u_j, c)$ as

$$\big|h_{F \cdot n_j^k}(u_k, u_j) - h_{F \cdot n_j^k}(u_k, c)\big| - \big|h_{F \cdot n_j^k}(u_k, c) - h_{F \cdot n_j^k}(c, c)\big|,$$

and observe that

$$|A_{j,k}| - |B_{j,k}| = \int_{S_{j,k}} E_j^k\big(u_k^{n+1}, u_j^{n+1}, c\big) \, d\sigma$$

and

$$E_j^k(u, u, c) = \mathrm{sgn}(u - c)\big(F(u) - F(c)\big) \cdot n_j^k.$$

The discrete entropy inequality and a lengthy calculation will show that

$$E_\phi(u^\delta) \leqslant k\big(\|\nabla \cdot \phi\|_\infty + \|\phi_t\|_\infty\big) \cdot (\delta + \Delta t)$$

$$\cdot \Bigg(\sum_{j, n \leqslant N} \Big(|u_j^{n+1} - u_j^n| \, |\Omega_j| \tag{3.6}$$

$$+ \Delta t^n \int_{S_{j,k}} \big|E_j^k\big(u_j^{n+1}, u_k^{n+1}, c\big) - E_j^k\big(u_j^{n+1}, u_j^{n+1}, c\big)\big| \, d\sigma \Bigg).$$

The inequality above, along with the result of Lemma 3.2 and the hypothesis of the proposition, will easily establish the desired result.

A sufficient condition to establish the compactness of a family of functions in $L^1_{\mathrm{loc}}(\mathbf{R}^d)$ is the bounded variation hypothesis of Proposition 3.1. As seen above, it also plays a key role in establishing the entropy error. It is well known that *if* the family of approximations $\{u^\delta\}_{\delta > 0}$ is uniformly bounded in L^∞ and BV on $[0, T]$, the previous proposition would imply that as $\delta \downarrow 0$, $u^\delta \to u$ in $L^\infty([0, T], L^1_{\mathrm{loc}})$, where u is the unique entropy solution of the scalar problem (1.1). This follows from the Kruzkov uniqueness theorem and the bounded convergence theorem. See [6, 1, or 14] for additional details. The bounded variation assumption is unfortunately a delicate matter for arbitrary partitions of $\mathbf{R}^d$. However, additional restrictions on the partitioning allows us to state two results.

THEOREM 3.1. *Suppose Ω_j is given by*

$$\Omega_{j_1,\ldots,j_d} = \left[x^1_{j_1}, x^1_{j_1+1}\right) \times \cdots \times \left[x^d_{j_d}, x^d_{j_d+1}\right).$$

Then

$$\left\|u^\delta(t)\right\|_{\mathrm{BV}} \leqslant \|u_0\|_{\mathrm{BV}},$$

where u_0 is the initial data of (1.1).

The proof of Theorem 3.1 can be found in [1] for a uniform mesh and in [14] for a nonuniform mesh. For more general partitions, it will be shown in a forthcoming paper [16] that the variation of approximations from (3.1) may indeed grow with t. The next theorem, however, gives a favorable estimate for a case more general than the one treated above. Its proof will appear in [16].

THEOREM 3.2. *Let T: $\mathbf{R}^d \to \mathbf{R}^d$ be a smooth bijective transformation of the "x" coordinate system into the "s" coordinate system. Suppose that this transformation satisfies*

$$\sum_{i,j,k} \int_{\mathbf{R}^d} \left| \frac{\partial^2 Ti}{\partial x_j \, \partial x_k} \right| dx \leqslant c.$$

Then, if $\bigcup_j \Omega_j$ is given by a cartesian product of the partitioned coordinate axes in the "s" variable, we have

$$\left\|u^\delta(t)\right\|_{\mathrm{BV}} \leqslant \mathrm{const}\, \exp(ct)\|u_0\|_{\mathrm{BV}}.$$

We should remark that if the sides of Ω_j are piecewise planar and its vertices satisfy the condition above, the variation estimate above is still valid.

The next section will deal with nonuniform time approximations obtained from the finite volume method. As has been true so far, no proofs will be given.

4. Time discretization. A notion fundamental to our time discretization techniques is to integrate over space-time. The basic ideas behind the finite volume approach can then be modified appropriately keeping this idea in mind. To simplify matters, we will restrict our attention to the one-space-dimensional Cauchy problem

$$\frac{\partial u}{\partial t} + \frac{\partial}{\partial x} f(u) = 0, \qquad u(x,0) = u_0(x). \tag{4.1}$$

Two types of time discretization are discussed below. First, a technique that allows local time increments is developed, i.e. time increments that may vary with spatial location. A second technique is developed that allows the spatial grid to vary in time, i.e. a moving grid. Either or both of these techniques can be incorporated into an adaptive code yielding greater computational efficiency as well as improved solution resolution. See [13, 5, or 15] for additional details.

To begin, let Ω be a regular subset of *space-time* $\mathbf{R} \times \mathbf{R}^+$. Integrating (4.1) over Ω and using the divergence theorem, we find that the exact solution of (4.1) satisfies

$$\int_{\partial\Omega} (f(\gamma^- u), \gamma^- u) \cdot n \, d\sigma = 0,$$

where here $n = (n_x, n_t)$. As in §2, formally replace $(f(u), u) \cdot n$ with $h_{F \cdot n}(\gamma^+ u, \gamma^- u)$, again requiring that $h_{F \cdot n}$ satisfy

$$h_{F \cdot n}(u, u) = (f(u), u) \cdot n, \tag{4.2}$$

$$h_{F \cdot (-n)}(u_1, u_2) = -h_{F \cdot n}(u_2, u_1). \tag{4.3}$$

Above, $F(u)$ is to represent the space-time vector $(f(u), u)$. The manner in which $h_{F \cdot e_t}$ is defined determines the type of scheme derived. (e_t will denote the unit vector in the positive time direction.)

To derive the local time discretization scheme, partition the space axis into a union of disjoint intervals, $\mathbf{R} = \cup_j \mathscr{I}_j$, where $\mathscr{I}_j = [x_j, x_{j+1})$. At each time level t^n, decompose this partition into two subsets $\cup_{j \in \mathscr{C}^n} \mathscr{I}_j$ and $\cup_{j \notin \mathscr{C}^n} \mathscr{I}_j$, where $\mathscr{C}^n$ is any subset of the integers (possibly dependent on n). The time increment $[t^n, t^{n+1})$ is associated with those j's belonging to $\mathscr{C}^n$. Otherwise, we partition $[t^n, t^{n+1})$ into $\cup_{l=0}^{M-1}[t^{n+\eta_l}, t^{n+\eta_{l+1}})$, where $t^{n+\eta_l}$ is defined below; associate these time increments to those j's not belonging to $\mathscr{C}^n$.

Let $\{\sigma_k\}_{k=1}^M$ be a sequence of positive numbers such that $\sum_{k=1}^M \sigma_k = 1$. Define $\eta_l = \sigma_1 + \cdots + \sigma_l$ with $\eta_0 \equiv 0$. We now define $t^{n+\eta_{l+1}} = t^{n+\eta_l} + \sigma_{l+1} \Delta t^n$.

On the cell $[x_j, x_{j+1}) \times [t^{n+\eta_l}, t^{n+\eta_{l+1}})$, $u^\delta(x, t)$ is defined in the usual way. The numerical flux $h_{F \cdot n}$ is defined on the top and bottom of each cell as $h_{F \cdot e_t}(u_1, u_2) = u_1$; this generates an *explicit* algorithm. On the sides, let $h_{F \cdot e_x}(u_1, u_2) = h_f(u_1, u_2)$, where h_f is any one of numerous two-point numerical flux functions consistent with $f(u)$.

The finite volume approach yields a differencing technique that advances approximations from time level t^n to t^{n+1} through time increments Δt^n when $j \in \mathscr{C}^n$ and fractions of Δt^n when $j \notin \mathscr{C}^n$. The technique obtained can be cast into a predictor-corrector type form.

The predictor is as follows:
For $k = 1, \ldots, M - 1$,

$$u_j(t^{n+\eta_k}) = \begin{cases} u_j(t^n), & j \in \mathscr{C}^n, \\ u_j(t^n) - \lambda_j^n \sum_{l=0}^{k-1} \sigma_{l+1} \Delta_+ h_f\big(u_j(t^{n+\eta_l}), u_{j-1}(t^{n+\eta_l})\big), \end{cases} \tag{4.4}$$

$$j \notin \mathscr{C}^n,$$

where λ_j^n will be defined as $\Delta t^n / \Delta x_j$ and Δ_+ is the forward difference operator. The corrector is

$$u_j(t^{n+1}) = u_j(t^n) - \lambda_j^n \sum_{l=0}^{M-1} \sigma_{l+1} \Delta_+ h_f\left(u_j(t^{n+\eta_l}), u_{j-1}(t^{n+\eta_l})\right).$$

$$(4.5)$$

We note that if $j-1$, j and $j+1$ all belong to $\mathscr{C}^n$, the algorithm reduces to

$$u_j(t^{n+1}) = u_j(t^n) - \lambda_j^n \Delta_+ h_f\left(u_j(t^n), u_{j-1}(t^n)\right).$$

Furthermore, for j not belonging to $\mathscr{C}^n$, the algorithm may be written inductively as

$$u_j(t^{n+\eta_{k+1}}) = u_j(t^{n+\eta_k}) - \lambda_j^n \sigma_{k+1} \Delta_+ h\left(u_j(t^{n+\eta_k}), u_{j-1}(t^{n+\eta_k})\right),$$

for $k = 0, \ldots, M - 1$. Thus, the necessary computer programming is quite simple. Values of u_j at the same time level depend only on the values of u_{j-1}, u_j and u_{j+1} at the previous time level, except when j belongs to $\mathscr{C}^n$ and either $j-1$ or $j+1$ does not. We call such points x_j interface points. For these points, we must store the associated neighboring values of u_j at all $M-1$ intermediate time levels so that u_j may be advanced from t^n to t^{n+1}.

Again for the scalar version of (4.1), the previous algorithm (4.4), (4.5) can be justified when h_f is a monotone numerical flux function. A local CFL type restriction appears in the following

THEOREM 4.1. *Let* $u^\delta(x, t)$ *be obtained from the algorithm* (4.4), (4.5). *Further assume that a* local *CFL type restriction is satisfied,*

$$\Lambda_j^{n+\eta_k}\left[\left|h_f(u, v_1) - h_f(u, v_2)\right| + \left|h_f(v_1, w) - h_f(v_2, w)\right|\right] \leqslant |v_1 - v_2|,$$

for all u, w, v_1, v_2 *between the values of* $u_{j+1}^{n+\eta_k}$, $u_j^{n+\eta_k}$, $u_{j-1}^{n+\eta_k}$ *and* $\Lambda_j^{n+\eta_k}$ *is defined by*

$$\Lambda_j^{n+\eta_k} = \begin{cases} \Delta t^n / \Delta x_j & \text{if } j \text{ or } j \pm 1 \in \mathscr{C}^n, \\ \sigma_{k+1} \Delta t^n / \Delta x_j & \text{otherwise}. \end{cases}$$

Then the $\lim_{\delta \downarrow 0} u^\delta(x, t)$ *exists in* $L^\infty([0, T], L^1_{\text{loc}}(\mathbf{R}))$ *and the limit is the entropy satisfying solution of* (4.1).

The proof of Theorem 4.1 can be found in [13].

We now show how to modify standard two-point numerical flux functions to incorporate a spatial grid that can vary with time.

Space-time is partitioned into strips $\cup_n \mathbf{R} \times [t^n, t^{n+1})$. Each strip is decomposed into pairwise disjoint trapezoids Ω_j^n. Denote the vertex coordinates of Ω_j^n by (x_j^n, t^n), (x_{j+1}^n, t^n), (ξ_j^n, t^{n+1}) and (ξ_{j+1}^n, t^{n+1}), and the line segment through the points (x_j^n, t^n) and (ξ_j^n, t^{n+1}) by S_j^n. The side S_{j+1}^n is given in a similar manner. See Figure 1.

$u^\delta(x, t)$ is defined in the usual way and to generate an *implicit* time differencing algorithm, $h_{F\cdot e_t}$ is given by $h_{F\cdot e_t}(u_1, u_2) = u_2$. Applying the finite volume method, we arrive at an implicit difference scheme

$$u_j^n \Delta \xi_j^n - \int_{x_j^n}^{x_{j+1}^n} u^\delta(x, t^n - 0)\, dx$$

$$+ \int_{S_{j+1}^n} h_{F\cdot n}(u_{j+1}^n, u_j^n)\, d\sigma + \int_{S_j^n} h_{F\cdot n}(u_{j-1}^n, u_j^n)\, d\sigma = 0. \tag{4.6}$$

Using (4.3), (4.6) may be cast into the more familiar conservative form. Define the local grid velocity by

$$\dot{X}_j = (\xi_j - x_j)/\Delta t.$$

A simple calculation will show that

$$|S_j| = \Delta t \left(1 + (\dot{X}_j)^2\right)^{1/2} \quad \text{and} \quad n = \pm \frac{\Delta t}{|S_j|}(-\dot{X}_j e_t + e_x).$$

Putting these together, (4.6) is written in conservation form as

$$u_j^n \Delta \xi_j^n + \Delta_+ |S_j^n| h_{\bar{j}}(u_j^n, u_{j-1}^n) = a_j^n \Delta x_j^n, \tag{4.7}$$

where

$$a_j^n = \frac{1}{\Delta x_j^n} \int_{x_j^n}^{x_{j+1}^n} u^\delta(x, t^n - 0)\, dx.$$

In (4.7), $h_{\bar{j}}$ represents any two-point numerical flux consistent with

$$\frac{f(u) - \dot{X}_j^n u}{\left(1 + (\dot{X}_j^n)^2\right)^{1/2}}.$$

Of course, when $\dot{X}_j = 0$, (4.7) reduces to the standard implicit method.

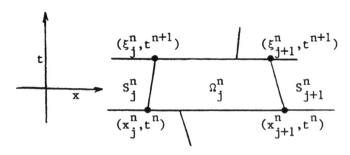

FIGURE 1

For the scalar equation, (4.7) can be justified when monotone numerical flux functions are employed. The proof of the following theorem can be found in [15].

THEOREM 4.2. *Suppose that the initial data of* (4.1) *are in* $L^{\infty} \cap$ BV. *Further, suppose that during mesh refinement the following conditions are maintained*:

(i) $x_{j+1}^n > x_j^n$, $\xi_{j+1}^n > \xi_j^n$,

(ii) $|\dot{X}_j^n| \leqslant K$.

Then $\lim_{\delta \downarrow 0} u^{\delta}(x, t)$ *exists in* L^{∞} $([0, T], L_{\text{loc}}^1)$ *and the limit is the entropy satisfying solution of* (4.1).

Other moving grid techniques are addressed in [15], explicit methods, etc. The reader is encouraged to see [5 and 15] for a broader treatment of this subject.

REFERENCES

1. M. G. Crandall and A. Majda, *Monotone difference approximations for scalar conservation laws*, Math. Comp. **34** (1980), 1–22.

2. B. Enguist and S. Osher, *One sided difference approximations for nonlinear conservation laws*, Math. Comp. **36** (1981), 321–351.

3. S. K. Godunov, A. V. Zabrodin and G. P. Prokopov, U.S.S.R. Comput. Math. and Math. Phys. **1** (1962), 1187–1219.

4. A. Harten and P. D. Lax, *A random choice finite-difference scheme for hyperbolic conservation laws*, SIAM J. Numer. Anal. **18** (1981), 289–315.

5. A. Harten and J. M. Hyman, *Self-adjusting grid methods for one-dimensional hyperbolic conservation laws*, J. Comput. Phys. **50** (1983), 235–269.

6. S. N. Kruzkov, *First order quasi-linear equations in several independent variables*, Math. USSR-Sb. **10** (1970), 217–243.

7. P. D. Lax, *Shock waves and entropy*, Contributions to Nonlinear Functional Analysis (E. H. Zarantonello, ed.), Academic Press, New York, 1971, pp. 603–634.

8. _____, *Hyperbolic systems of conservation laws and the mathematical theory of shock waves*, SIAM Regional Conf. Ser. Lectures in Applied Math., Vol. 11, 1972.

9. P. D. Lax and B. Wendroff, *Systems of conservation laws*, Comm. Pure Appl. Math. **13** (1960), 217–237.

10. O. A. Olenik, *Discontinuous solutions of nonlinear differential equations*, Uspekhi Mat. Nauk **31** (1957), 3–73; Amer. Math. Soc. Transl. (2) **26** (1963), 95–172.

11. S. Osher, *Numerical solution of singular perturbation problems and hyperbolic systems of conservation laws*, North-Holland Math. Stud., No. 47, North-Holland, Amsterdam and New York, 1981, pp. 179–205.

12. S. Osher and S. Chakravarthy, *Upwind schemes and boundary conditions with applications to Euler equations in general geometries*, J. Comput. Phys. **50** (1983), 447–481.

13. S. Osher and R. Sanders, *Numerical approximations to nonlinear conservation laws with locally varying time and space grids*, Math. Comp. **41** (1983), 321–336.

14. R. Sanders, *On convergence of monotone finite difference schemes with variable spatial differencing*, Math. Comp. **40** (1983), 91–106.

15. _____, *The moving grid method for nonlinear hyperbolic conservation laws*, SIAM J. Numer. Anal. (to appear).

16. _____, *The finite volume method for nonlinear hyperbolic conservation laws*, in preparation.

17. J. L. Steger and R. F. Warming, *Flux vector splitting of the inviscid gas dynamics equations with applications to finite difference methods*, NASA Technical Memorandum 78605, 1979.

DEPARTMENT OF MATHEMATICS, UNIVERSITY OF SOUTHERN CALIFORNIA, LOS ANGELES, CALIFORNIA 90089 - 0781

Lectures in Applied Mathematics
Volume **22**, 1985

Conforming Finite Element Methods for Incompressible and Nearly Incompressible Continua[1]

L. R. Scott and M. Vogelius

1. Introduction. We shall be interested here in finite element discretizations of problems involving an incompressibility condition. As model problems we consider the Stokes equations for the flow of a viscous, incompressible fluid and the equations of linear plane-strain elasticity for the deformation of an isotropic, nearly incompressible solid. In both cases the incompressibility condition takes the form of a divergence constraint. Although this is the most simple formulation, the proper understanding of how an approximate method satisfies the constraint represents an important step towards the understanding of more complicated situations, involving e.g. the Navier-Stokes equations or the equations of nonlinear elasticity. The finite element methods we study have the property that the approximations to the velocities, respecively to the displacements, are continuous; such methods are generally referred to as conforming.

2. The Stokes equations. Let Ω be a bounded polygonal domain in the plane, and let $\mathbf{U}$ and P solve

$$
\begin{aligned}
-\Delta \mathbf{U} + \nabla P &= \mathbf{F} &&\text{in } \Omega, \\
\nabla \cdot \mathbf{U} &= 0 &&\text{in } \Omega, \quad \text{and} \\
\mathbf{U} &= 0 &&\text{on } \partial\Omega.
\end{aligned}
\tag{2.1}
$$

1980 *Mathematics Subject Classification*. Primary 65N30, 76D05.

[1] This work was partially supported by NSF grant MCS 83-03242 (LRS) and ONR contract N00014-77-C-0623 (MV).

Here $\mathbf{U} = (U_1, U_2)$ represents the fluid velocities and P the pressure; the viscosity has been set to 1. To simplify the exposition we are assuming homogeneous boundary data on $\partial\Omega$. For the linear problem (2.1) we can convert inhomogeneous boundary data into an external force term $\mathbf{F}$. However, for a nonlinear problem one must deal directly with the inhomogeneous boundary data; cf. Gunzburger and Peterson [17]. Other possible boundary conditions could involve the normal fluid stresses

$$\sum_{j=1}^{2} \left(\frac{\partial U_i}{\partial x_j} + \frac{\partial U_j}{\partial x_i} \right) n_j - Pn_i, \qquad i = 1, 2,$$

but since stress boundary conditions are physically much more frequent when dealing with solids, we shall reserve these for our formulation of the boundary value problem for the equations of elasticity. For regularity results concerning the solution to (2.1) on a polygonal domain, see Kellogg and Osborn [20] and Osborn [29]. Note that P is determined only up to an additive constant.

3. The equations of linear elasticity. As before, let Ω be a bounded polygonal domain in the plane and consider the problem

$$-\Delta\mathbf{U} - \frac{1}{1 - 2\nu} \nabla(\nabla \cdot \mathbf{U}) = \mathbf{F} \quad \text{in } \Omega, \text{ and}$$

$$\sum_{j=1}^{2} \varepsilon_{ij}(\mathbf{U})n_j + \frac{\nu}{1 - 2\nu} \nabla \cdot \mathbf{U} n_i = g_i \quad \text{on } \partial\Omega, \, i = 1, 2,$$

(3.1)

$\mathbf{n} = (n_1, n_2)$ here denotes the outward unit normal to Ω, and $\varepsilon_{ij}(\mathbf{U})$ is the usual symmetric strain tensor

$$\varepsilon_{ij}(\mathbf{U}) = \frac{1}{2} \left(\frac{\partial U_i}{\partial x_j} + \frac{\partial U_j}{\partial x_i} \right).$$

The equations (3.1) are the equations of isotropic, plane-strain linear elasticity corresponding to the domain Ω. $0 < \nu < \frac{1}{2}$ is a material-dependent constant, the so-called Poisson's ratio, which describes the compressibility. The other constant, the shear modulus, that is needed in order to characterize fully an isotropic material has been absorbed into the external load $\mathbf{F}$ and boundary load $\mathbf{g}$. If $\mathbf{U}$ is a solution to (3.1) then the vector

$$\tilde{\mathbf{U}} = (U_1, U_2, 0)$$

solves the equations of 3-dimensional isotropic elasticity on the domain $\Omega \times \mathbb{R}$, with external load $\tilde{\mathbf{F}} = (\mathbf{F}, 0)$, independent of x_3, and boundary load $\tilde{\mathbf{g}} = (\mathbf{g}, 0)$, independent of x_3, on the vertical boundary $\partial\Omega \times \mathbb{R}$.

This is the reason for the notion of "plane-strain" in connection with (3.1). Values of ν near $\frac{1}{2}$ correspond to a nearly incompressible material.

Both the problem (2.1) and the problem (3.1) can (on a simply connected domain) be reduced to the solution of the biharmonic equation with Dirichlet boundary conditions. For (2.1) this follows through the introduction of a stream function such that $\mathbf{U} = \nabla \times \Psi$; the corresponding boundary conditions for Ψ are homogeneous Dirichlet. For (3.1) one first subtracts a particular solution corresponding to the external load $\mathbf{F}$ so that the resulting system has a vanishing external load. For such a system one may introduce an Airy stress function Φ ($\Delta^2 \Phi = 0$) so that

$$\left(\frac{\partial}{\partial x_1} \right)^2 \Phi = \sigma_{22}, \qquad -\frac{\partial^2}{\partial x_1 \partial x_2} \Phi = \sigma_{12}, \qquad \left(\frac{\partial}{\partial x_2} \right)^2 \Phi = \sigma_{11},$$

where

$$\sigma_{ij} = \frac{E}{1 + \nu} \left(\varepsilon_{ij}(\mathbf{U}) + \delta_{ij} \frac{\nu}{1 - 2\nu} \nabla \cdot \mathbf{U} \right)$$

denotes the stresses corresponding to the displacement $\mathbf{U}$ (E is the so-called Young's modulus). The boundary conditions for Φ are in this case (inhomogeneous) Dirichlet.

Regularity results and *a priori* estimates for solutions to (3.1) (as well as (2.1)) may thus be derived from the properties of solutions to the biharmonic equation on a polygonal domain (cf. Grisvard [15]).

4. Variational formulations. In order to introduce finite element discretizations of the equations (2.1) and (3.1) we have to cast them in a variational (or weak) form. This is done following the standard 3-step recipe:

(1) multiply each differential equation by a suitable test function,
(2) integrate the result over Ω,
(3) integrate by parts (to taste).

By multiplication of the first equation in (2.1) by $\mathbf{v}$ (vanishing on $\partial\Omega$) the above three steps lead to

$$a(\mathbf{U}, \mathbf{v}) + b(\mathbf{v}, P) = (\mathbf{F}, \mathbf{v}),$$

where $(\mathbf{F}, \mathbf{v})$ is the usual $[L_2(\Omega)]^2$ inner product,

$$a(\mathbf{U}, \mathbf{v}) = 2 \int_\Omega \sum_{i,j} \varepsilon_{ij}(\mathbf{U}) \varepsilon_{ij}(\mathbf{v}) \, d\mathbf{x} \quad \text{and} \tag{4.1}$$

$$b(\mathbf{v}, P) = -\int_\Omega \nabla \cdot \mathbf{v} P \, d\mathbf{x}. \tag{4.2}$$

Multiplying the second equation in (2.1) by a suitable function q and integrating over Ω, we get $b(\mathbf{U}, q) = 0$. The appropriate spaces of "test" functions **v** and q are given by

$$\mathring{H}^1(\Omega) \times \mathring{H}^1(\Omega) \quad \text{and} \quad L_2(\Omega). \qquad (4.3)$$

Here, $\mathring{H}^1(\Omega)$ is the standard Sobolev space of functions whose gradients are square integrable and whose traces vanish on $\partial\Omega$. With these spaces our "variational form" of (2.1) is

Find $\mathbf{U} \in [\mathring{H}^1(\Omega)]^2$ and $P \in L_2(\Omega)$ such that

$$a(\mathbf{U}, \mathbf{v}) + b(\mathbf{v}, P) = (\mathbf{F}, \mathbf{v}) \quad \text{for all } \mathbf{v} \in \left[\mathring{H}^1(\Omega)\right]^2,$$
$$b(\mathbf{U}, q) = 0 \qquad \text{for all } q \in L_2(\Omega). \qquad (4.4)$$

In order to find the variational formulation of (3.1) we multiply the first equation by $\mathbf{v} \in [H^1(\Omega)]^2$ (not vanishing on $\partial\Omega$) and integrate by parts. Because of the form of the boundary conditions this leads to

$$a(\mathbf{U}, \mathbf{v}) + b\left(\mathbf{v}, -\frac{2\nu}{1-2\nu} \nabla \cdot \mathbf{U}\right) = (\mathbf{F}, \mathbf{v}) + 2\langle \mathbf{g}, \mathbf{v}\rangle,$$

where $\langle \cdot, \cdot \rangle$ is the $[L_2(\partial\Omega)]^2$ inner product. The differential equation (3.1) therefore has the weak form

Find $\mathbf{U} \in [H^1(\Omega)]^2$ such that for all $\mathbf{v} \in [H^1(\Omega)]^2$

$$a(\mathbf{U}, \mathbf{v}) + b\left(\mathbf{v}, -\frac{2\nu}{1-2\nu} \nabla \cdot \mathbf{U}\right) = (\mathbf{F}, \mathbf{v}) + 2\langle \mathbf{g}, \mathbf{v}\rangle. \qquad (4.5)$$

This may be rewritten as

Find $\mathbf{U} \in [H^1(\Omega)]^2$ and $P \in L_2(\Omega)$ such that

$$a(\mathbf{U}, \mathbf{v}) + b(\mathbf{v}, P) = (\mathbf{F}, \mathbf{v}) + 2\langle \mathbf{g}, \mathbf{v}\rangle \quad \text{for all } \mathbf{v} \in \left[H^1(\Omega)\right]^2,$$
$$-\frac{1-2\nu}{2\nu}(P, q) + b(\mathbf{U}, q) = 0 \qquad \text{for all } q \in L_2(\Omega).$$

Setting $\nu = \frac{1}{2}$ this gives "formally" a Stokes equation of the form (4.4) only, of course, with different boundary conditions. Indeed the equations of elasticity for ν near $\frac{1}{2}$ may be viewed as a penalized version of a Stokes problem (for more details, see e.g. Temam [35]). Note that the equation (4.5) only has a solution provided the loads $\mathbf{F}$ and $\mathbf{g}$ are statically admissible, *i.e.* provided

$$(\mathbf{F}, \mathbf{R}) + 2\langle \mathbf{g}, \mathbf{R}\rangle = 0$$

for any rigid motion $\mathbf{R}$. The solution $\mathbf{U}$ is also only determined modulo a rigid motion; whenever we discuss the problem (4.5) we shall avoid this nonuniqueness by thinking of functions as equivalence classes modulo rigid motions (remember, a rigid motion is one for which $\varepsilon_{ij}(\mathbf{R}) = 0$, or $\mathbf{R}(x_1, x_2) = (-\gamma x_2 + \alpha, \gamma x_1 + \beta)$). In this paper we shall always assume that Ω is connected and for the Stokes equations we make P unique by imposing

$$\int_\Omega P\,d\mathbf{x} = 0.$$

This replaces $L_2(\Omega)$ in (4.4) by $\mathring{L}_2(\Omega) = L_2(\Omega) \cap \{q: \int_\Omega q\,d\mathbf{x} = 0\}$.

5. Stability of finite element approximations. Given a variational formulation such as (4.4) or (4.5) the finite element method consists of choosing finite-dimensional subspaces

$$\mathring{V}_h \subseteq \left[\mathring{H}^1(\Omega)\right]^2, \quad \Pi_h \subseteq \mathring{L}_2(\Omega), \quad \text{or} \quad V_h \subseteq \left[H^1(\Omega)\right]^2,$$

and replacing everywhere the infinite-dimensional spaces by their finite-dimensional counterparts.[2] The spaces V_h, $\mathring{V}_h$ and Π_h are typically made up of piecewise polynomial functions on some triangulation Σ_h, h denoting the mesh size. For the Stokes problem the discrete version thus becomes

Find $\mathbf{U}_h \in \mathring{V}_h$ and $P_h \in \Pi_h$ such that

$$
\begin{aligned}
a(\mathbf{U}_h, \mathbf{v}) + b(\mathbf{v}, P_h) &= (\mathbf{F}, \mathbf{v}) \quad \text{for all } \mathbf{v} \in \mathring{V}_h, \\
b(\mathbf{U}_h, q) &= 0 \qquad\quad \text{for all } q \in \Pi_h,
\end{aligned}
\tag{5.1}
$$

and for the equations of elasticity (4.5) it reads

Find $\mathbf{U}_h \in V_h$ such that

$$a(\mathbf{U}_h, \mathbf{v}) + b\left(\mathbf{v}, -\frac{2\nu}{1-2\nu}\nabla\cdot\mathbf{U}_h\right) = (\mathbf{F}, \mathbf{v}) + 2\langle \mathbf{g}, \mathbf{v}\rangle \quad \text{for all } \mathbf{v} \in V_h.
\tag{5.2}$$

As an example we consider a uniform triangulation which locally is as shown in Figure 1. Perhaps the simplest finite-dimensional subspaces are

$$V_h = \left[C^0 \text{ piecewise linear functions}\right]^2,$$

or

$$\mathring{V}_h = \left[C^0 \text{ piecewise linear functions that vanish on } \partial\Omega\right]^2 \tag{5.3}$$

and correspondingly

$$\Pi_h = \text{piecewise constants having integral zero (over } \Omega).$$

One reason why finite element methods for the equations of elasticity (near incompressibility), or the Stokes equations, are intriguing is that *the*

---

[2] Indeed V_h should be a subset of $[H^1(\Omega)]^2/\{\text{Rigid motions}\}$, but this is not explicitly mentioned; our convention is to identify a function with its equivalence class modulo rigid motions, whenever appropriate.

choices of spaces (5.3) *do not work for the mesh in Figure* 1! In the case of the discrete equations of elasticity (5.2) the relative error as $\nu \to \frac{1}{2}$ for fixed h approaches a constant, which is bounded away from 0 independently of h. For the discrete Stokes equation (5.1), $\mathbf{U}_h \equiv 0$ for any h. Below we give an explanation of this phenomenon in the case of the Stokes equations. Since the Stokes equations are the "limit" as $\nu \to \frac{1}{2}$ of the equations of elasticity this also intuitively explains the lack of uniformity in ν of the accuracy of the approximation (5.2) to the equations of elasticity. (Note, however, the difference in our boundary conditions.)

The second equation in (5.1) requires that $\mathbf{U}_h$ lies in the subspace

$$Z_h = \{\mathbf{v} \in \mathring{V}_h \colon b(\mathbf{v}, q) = 0 \ \forall q \in \Pi_h\}, \tag{5.4}$$

and part of (5.1) may thus be restated as

Find $\mathbf{U}_h \in Z_h$ such that

$$a(\mathbf{U}_h, \mathbf{v}) = (\mathbf{F}, \mathbf{v}) \quad \text{for all } \mathbf{v} \in Z_h. \tag{5.5}$$

The reason for the deficiency of the choice of spaces $\mathring{V}_h$ and Π_h, given by (5.3) is that the corresponding Z_h on the mesh Σ_h^1 consists of the zero element only. This can be seen as follows: we first observe that

$Z_h = \nabla \times \{C^1$ piecewise quadratics all of whose first

derivatives vanish on $\partial\Omega\}$,

with the curl operator $\nabla \times$ given by

$$\nabla \times \psi = \left(\frac{\partial}{\partial x_2}\psi, \ -\frac{\partial}{\partial x_1}\psi\right).$$

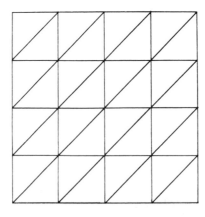

FIGURE 1. Uniform triangulation of size h, Σ_h^1.

A C^1 piecewise quadratic whose first derivatives vanish on $\partial\Omega$ can be extended by constants onto $\mathbb{R}^2 \setminus \Omega$. By subtracting the constant which is attained in the unbounded component of $\mathbb{R}^2 \setminus \Omega$, we thus obtain a C^1 piecewise quadratic with compact support. On the mesh shown in Figure 1 there is only one piecewise quadratic with compact support (cf. Morgan and Scott [26], Chui and Wang [7]), namely, the constant 0! Consequently, it follows that

$$Z_h = \nabla \times \{\text{constant function}\} = \{0\}.$$

The formulation (5.5) mimics the following equivalent version of the Stokes problem

Find $\mathbf{U} \in Z$ such that

$$a(\mathbf{U}, \mathbf{v}) = (\mathbf{F}, \mathbf{v}) \quad \text{for all } \mathbf{v} \in Z, \tag{5.6}$$

where Z denotes the subspace

$$Z = \left\{ \mathbf{v} \in \left[\mathring{H}^1(\Omega) \right]^2 : b(\mathbf{v}, q) = 0 \; \forall q \in \mathring{L}_2(\Omega) \right\}$$
$$= \left\{ \mathbf{v} \in \left[\mathring{H}^1(\Omega) \right]^2 : \nabla \cdot \mathbf{v} = 0 \right\}.$$

Note that (5.6) represents the well-known Hodge decomposition (cf. Temam [35]). The space Z_h, as defined in (5.4), is of course not necessarily contained in Z, but in many interesting cases this inclusion holds, i.e. $Z_h = \mathring{V}_h \cap Z$. Since $a(\cdot, \cdot)$ is a positive definite symmetric form that coerces the H^1-norm (on $\mathring{H}^1(\Omega)$) it follows in this case that

$$\|\mathbf{U} - \mathbf{U}_h\|_{H^1(\Omega)} \leqslant C \inf_{\mathbf{z} \in Z_h} \|\mathbf{U} - \mathbf{z}\|_{H^1(\Omega)}.$$

We would thus obtain a quasioptimal velocity approximation $\mathbf{U}_h$ if the following condition were to hold:

For any $\mathbf{U} \in Z$,

$$\inf_{\mathbf{z} \in Z_h} \|\mathbf{U} - \mathbf{z}\|_{H^1(\Omega)} \leqslant C \inf_{\mathbf{v} \in \mathring{V}_h} \|\mathbf{U} - \mathbf{v}\|_{H^1(\Omega)}, \tag{5.7}$$

with C independent of $\mathbf{U}$ and h.

DEFINITION I. A family of closed (not necessarily finite-dimensional) subspaces $W_h \subseteq [H^1(\Omega)]^2$ is called divergence-stable if
  (i) the spaces $\nabla \cdot W_h$ are closed in $L_2(\Omega)$,
  (ii) $\exists c > 0$, independent of h, such that

$$\sup_{\mathbf{w} \in W_h \setminus \{0\}} \frac{b(\mathbf{w}, q)}{\|\mathbf{w}\|_{H^1(\Omega)}} \geqslant c \|q\|_{L_2(\Omega)} \quad \text{for all } q \in \nabla \cdot W_h.$$

REMARK 5.1. The definition of divergence-stability as given above is equivalent to the requirement that there exists a uniformly bounded maximal right inverse for the divergence operator on the spaces W_h, i.e. there exists a family of linear operators $\mathscr{L}_h \colon \nabla \cdot W_h \to W_h$ such that

(i) $\nabla \cdot (\mathscr{L}_h q) = q,\ \forall q \in \nabla \cdot W_h$, and

(ii) $\|\mathscr{L}_h q\|_{H^1(\Omega)} \leqslant C\|q\|_{L_2(\Omega)}$,

with a constant C that is independent of h and q.

The condition (5.7) and the concept of divergence-stability are intimately related as shown by the following result.

PROPOSITION 5.1. *The spaces* $Z_h = \mathring{V}_h \cap Z$ *and* $\mathring{V}_h$ *satisfy* (5.7) *for any* $\mathbf{U} \in Z$, *with a constant* C *that is independent of* h *and* $\mathbf{U}$, *if and only if* $\mathring{V}_h$ *is divergence-stable.*

PROOF. Assume that $Z_h = \mathring{V}_h \cap Z$ and $\mathring{V}_h$ satisfy (5.7). For any $q \in \nabla \cdot \mathring{V}_h$ there exists $\mathbf{v} \in [\mathring{H}^1(\Omega)]^2$, such that

$$\nabla \cdot \mathbf{v} = q \quad \text{and} \quad \|\mathbf{v}\|_{H^1(\Omega)} \leqslant C\|q\|_{L_2(\Omega)}$$

(cf. Temam [35]).

Since $q \in \nabla \cdot \mathring{V}_h$ there also exists $\mathbf{v}_h \in \mathring{V}_h$ such that $\nabla \cdot \mathbf{v}_h = q$. Hence

$$\mathbf{v} - \mathbf{v}_h \in Z,$$

and therefore there exists $\mathbf{z}_h \in Z_h$ with

$$\begin{aligned}
\|\mathbf{v} - \mathbf{v}_h - \mathbf{z}_h\|_{H^1(\Omega)} &\leqslant C \inf_{\mathbf{w} \in \mathring{V}_h} \|\mathbf{v} - \mathbf{v}_h - \mathbf{w}\|_{H^1(\Omega)} \\
&\leqslant C\|\mathbf{v}\|_{H^1(\Omega)}.
\end{aligned} \tag{5.8}$$

The estimate (5.8) immediately leads to

$$\|\mathbf{v}_h + \mathbf{z}_h\|_{H^1(\Omega)} \leqslant C\|\mathbf{v}\|_{H^1(\Omega)} \leqslant C\|q\|_{L_2(\Omega)}. \tag{5.9}$$

Since also

$$\nabla \cdot (\mathbf{v}_h + \mathbf{z}_h) = q, \tag{5.10}$$

the field $\mathbf{w}_h = -(\mathbf{v}_h + \mathbf{z}_h) \in \mathring{V}_h$ satisfies

$$\frac{b(\mathbf{w}_h, q)}{\|\mathbf{w}_h\|_{H^1(\Omega)}} = \frac{\|q\|_{L_2(\Omega)}^2}{\|\mathbf{w}_h\|_{H^1(\Omega)}} \geqslant c\|q\|_{L_2(\Omega)}. \tag{5.11}$$

As a consequence of (5.9)–(5.11) we conclude that $\mathring{V}_h$ is divergence-stable.

The proof in the opposite direction is a simple consequence of the results of Babuška [2] and Brezzi [6]; for completeness we include the details of a proof here. For each $\mathbf{v} \in \mathring{V}_h$, let $\mathbf{z}_h \in Z_h = \mathring{V}_h \cap Z$ be its orthogonal projection with respect to the inner-product $a(\cdot, \cdot)$, and define $q_h \in \nabla \cdot \mathring{V}_h$ via

$$b(\mathbf{w}, q_h) = a(\mathbf{v} - \mathbf{z}_h, \mathbf{w}) \quad \text{for all } \mathbf{w} \in \mathring{V}_h.$$

Since $\mathring{V}_h$ is divergence-stable, q_h is well defined and satisfies

$$\|q_h\|_{L_2(\Omega)} \leqslant C\|\mathbf{v} - \mathbf{z}_h\|_{H^1(\Omega)}.$$

Now by definition of $\mathbf{z}_h$ and q_h,

$$
\begin{aligned}
c\|\mathbf{v} - \mathbf{z}_h\|_{H^1(\Omega)}^2 &\leqslant a(\mathbf{v} - \mathbf{z}_h, \mathbf{v} - \mathbf{z}_h) = a(\mathbf{v} - \mathbf{z}_h, \mathbf{v}) \\
&= b(\mathbf{v}, q_h) = b(\mathbf{v} - \mathbf{U}, q_h) \\
&\leqslant C\|\mathbf{v} - \mathbf{U}\|_{H^1(\Omega)}\|q_h\|_{L_2(\Omega)} \\
&\leqslant C\|\mathbf{v} - \mathbf{U}\|_{H^1(\Omega)}\|\mathbf{v} - \mathbf{z}_h\|_{H^1(\Omega)}
\end{aligned}
$$

for any $\mathbf{U} \in Z$. Thus we have, for any $\mathbf{U} \in Z$,

$$\|\mathbf{v} - \mathbf{z}_h\|_{H^1(\Omega)} \leqslant C\|\mathbf{v} - \mathbf{U}\|_{H^1(\Omega)}.$$

By the triangle inequality, this means that, for $\mathbf{U} \in Z$ and $\mathbf{v} \in \mathring{V}_h$,

$$\|\mathbf{U} - \mathbf{z}_h\|_{H^1(\Omega)} \leqslant \|\mathbf{U} - \mathbf{v}\|_{H^1(\Omega)} + \|\mathbf{v} - \mathbf{z}_h\|_{H^1(\Omega)} \leqslant C\|\mathbf{U} - \mathbf{v}\|_{H^1(\Omega)}, \tag{5.12}$$

provided $\mathbf{z}_h$ is the projection (relative to the inner product $a(\cdot, \cdot)$) of $\mathbf{v}$ onto $Z_h = \mathring{V}_h \cap Z$. The estimate (5.12) clearly implies (5.7). (Note that at no point in this proof did we use that $\mathring{V}_h$ is finite dimensional.)

REMARK 5.2. Proposition 5.1 *does not* assert that divergence-stability is always necessary in order to get optimal-rate velocity approximations. Indeed optimal-rate velocity approximations are in certain cases achieved with piecewise polynomial spaces $\mathring{V}_h$ that are not divergence-stable (cf. Remark 6.1).

We now suppose that $\mathbf{U}_h \in Z_h$ is known through the solution of (5.5), and we consider determining $P_h \in \Pi_h$ from the first equation in (5.1)

$$b(\mathbf{v}, P_h) = (\mathbf{F}, \mathbf{v}) - a(\mathbf{U}_h, \mathbf{v}) \quad \text{for all } \mathbf{v} \in \mathring{V}_h. \tag{5.13}$$

In order for P_h to be unique it is necessary and sufficient that there exists $c > 0$ such that

$$\sup_{\mathbf{v} \in \mathring{V}_h \backslash \{0\}} \frac{b(\mathbf{v}, q)}{\|\mathbf{v}\|_{H^1(\Omega)}} \geqslant c\|q\|_{L_2(\Omega)} \tag{5.14}$$

for all $q \in \Pi_h$. Furthermore, if we want $\|P_h\|_{L_2(\Omega)}$ uniformly bounded by $C(\|P\|_{L_2(\Omega)} + \|\mathbf{U}\|_{H^1(\Omega)})$ independently of h, then it is natural to require that the constant c in (5.14) is independent of h.

We note that, although it is mathematically simple to describe, it is in practice not always obvious how to use the equations (5.5) and (5.13) to solve for $\mathbf{U}_h$ and P_h (cf. Glowinski [14]).

DEFINITION II. Let W_h and Q_h be two families of closed subspaces of $[H^1(\Omega)]^2$ and $L_2(\Omega)$, respectively. The family W_h is said to be divergence-stable relative to Q_h if $\exists c > 0$, independent of h, such that

$$\sup_{\mathbf{w} \in W_h \setminus \{0\}} \frac{b(\mathbf{w}, q)}{\|\mathbf{w}\|_{H^1(\Omega)}} \geqslant c\|q\|_{L_2(\Omega)}$$

for all $q \in Q_h$.

REMARK 5.3. By comparison of the Definitions I and II, we note that a family of closed subspaces $W_h \subseteq [H^1(\Omega)]^2$ is divergence-stable if and only if it is divergence-stable relative to the family $\nabla \cdot W_h$.

Using the concepts of divergence-stability introduced in Definitions I and II, the question of stability of finite element approximations to the Stokes equations (2.1) and the equations of elasticity (3.1) is well understood.

PROPOSITION 5.2. *Suppose that the family of spaces $\mathring{V}_h \subseteq [\mathring{H}^1(\Omega)]^2$ is divergence-stable relative to the family $\Pi_h \subseteq \mathring{L}_2(\Omega)$. Let $(\mathbf{U}, P) \in [\mathring{H}^1]^2 \times \mathring{L}_2$ and $(\mathbf{U}_h, P_h) \in \mathring{V}_h \times \Pi_h$ denote the solutions to (4.4) and (5.1), respectively; then*

$$\|\mathbf{U} - \mathbf{U}_h\|_{H^1(\Omega)} + \|P - P_h\|_{L_2(\Omega)}$$

$$\leqslant C\left(\inf_{\mathbf{v} \in \mathring{V}_h} \|\mathbf{U} - \mathbf{v}\|_{H^1(\Omega)} + \inf_{q \in \Pi_h} \|P - q\|_{L_2(\Omega)} \right),$$

*and C is independent of h (and **F**).*

Proposition 5.2 follows directly from the results of Babuška [2] or Brezzi [6]. It reduces the question of convergence to one of approximation theory, and this inherently leads to contradictory requirements: on the one hand the spaces Π_h should have good approximation properties —on the other hand they should not be too big, since that would jeopardize stability. One very natural choice of Π_h seems to be $\Pi_h = \nabla \cdot \mathring{V}_h$.

PROPOSITION 5.3. *Suppose that the family of spaces $\mathring{V}_h \subseteq [\mathring{H}^1(\Omega)]^2$ is divergence-stable, and choose $\Pi_h = \nabla \cdot \mathring{V}_h$. Let $(\mathbf{U}, P) \in [\mathring{H}^1]^2 \times \mathring{L}_2$ and $(\mathbf{U}_h, P_h) \in \mathring{V}_h \times \Pi_h$ be the solutions to (4.4) and (5.1), respectively, then*
  (i)

$$\|\mathbf{U} - \mathbf{U}_h\|_{H^1(\Omega)} \leqslant C \inf_{\mathbf{v} \in \mathring{V}_h} \|\mathbf{U} - \mathbf{v}\|_{H^1(\Omega)},$$

  (ii)

$$\|P - P_h\|_{L_2(\Omega)} \leqslant C\left(\inf_{\mathbf{v} \in \mathring{V}_h} \|\mathbf{U} - \mathbf{v}\|_{H^1(\Omega)} + \inf_{q \in \Pi_h} \|P - q\|_{L_2(\Omega)} \right),$$

*and C is independent of h (and **F**).*

PROOF. This follows immediately from Proposition 5.1, and 5.2, and the analysis leading to Definition I.

In the following we shall denote by $\mathbf{U}^\nu \in [H^1]^2$, $\mathbf{U}_h^\nu \in V_h$, the solutions to (4.5) and (5.2), thus emphasizing the dependence on $0 < \nu < \frac{1}{2}$ of the solution to the problem of elasticity.

PROPOSITION 5.4. *Suppose that the family of spaces $V_h \subseteq [H^1(\Omega)]^2$ is divergence-stable. Then*

$$\|\mathbf{U}^\nu - \mathbf{U}_h^\nu\|_{H^1(\Omega)} + \frac{1}{1 - 2\nu}\left\|\nabla \cdot (\mathbf{U}^\nu - \mathbf{U}_h^\nu)\right\|_{L_2(\Omega)}$$

$$\leqslant C\left(\inf_{\mathbf{v} \in V_h}\|\mathbf{U}^\nu - \mathbf{v}\|_{H^1(\Omega)} + \inf_{q \in \nabla \cdot V_h}\left\|\frac{1}{1 - 2\nu}\nabla \cdot \mathbf{U}^\nu - q\right\|_{L_2(\Omega)}\right),$$

with a constant C that is independent of h and ν (and $\mathbf{F}$ and $\mathbf{g}$).

PROOF. For a fixed $0 < \nu < \frac{1}{2}$ this estimate with a constant C_ν, possibly depending on ν, follows directly from the results of Babuška [2]. The fact that C may be chosen independently of ν, provided the family V_h is divergence-stable, is e.g. proven in Vogelius [36].

Several ways to remedy the deficiency of the spaces in (5.3) have been suggested and analyzed in the literature. A nonconforming, piecewise linear, triangular element was proposed by Crouzeix and Raviart [10] which together with the Π_h in (5.3) yields optimal approximation. For computational experience with this element, see Pritchard, Renardy and Scott [31] and references therein. Increasing Z_h can also be achieved by choosing a special triangulation, as shown in Figure 2, although this is by no means obvious. The mesh consists of squares with diagonals drawn in; the intersections of the diagonals are special examples of what we shall denote "singular vertices" (the exact definition of a singular vertex will be given in the next section). It was discovered by Powell [30] that the space of C^1 piecewise quadratics on the mesh in Figure 2 has good approximation properties—there is roughly one degree of freedom per

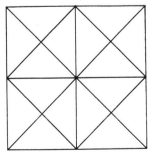

FIGURE 2. Special mesh Σ_h^2, with "singular vertices".

singular vertex. One must reduce Π_h in (5.3) appropriately at singular vertices; cf. later in this paper. For computational experience with, and analysis of, this method, see Fix, Gunzburger and Nicolaides [13], Malkus and Olsen [22], Mercier [24] and references therein.

From Definition II it follows that a possible method to achieve divergence-stability of W_h relative to Q_h (if it is not already there) is to increase W_h, or decrease Q_h (or both). The well-known Hood-Taylor [18] element is obtained from continuous piecewise quadratic velocities, piecewise linear pressures exactly this way by requiring the pressures also to be continuous. The element analyzed by Arnold, Brezzi and Fortin [1] has a similar flavor: for velocities the continuous piecewise linears are enriched by cubic bubble functions (the pressures however are taken to be continuous piecewise linear).

In the context of the equations of elasticity, it has been proposed to use "reduced integration" on the term involving b, while maintaining simple choices for both the meshes and V_h. We refer to Zienkiewicz [38], Malkus and Hughes [21] and references therein for computational experience with reduced integration methods, and an account of their relation to mixed methods.

An obvious question at this point seems to be:

> are the problems that we encountered with conforming piecewise linear elements also present for conforming piecewise polynomials of degree $p + 1$, $p \geqslant 1$?

This question is the focus of the next two sections.

6. Divergence-stability of high-order conforming spaces. Let us consider the spaces

$$V_h = \left[\mathscr{P}_h^{[p+1],0} \right]^2 = \left[C^0 \text{ piecewise polynomials of degree } \leqslant p + 1 \right]^2,$$
$$\mathring{V}_h = \left[\mathring{\mathscr{P}}_h^{[p+1],0} \right]^2 = \left\{ \mathbf{v} \in \left[\mathscr{P}_h^{[p+1],0} \right]^2 : \mathbf{v} = 0 \text{ on } \partial\Omega \right\}, \qquad (6.1)$$

on an arbitrary family of triangulations Σ_h of Ω. We have already seen that for $p = 0$ these spaces are not in general divergence-stable. Results due to de Boor and Höllig [11], and Jia [19] show that these spaces are not in general divergence-stable for $p = 1$ or 2; this lack of divergence-stability is well-documented by computational experience (cf. §7 for more details). We now turn our attention to the case $p \geqslant 3$. In order to state rigorously a result concerning the divergence-stability of the choice (6.1) for $p \geqslant 3$ we need some notation. A *singular internal vertex* [25] is one where precisely four triangles meet through the intersection of two straight lines, as shown in Figure 3. A *singular boundary vertex* is a vertex on $\partial\Omega$ where $1 \leqslant k \leqslant 4$ triangles meet through the intersection of two straight lines. There are four such possibilities as shown in Figure 4.

THEOREM 6.1 [**32**]. *Suppose that* Σ_h *is quasi-uniform and that no nonsingular vertices degenerate towards singular as h tends to zero. Then for any fixed* $p \geqslant 3$ *the spaces* (6.1) *are divergence-stable* (*in the sense of Definition* I). *Furthermore, the constant c entering into Definition* I *is bounded from below by*

$$c'p^{-K}, c' > 0,$$

with c' *and* K *independent of* h *and* p.

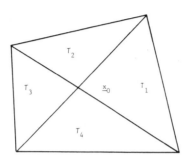

FIGURE 3. Singular internal vertex.

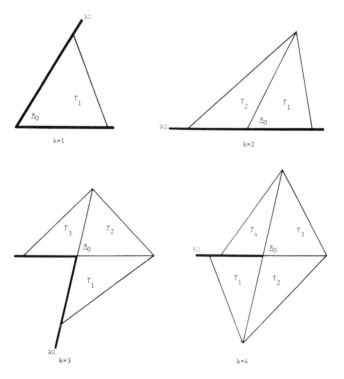

FIGURE 4. Singular boundary vertices.

REMARK 6.1. By quasi-uniform we mean that each triangle $\mathcal{T} \in \Sigma_h$ contains a ball of radius ρh, where $\rho > 0$ is independent of h. Whether this restriction is essential to guarantee divergence-stability of the spaces (6.1) for $p \geqslant 3$, we do not know. It is easy to see that the spaces (6.1) cannot be divergence-stable if a nonsingular vertex tends towards becoming singular (as seen later the dimension of the space $\nabla \cdot V_h$ or $\nabla \cdot \overset{\circ}{V}_h$ decreases by one when a vertex becomes singular). This does not imply that we will observe suboptimal convergence-rates when discretizing the Stokes problem or the equations of elasticity, $\nu \sim \frac{1}{2}$, using a mesh with nearly-singular (nonsingular) vertices, however, the ratio between the error in the finite element approximation of the velocities (or displacements) and the error of the best (H^1-) approximation may be arbitrarily large depending on the data. As stated in Theorem 6.1 the lower bound in the estimate of divergence-stability may approach zero algebraically in p^{-1} as $p \uparrow \infty$; we do not know whether this is indeed the case or whether the lower bound is uniform in p also. Note that divergence stability of V_h requires only that nonsingular *internal* vertices do not degenerate to singularity as $h \rightarrow 0$.

For the proof of the theorem, as well as more details concerning the results, we refer to Scott and Vogelius [32] and Vogelius [37]. In case of the equations of elasticity (4.5), Theorem 6.1 in combination with Proposition 5.4 leads to the fact that the spaces (6.1) give quasioptimal finite element approximations to

$$\left(\mathbf{U}^\nu, \frac{1}{1 - 2\nu} \nabla \cdot \mathbf{U}^\nu \right) \quad \text{for fixed } p \geqslant 3.$$

The question of convergence properties thus becomes one of approximation theory, and in addition to understanding the approximation properties of V_h, it is important to understand the character of the spaces $\nabla \cdot V_h$. The fact that the lower bound in the stability estimate approaches zero at most algebraically in p^{-1} as $p \uparrow \infty$ was used in Vogelius [36] to prove "almost" optimal convergence rates for the so-called p-version of the finite element method. Using an interpolation trick it was shown (on a smooth domain, with curved elements at the boundary) that the p-version converges at "almost" optimal rates in the energy norm, uniformly with respect to Poisson's ratio. "Almost" optimal here means that it converges at any rate strictly less than optimal. A slight variation of the present result (permitting for curved elements at the boundary) could likewise be used to verify a similar result for combinations of the h- and p-versions (*i.e.* when simultaneously changing mesh size and degree of the polynomials).

For the case of the Stokes problem it is clear from both Propositions 5.2 and 5.3 that, in addition to divergence-stability and good approximation properties of the spaces $\mathring{V}_h$, it is important to have good approximation properties of the corresponding pressure spaces Π_h. Since Theorem 6.1 guarantees divergence-stability relative to *any* subspace of $\nabla \cdot \mathring{V}_h$, this naturally leads to the question of finding a characterization of the spaces $\nabla \cdot \mathring{V}_h$. Part of this characterization (its necessity) is implicit in the work of Nagtegaal, Parks and Rice [27] and that of Mercier [23].

Let $\mathbf{x}_0$ be an internal vertex where four triangles meet with the common edges lying on either the x_1-axis or the x_2-axis, as shown in Figure 5.

Let $\mathbf{v} \in V_h$ and set $\mathbf{v}^i = \mathbf{v}|_{\mathscr{T}_i}$, $i = 1, \ldots, 4$. Since $\mathbf{v}$ is continuous, $v_1^1 - v_1^4$ vanishes identically on the edge e_1; consequently,

$$\frac{\partial}{\partial x_1} v_1^1 = \frac{\partial}{\partial x_1} v_1^4 \quad \text{on } e_1. \tag{6.2a}$$

Similarly, we get

$$\frac{\partial}{\partial x_1} v_1^3 = \frac{\partial}{\partial x_1} v_1^2 \quad \text{on } e_3, \tag{6.2b}$$

$$\frac{\partial}{\partial x_2} v_2^1 = \frac{\partial}{\partial x_2} v_2^2 \quad \text{on } e_2, \tag{6.2c}$$

and

$$\frac{\partial}{\partial x_2} v_2^3 = \frac{\partial}{\partial x_2} v_2^4 \quad \text{on } e_4. \tag{6.2d}$$

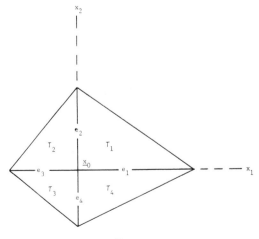

FIGURE 5.

The vertex $\mathbf{x}_0$ is common to all the edges e_1 through e_4 and by summation of (6.2a)–(6.2d) at $\mathbf{x}_0$ we get

$$\nabla \cdot \mathbf{v}^1(\mathbf{x}_0) + \nabla \cdot \mathbf{v}^3(\mathbf{x}_0) = \nabla \cdot \mathbf{v}^2(\mathbf{x}_0) + \nabla \cdot \mathbf{v}^4(\mathbf{x}_0), \quad (6.3)$$

which, by introduction of $\phi = \nabla \cdot \mathbf{v}$, may be restated

$$\sum_{i=1}^{4} (-1)^i \phi|_{\mathcal{T}_i}(\mathbf{x}_0) = 0. \quad (6.4)$$

The chain rule of differentiation gives that

$$\nabla \cdot (T^{-1}\mathbf{v} \circ T)(\mathbf{x}) = (\nabla \cdot \mathbf{v})(T\mathbf{x}) \quad (6.5)$$

for any invertible affine map $T: \mathbb{R}^2 \to \mathbb{R}^2$. If $\mathbf{v} \in V_h$ then so is $T^{-1}\mathbf{v}(T\mathbf{x})$ (on the corresponding triangulation), and it thus follows immediately from (6.5) that (6.3) (or (6.4)) must be satisfied at all internal vertices which can be obtained by an affine transformation of one as in Figure 5. These vertices are exactly the singular internal vertices.

This simple calculation thus shows that for any $p \geqslant 0$

$$\nabla \cdot V_h \subseteq \mathscr{P}_h^{[p],-1} = \{(\text{discontinuous}) \text{ piecewise polynomials } \phi \text{ of}$$

degree $\leqslant p$ that satisfy (6.4) at singular internal vertices}.

$$(6.6)$$

For the spaces $\mathring{V}_h$, with homogeneous Dirichlet boundary conditions, we find two additional sets of constraints, namely with $\phi = \nabla \cdot \mathbf{v}$:

$$\sum_{i=1}^{k} (-1)^i \phi|_{\mathcal{T}_i}(\mathbf{x}_0) = 0, \quad (6.7)$$

at any singular boundary vertex $\mathbf{x}_0$, where $1 \leqslant k \leqslant 4$ triangles meet as in Figure 4, and

$$\int_{\Omega} \phi \, d\mathbf{x} = 0. \quad (6.8)$$

If we define

$$\tilde{\mathscr{P}}_h^{[p],-1} = \{(\text{discontinuous}) \text{ piecewise polynomials } \phi \text{ of}$$
degree $\leqslant p$, that satisfy (6.4) at singular internal vertices,
(6.7) at singular boundary vertices and, furthermore,
satisfy (6.8)\},

then this shows that for any $p \geqslant 0$

$$\nabla \cdot \mathring{V}_h \subseteq \tilde{\mathscr{P}}_h^{[p],-1}. \quad (6.9)$$

It was shown in Vogelius [37] that (6.6) is indeed an equality provided $p \geqslant 3$. By a similar combinatorial argument it is proven in Scott and Vogelius [32] that (6.9) is also an equality for any $p \geqslant 3$ (remember, Ω is connected). This characterizes the spaces $\nabla \cdot V_h$ and $\nabla \cdot \mathring{V}_h$ on an arbitrary triangulation for any $p \geqslant 3$.

REMARK 6.2. From the characterization of $\nabla \cdot \mathring{V}_h$ given above it follows that the space

{continuous piecewise polynomials of degree $\leq p$ that satisfy (6.8)}

$$(6.10)$$

is a subspace of $\nabla \cdot \mathring{V}_h$ provided $p \geq 3$. Selecting the pressure space Π_h to be as in (6.10), $p \geq 3$, generalizes the methods studied by Hood and Taylor [18] for $p = 1$ and 2. Theorem 6.1 shows that these generalized Hood-Taylor elements, $p \geq 3$, lead to quasioptimal approximation. Bercovier and Pironneau [3] proved the same to be the case for the Hood-Taylor element for $p = 1$. It would be natural to conjecture that the case $p = 2$ also leads to quasioptimal approximations.

REMARK 6.3. If one uses the approach (5.5) to compute the finite element approximation to the velocity U in the Stokes problem, then it is important to have a local basis for Z_h. Assuming that Π_h has been selected so that $Z_h \subseteq Z$, it then follows that, for all $p \geq 0$,

$$Z_h = \{\mathbf{v} \in \mathring{V}_h : \nabla \cdot \mathbf{v} = 0\} = \nabla \times \mathring{\mathscr{P}}_h^{[p+2],1}, \qquad (6.11)$$

where

$$\mathring{\mathscr{P}}_h^{[p+2],1} = \{C^1 \text{ piecewise polynomials of degree} \leq p + 2$$

all of whose first derivatives vanish at the boundary}.

Note that elements of $\mathring{\mathscr{P}}_h^{[p+2],1}$ are constant on each connected component of $\partial\Omega$, however the constants need not be the same; if $\partial\Omega$ is connected then we shall always pick the constant to be zero for functions in $\mathring{\mathscr{P}}_h^{[p+2],1}$.

A basis for $\mathscr{P}_h^{[p+2],1}$, $p \geq 3$ (with no boundary conditions), was constructed in Morgan and Scott [25]; the corresponding basis for $\mathring{\mathscr{P}}_h^{[p+2],1}$, $p \geq 3$, is not altogether obvious, but is described in Scott and Vogelius [32]. The finite element method corresponding to the spaces $\mathring{V}_h$ and $\Pi_h = \nabla \cdot \mathring{V}_h$ in the case $p = 3$ is currently being tested numerically for the full (nonlinear, time-dependent) Navier-Stokes equations.

7. Results concerning lower-degree spaces. There are really three aspects of the finite element approximation to the Stokes equations or the equations of elasticity addressed in the previous sections.

(a) Bounds for a maximal right inverse for the divergence operator on V_h and $\mathring{V}_h$, or estimates for the divergence-stability of these spaces relative to families of pressure spaces Π_h.

(b) Characterization of the spaces $\nabla \cdot V_h$ and $\nabla \cdot \mathring{V}_h$.

(c) Determination of the approximation properties of the spaces $\mathring{V}_h \cap \{\mathbf{v} : \nabla \cdot \mathbf{v} = 0\}$ (or $V_h \cap \{\mathbf{v} : \nabla \cdot \mathbf{v} = 0\}$).

For the equations of elasticity (4.5) we are as indicted by Proposition 5.4 mainly concerned with (a) and (b) whereas the aspect (c) is also of importance when solving the Stokes problem (4.4). As pointed out in §§5 and 6 (particularly in Remarks 5.2 and 6.1), (c) may be of independent interest even though (a) and (b) do not have satisfactory answers. We shall now discuss the aspects (a)–(c) for the spaces

$$V_h = \left[\mathscr{P}_h^{[p+1],0} \right]^2 \quad \text{and} \quad \mathring{V}_h = \left[\mathring{\mathscr{P}}_h^{[p+1],0} \right]^2, \quad p = 0, 1, 2. \quad (7.1)$$

We know of very few results that are valid on a quite arbitrary triangulation for these low-degree spaces; as a consequence we shall restrict our attention to the triangulations, a local picture of which are shown in Figures 1 and 2 (we denote these by Σ_h^1 and Σ_h^2, respectively). On these triangulations, the dimension formula conjectured by Strang [33]

$$\dim\left(\mathscr{P}_h^{[p+2],1} \right) = \tfrac{1}{2}(p+3)(p+4)T - (2p+5)E_0 + 3V_0 + \sigma_0 \tag{7.2}$$

is known to hold also for $p = 0$, 1 and 2 (cf. Morgan and Scott [26]). Here T denotes the number of triangles, E_0 is the number of internal edges, V_0 is the number of internal vertices, and σ_0 is the number of singular internal vertices. On a simply connected domain, the null space of the divergence operator acting on $V_h = [\mathscr{P}_h^{[p+1],0}]^2$ is the curl of $\mathscr{P}_h^{[p+2],1}$; it thus follows as in Vogelius [37] (or Scott and Vogelius [32]) that, whenever (7.2) holds, $\nabla \cdot V_h$ must have the same dimension $\tfrac{1}{2}(p+2)(p+1)T - \sigma_0$ as the space $\mathscr{P}_h^{[p],-1}$ (cf. (6.6)). Since $\nabla \cdot V_h \subseteq \mathscr{P}_h^{[p],-1}$, we conclude that

$$\nabla \cdot V_h = \mathscr{P}_h^{[p],-1} \tag{7.3}$$

whenever (7.2) holds for a given triangulation, and in particular for $p = 0$, 1 and 2 on the triangulations Σ_h^1 and Σ_h^2. (For $p = 2$ we could indeed have derived (7.2) for much more general triangulations; cf. Morgan and Scott [26].) By a hole-filling procedure we can extend our argument to verify (7.3) even though the domain is not simply connected. The characterization (7.3) for the case $p = 0$ and the triangulation Σ_h^2 was also noted by Fix, Gunzburger and Nicolaides (cf. [13]).

In contrast the relation between the spaces $\nabla \cdot \mathring{V}_h$ and $\mathring{\mathscr{P}}_h^{[p],-1}$ is not nearly as simple for $p = 0$, 1 and 2 as the characterization given in the previous section, for $p \geqslant 3$, might lead one to believe. Let us start by considering the piecewise linear case, $p = 0$. If σ denotes the total number of singular vertices, including singular boundary vertices, then we have for the triangulations Σ_h^1 and Σ_h^2

$$\dim \mathring{V}_h = 2V_0 \quad \text{and} \quad \dim \mathring{\mathscr{P}}_h^{[0],-1} = T - \sigma - 1$$

(excluding the trivial case of a rectangle divided into two triangles by the diagonal, when both $\mathring{V}_h$ and $\tilde{\mathscr{P}}_h^{[0],-1}$ consist of 0 only). Using the relations

$$E - E_0 = V - V_0, \qquad E + E_0 = 3T$$

and Euler's formula

$$T + V - E = 1$$

(assuming Ω is simply connected), we get

$$\dim \tilde{\mathscr{P}}_h^{[0],-1} - \dim \mathring{V}_h = (E - E_0) - \sigma - 3; \qquad (7.4)$$

T, E_0, V_0 are as before, and E denotes the total number of edges, V the total number of vertices. From (7.4) we immediately conclude that if $\sigma < E - E_0 - 3$, then $\dim \mathring{V}_h < \dim \tilde{\mathscr{P}}_h^{[0],-1}$ and consequently $\nabla \cdot \mathring{V}_h$ is a *proper* subspace of $\tilde{\mathscr{P}}_h^{[0],-1}$ (see Nagtegaal, Parks and Rice [27] and Malkus and Hughes [21] for a similar constraint-counting method). If Ω is a rectangle and Σ_h^1 is used for a triangulation, then there are exactly 2 singular vertices (one in the upper-left and one in the lower-right corner of Ω); thus $\nabla \cdot \mathring{V}_h$ is a *proper* subspace of $\tilde{\mathscr{P}}_h^{[0],-1}$ except for the trivial case that Ω is divided into only two triangles. Since $\mathring{V}_h \cap Z = \{\mathbf{v} \in \mathring{V}_h: \nabla \cdot \mathbf{v} = 0\} = \{0\}$ on the mesh Σ_h^1, it follows that $\nabla \cdot$ is injective on $\mathring{V}_h$ and hence we get from (7.4) that

$$\dim \tilde{\mathscr{P}}_h^{[0],-1} - \dim \nabla \cdot \mathring{V}_h = (E - E_0) - 5$$

on a rectangle triangulated by the mesh Σ_h^1 (excluding the trivial case of only 2 triangles).

In the case that Ω is a rectangle divided into triangles by the mesh Σ_h^2 then the dimension of $\mathring{\mathscr{P}}_h^{[2],1} = \mathscr{P}_h^{[2],1} \cap \{\psi: \psi = \partial\psi/\partial n = 0 \text{ on } \partial\Omega\}$ has been calculated by Chui, Schumaker and Wang [8] to be

$$\dim \mathring{\mathscr{P}}_h^{[2],1} = \sigma - (E - E_0) + 4,$$

provided Ω has at least two boundary edges on each side. As a consequence of this and (6.11), it follows that

$$\dim \mathring{V}_h = \dim \nabla \cdot \mathring{V}_h + \sigma - (E - E_0) + 4,$$

and thus, using (7.4), we find that

$$\dim \tilde{\mathscr{P}}_h^{[0],-1} - \dim \nabla \cdot \mathring{V}_h = 1 \qquad (7.5)$$

for the choice (7.1), with $p = 0$, on this mesh Σ_h^2.

Now consider the case of piecewise quadratic fields, i.e. the choice $\mathring{V}_h = (\mathring{\mathscr{P}}_h^{[2],0})^2$ on the mesh Σ_h^1. The dimension of $\mathring{\mathscr{P}}_h^{[3],1} = \mathscr{P}_h^{[3],1} \cap \{\psi: \psi = \partial\psi/\partial n = 0 \text{ on } \partial\Omega\}$ on a rectangle triangulated by this mesh has been determined in Chui, Schumaker and Wang [9], and via calculations similar to those just given we find that

$$\dim \tilde{\mathscr{P}}_h^{[1],-1} - \dim \nabla \cdot \mathring{V}_h = 3. \qquad (7.6)$$

Note that this agrees with the results of Malkus and Olsen [**22**]. For the space $\mathring{V}_h = (\mathring{\mathscr{P}}_h^{[2],0})^2$ on Σ_h^2 the situation is different. Mercier [**23**] observed that in this case $\mathring{\mathscr{P}}_h^{[3],1}$ has a local basis given by the Fraeijs de Veubeke-Sander cubic macroelement. Using calculations similar to those above one can show that

$$\nabla \cdot \mathring{V}_h = \mathring{\mathscr{P}}_h^{[1],-1} \tag{7.7}$$

on a rectangle triangulated by the mesh Σ_h^2 ($p = 1$).

For the case $p = 2$, that is piecewise cubic fields, $\mathring{V}_h = (\mathring{\mathscr{P}}_h^{[3],0})^2$, with homogeneous boundary conditions on a rectangle triangulated by the mesh Σ_h^2, one can use the local basis for $\mathring{\mathscr{P}}_h^{[4],1}$, given via macroelements in Douglas et al. [**12**], to show that

$$\nabla \cdot \mathring{V}_h = \mathring{\mathscr{P}}_h^{[2],-1}. \tag{7.8}$$

To the best of our knowledge, no result is known for the mesh Σ_h^1 for $p = 2$, although it seems clear that one would not expect (7.8) to hold in general.

Whenever, as in (7.5) and (7.6), the codimension of $\nabla \cdot \mathring{V}_h$ in $\mathring{\mathscr{P}}_h^{[p],-1}$ is nonzero but independent of the mesh size, it is reasonable to conjecture that the orthogonal complements consist of global modes. For an explicit calculation of the global constraint on $\nabla \cdot \mathring{V}_h$ in the case corresponding to (7.5), see Olsen [**28**].

The previous discussion centered on the characterization of $\nabla \cdot V_h$ or $\nabla \cdot \mathring{V}_h$ for low-degree polynomials; we now turn to the aspects (a) and (c) listed at the beginning of this section. Proposition 5.1 shows that there is a close connection between approximation properties of the spaces $\mathring{V}_h \cap \{\mathbf{v}: \nabla \cdot \mathbf{v} = 0\}$ (or $V_h \cap \{\mathbf{v}: \nabla \cdot \mathbf{v} = 0\}$) and the divergence stability of $\mathring{V}_h$ (or V_h, respectively).

In the following we shall for simplicity assume that Ω is simply connected. Then $V_h \cap \{\mathbf{v}: \nabla \cdot \mathbf{v} = 0\} = \nabla \times \mathscr{P}_h^{[p+2],1}$ and $\mathring{V}_h \cap \{\mathbf{v}: \nabla \cdot \mathbf{v} = 0\} = \nabla \times \mathring{\mathscr{P}}_h^{[p+2],1}$, and this in combination with Proposition 5.1 shows that there is a close connection between approximation properties of $\mathscr{P}_h^{[p+2],1}$ and $\mathring{\mathscr{P}}_h^{[p+2],1}$ and divergence-stability. The next proposition elaborates more on that connection.

PROPOSITION 7.1. *Let $p = 0$, 1 or 2 and assume that there exist $\psi \in H^{p+3}(\Omega)$, $c > 0$, and $0 < \alpha \leqslant p + 3$ such that*

$$\inf\|\psi - \phi\|_{L_2(\Omega)} \geqslant ch^{p+3-\alpha},$$

where the inf is taken over $\phi \in \mathscr{P}_h^{[p+2],1}$ (for some quasi-uniform family of triangulations). Then

$$\inf_{q \in \nabla \cdot V_h \setminus \{0\}} \sup_{\mathbf{v} \in V_h \setminus \{0\}} \frac{b(\mathbf{v}, q)}{\|\mathbf{v}\|_{H^1(\Omega)} \|q\|_{L_2(\Omega)}} \leqslant Ch^\beta$$

for the subspaces $V_h = (\mathscr{P}_h^{[p+1],0})^2$, with $\beta = (p + 1)\alpha/(p + 3)$.

PROOF. By contradiction let us assume that

$$C_h = h^{-\beta} \inf_{q \in \nabla \cdot V_h \setminus \{0\}} \sup_{\mathbf{v} \in V_h \setminus \{0\}} \frac{b(\mathbf{v}, q)}{\|\mathbf{v}\|_{H^1(\Omega)} \|q\|_{L_2(\Omega)}}$$

is unbounded as $h \to 0$. Using the same argument as in the proof of Proposition 5.1, we get directly from the definition of C_h that

$$\inf_{\mathbf{z} \in V_h \cap \{\mathbf{v}: \nabla \cdot \mathbf{v} = 0\}} \|\mathbf{U} - \mathbf{z}\|_{H^1(\Omega)} \leqslant C C_h^{-1} h^{-\beta} \inf_{\mathbf{v} \in V_h} \|\mathbf{U} - \mathbf{v}\|_{H^1(\Omega)}, \qquad (7.9)$$

for any $\mathbf{U} \in [H^1(\Omega)]^2 \cap \{\mathbf{v}: \nabla \cdot \mathbf{v} = 0\}$. For any $\psi \in H^{p+3}(\Omega)$ we get that $\nabla \times \psi \in [H^{p+2}(\Omega)]^2 \cap \{\mathbf{v}: \nabla \cdot \mathbf{v} = 0\}$, and by insertion into (7.9) it follows that

$$\inf_{\phi \in \mathscr{P}_h^{[p+2],1}} \|\nabla \times \psi - \nabla \times \phi\|_{H^1(\Omega)} \leqslant C C_h^{-1} h^{-\beta} \inf_{\mathbf{v} \in V_h} \|\nabla \times \psi - \mathbf{v}\|_{H^1(\Omega)}$$

$$\leqslant C h^{p+1} C_h^{-1} h^{-\beta} \|\psi\|_{H^{p+3}(\Omega)}.$$

From this we conclude that for any $\psi \in H^{p+3}(\Omega)$

$$\inf_{\phi \in \mathscr{P}_h^{[p+2],1}} \|\psi - \phi\|_{H^2(\Omega)} \leqslant C C_h^{-1} h^{p+1-\beta} \|\psi\|_{H^{p+3}(\Omega)}.$$

Using the results of Bramble and Scott [5] (these results are valid if we require that all corners of Ω have interior angles $< 2\pi$) we thus get

$$\inf_{\phi \in \mathscr{P}_h^{[p+2],1}} \|\psi - \phi\|_{L_2(\Omega)} \leqslant C C_h^{-(p+3)/(p+1)} h^{p+3-\alpha} \|\psi\|_{H^{p+3}(\Omega)},$$

which, due to the fact that C_h is unbounded as $h \to 0$, produces a contradiction to the assumptions of this proposition.

Proposition 7.1 could just as easily have been verified for $H^{p+3}(\Omega) \cap \mathring{H}^2(\Omega)$, $\mathring{\mathscr{P}}_h^{[p+2],1} = \{\phi \in \mathscr{P}_h^{[p+2],1}: \phi = \partial\phi/\partial n = 0 \text{ at } \partial\Omega\}$ and $\mathring{V}_h = (\mathring{\mathscr{P}}_h^{[p+1],0})^2$.

In the light of Propositions 5.1 and 7.1, we shall center our discussion of the aspects (a) and (c) on known results concerning the approximation rates of $\mathscr{P}_h^{[p+2],1}$ and $\mathring{\mathscr{P}}_h^{[p+2],1}$ on the meshes Σ_h^1 and Σ_h^2.

First, suppose $p = 0$, the case of piecewise linear fields. On the mesh Σ_h^1 we know as previously stated that

$$\mathscr{P}_h^{[2],1} = \{0\};$$

this space obviously has no approximation rate associated to it and the assumptions of Proposition 7.1 are thus satisfied with $\alpha = 3$, i.e. the "best" constant in the "divergence-stability" estimate is $\leqslant Ch$; cf. Gunzburger and Nicolaides [16]. We do not anticipate that relaxing the boundary conditions would lead to any approximation rates for the spaces $\mathscr{P}_h^{[2],1}$ on the mesh Σ_h^1, though we do not know this for a fact. On the mesh Σ_h^2, Fix, Gunzburger and Nicolaides [13] give a construction of a uniformly bounded inverse for the divergence operator $(\nabla \cdot)^{-1}$: $\mathscr{P}_h^{[0]-1} \to V_h = [\mathscr{P}_h^{[1],0}]^2$, uniformly bounded, that is in $\mathscr{B}(H^{-1}; L_2)$ not $\mathscr{B}(L_2; H^1)$ as we are concerned with. Powell [30] and Mercier [23] both show that $\mathscr{P}_h^{[2],1}$ has optimal approximation rates on this mesh. Malkus and Olsen [22] *conjecture* that optimal approximation properties hold as well for $\mathring{\mathscr{P}}^{[2],1}$ on Σ_h^2, although this has not been proven. (Because of the local basis for $\mathring{\mathscr{P}}^{[2],1}$ given by Powell [30], one would have $\alpha = 1$ at worst.)

Now consider piecewise quadratic fields, *i.e.* $p = 1$. On the mesh Σ_h^1, de Boor and Höllig [11] show that approximation by $\mathscr{P}_h^{[3],1}$ is suboptimal by precisely one order of h in L_∞. This result (adapted to L_2) according to Proposition 7.1 leads to the conclusion that the "best" constant in the "divergence-stability" estimate for the piecewise quadratic spaces V_h is $\leqslant Ch^{1/2}$. Malkus and Olsen [22] report numerical evidence that the corresponding constant for the piecewise quadratic spaces $\mathring{V}_h$ (with boundary conditions) is $O(h)$ on the mesh Σ_h^1. On the mesh Σ_h^2 the situation is quite different, since both $\mathscr{P}_h^{[3],1}$ and $\mathring{\mathscr{P}}_h^{[3],1}$ are long known to have optimal approximation rates (cf. Mercier [23]); indeed, the recent results of Boland and Nicolaides [4] together with ther result (7.7) show that the corresponding spaces V_h and $\mathring{V}_h$ are divergence-stable on the mesh Σ_h^2 for $p = 1$. (For the macroelement, one takes simply the quadrilateral surrounding each singular vertex. Local stability is then guaranteed by (7.7) applied to each macroelement. The reason that these macroelements are locally stable in the sense of Boland and Nicolaides [4] is that the boundary vertices of each macroelement are nonsingular.)

For piecewise cubic fields ($p = 2$) on the mesh Σ_h^1, numerical experience (e.g. Szabo et al. [34]) had indicated that neither V_h nor $\mathring{V}_h$ was divergence-stable. Moreover, the results of Jia [19], similar to those of de Boor and Höllig [11] quoted above, also lead to this conclusion via Proposition 7.1. For the mesh Σ_h^2 and $p = 2$, the results of Douglas et al. [12] prove that (7.8) holds, and consequently that both V_h and $\mathring{V}_h$ are divergence-stable, in the same fashion as for the case $p = 1$ described previously. Finally, note that all the results for $p = 1$ and 2 for the mesh Σ_h^2 hold as well for a mesh based on the macroelement of Clough and Tocher; cf. Mercier [23].

To summarize the previous discussion:

(1) The characterization of the range of the divergence operator on spaces of continuous piecewise polynomials of degree at most $p + 1$, $p \geqslant 3$, given in §6, is widely valid also for $p \leqslant 2$, provided no boundary conditions are imposed. With boundary conditions imposed this characterization fails on the most natural triangulations.

(2) The divergence stability of the spaces $[\mathscr{P}^{[p+1],0}]^2$ or $[\mathring{\mathscr{P}}^{[p+1],0}]^2$ is intimately connected to the approximation properties of the spaces

$\mathcal{P}_h^{[p+2],1}$ or $\mathring{\mathcal{P}}_h^{[p+2],1}$. Since these spaces of C^1 piecewise polynomials have essentially one additional degree of freedom for each singular vertex, it is only natural that the spaces $[\mathcal{P}^{[p+1],0}]^2$ or $[\mathring{\mathcal{P}}^{[p+1],0}]^2$, $p \leqslant 2$, are much more likely to be divergence-stable, the more singular vertices the triangulation has. In this sense singular vertices are desirable when working with piecewise polynomials of degree $\leqslant 3$.

REFERENCES

1. Arnold, D. N., F. Brezzi and M. Fortin, *A stable finite element for the Stokes equations*, Calcolo (to appear).

2. Babuška, I., *Error-bounds for the finite element method*, Numer. Math. **16** (1971), 322–333.

3. Bercovier, M. and O. Pironneau, *Error estimates for the finite element method solution of the Stokes problem in the primitive variables*, Numer. Math. **33** (1979), 211–224.

4. Boland, J. M. and R. A. Nicolaides, *Stability of finite elements under divergence constraints*, SIAM J. Numer. Anal. **20** (1983), 722–731.

5. Bramble, J. H. and R. Scott, *Simultaneous approximation in scales of Banach spaces*, Math. Comp. **32** (1978), 947–954.

6. Brezzi, F., *On the existence, uniqueness and approximation of saddlepoint problems arising from Lagrangian multipliers*, RAIRO **8** (1974), 129–151.

7. Chui, C. K. and R.-H. Wang, *On smooth multivariate spline functions*, Math. Comp. (to appear).

8. Chui, C. K., L. L. Schumaker and R.-H. Wang, *On spaces of piecewise polynomials with boundary conditions. II. Type-1 triangulations*, Proc. 2nd Edmonton Conf. on Approximation Theory (to appear).

9. _____, *On spaces of piecewise polynomials with boundary conditions. III. Type-2 triangulations*, in Proc. 2nd Edmonton Conf. on Approximation Theory (to appear).

10. Crouzeix, M. and P.-A. Raviart, *Conforming and nonconforming finite element methods for solving the stationary Stokes equations*, RAIRO **7** (1973), 33–76.

11. de Boor, C. and K. Höllig, *Approximation order from bivariate C^1-cubics: A counterexample*, MRC Tech. Rep. No. 2389, (1982); Proc. Amer. Math. Soc. (to appear).

12. Douglas Jr., J., T. Dupont, P. Percell and R. Scott, *A family of C^1 finite elements with optimal approximation properties for various Galerkin methods for 2nd and 4th order problems*, RAIRO **13** (1979), 227–255.

13. Fix, G. J., M. D. Gunzburger and R. A. Nicolaides, *On mixed finite element methods for first-order elliptic systems*, Numer. Math. **37** (1981), 29–48.

14. Glowinski, R., *Numerical methods for nonlinear variational problems*, 2nd ed., Springer Ser. Comput. Phys., Springer-Verlag, Berlin and New York, 1984.

15. Grisvard, P., *Boundary value problems in non-smooth domains*. Part 2. Université de Nice, France, 1981.

16. Gunzburger, M. D. and R. A. Nicolaides, *The computational accuracy of some finite element methods for incompressible viscous flow problems*, ICASE Report No. 81-7, 1981.

17. Gunzburger, M. D. and J. S. Peterson, *On conforming finite element methods for the inhomogeneous stationary Navier-Stokes equations*, Numer. Math. **42** (1983), 173–194.

18. Hood, P. and C. Taylor, *Navier-Stokes equations via mixed interpolation*, Finite Elements in Flow Problems (Oden et al., eds.), Univ. of Alabama Press, Huntsville, Alabama, 1974, pp. 121–132.

19. Jia, R., *Approximation by smooth bivariate splines on a three-direction mesh*, MRC Tech. Rep. No. 2494, 1983.

20. Kellogg, R. B. and J. E. Osborn, *A regularity result for the Stokes problem in a convex polygon*, J. Funct. Anal. **21** (1976), 397–431.

21. Malkus, D. S. and T. J. R. Hughes, *Mixed finite element methods—reduced and selective integration techniques: a unification of concepts*, Comput. Methods Appl. Mech. Engrg., **15** (1978), 63–81.

22. Malkus, D. S. and E. T. Olsen, *Obtaining error estimates for optimally constrained incompressible finite elements*, Comput. Methods Appl. Mech. Engrg. (to appear).

23. Mercier, B., *A conforming finite element method for two dimensional, incompressible elasticity*, Internat. J. Numer. Methods Engrg. **14** (1979), 942–945.

24. _____ , *Lectures on topics in finite element solution of elliptic problems*, Springer-Verlag, Berlin and New York, 1979.

25. Morgan, J. and R. Scott, *A nodal basis for C^1 piecewise polynomials of degree $n \geq 5$*, Math. Comp. **29** (1975), 736–740.

26. _____ , *The dimension of the space of C^1 piecewise polynomials*, unpublished manuscript, 1975.

27. Nagtegaal, J. C., D. M. Parks and J. R. Rice, *On numerically accurate finite element solutions in the fully plastic range*, Comput. Methods Appl. Mech. Engrg. **4** (1974), 153–177.

28. Olsen, E. T., *Stable finite elements for non-Newtonian flows: first-order elements which fail the LBB condition*, Ph.D. Thesis, Illinois Institute of Technology, 1983.

29. Osborn, J. E., *Regularity of solutions of the Stokes problem in a polygonal domain*, Numerical Solution of Partial Differential Equations. III (B. Hubbard, ed.), Academic Press, New York, 1976, pp. 393–411.

30. Powell, M. J. D., *Piecewise quadratic surface fitting for contour plotting*, Software for Numerical Mathematics (D. J. Evans, ed.), Academic Press, New York, 1974, pp. 253–271.

31. Pritchard, W. G., Y. Renardy and L. R. Scott, *Tests of a numerical method for viscous, incompressible flow*. I. *Fixed-domain problems* (to appear).

32. Scott, L. R. and M. Vogelius, *Norm estimates for a maximal right inverse of the divergence operator in spaces of piecewise polynomials*, Tech. Note BN-1013, Inst. Phys. Sci. & Tech., Univ. of Maryland, 1983.

33. Strang, G., *Piecewise polynomials and the finite element method*, Bull. Amer. Math. Soc. **79** (1973), 1128–1137.

34. Szabo, B. A., P. K. Basu, D. A. Dunavant and D. Vasilopoulos, *Adaptive finite element technology in integrated design and analysis*, Report WU/CCM-81/1, Washington Univ., St. Louis, 1981.

35. Temam, R., *Navier-Stokes equations*, North-Holland, Amsterdam, 1977.

36. Vogelius, M., *An analysis of the p-version of the finite element method for nearly incompressible materials. Uniformly valid, optimal error estimates*, Numer. Math. **41** (1983), 39–53.

37. _____ , *A right-inverse for the divergence operator in spaces of piecewise polynomials. Application to the p-version of the finite element method*, Numer. Math. **41** (1983), 19–37.

38. Zienkiewicz, O. C., *The finite element method in engineering science*, McGraw-Hill, New York, 1971.

DEPARTMENT OF MATHEMATICS, UNIVERSITY OF MICHIGAN, ANN ARBOR, MICHIGAN 48109

DEPARTMENT OF MATHEMATICS AND INSTITUTE FOR PHYSICAL SCIENCE AND TECHNOLOGY, UNIVERSITY OF MARYLAND, COLLEGE PARK, MARYLAND 20742

Lectures in Applied Mathematics
Volume **22**, 1985

Vortex Methods and Turbulent Combustion[1]

J. A. Sethian[2]

ABSTRACT. We describe a series of numerical experiments using random vortex element techniques coupled to a flame propagation algorithm based on Huyghens' principle to model turbulent combustion. We solve the equations of zero Mach number combustion for the problem of a flame propagating in a swirling flow inside a closed vessel. We analyze the competing effects of viscosity, exothermicity, boundary conditions and pressure on the rate of combustion in the vessel.

A particularly challenging problem in the study of turbulent combustion within an internal combustion engine is the interaction between hydrodynamic turbulence and the propagation of a flame. The more reactants reached by the flame, the more energy released and the less unburnt fuel expelled at the end of a stroke. At high Reynolds numbers, turbulent eddies and recirculation zones form which affect the position of the flame and the distribution of unburnt fuel available for combustion. Conversely, exothermic effects along the flame front influence the fluid motion.

As one might expect, the full set of equations that describe the above phenomenon is highly complex; the equations are usually simplified in

[1] This work was supported in part by the Director, Office of Energy Research, Office of Basic Energy Sciences, Engineering, Mathematical and Geosciences Division of the U.S. Department of Energy under contract DE-AC03-76SF00098.
[2] Supported by a National Science Foundation Mathematical Sciences Postdoctoral Fellowship.
1980 *Mathematics Subject Classification*. Primary 76D05, 76N99; Secondary 35A40.

such a way as to highlight a particular aspect of the combustion process; see [3, 12, 18]. For example, most partial differential equation models of turbulent flow are based on a formulation of the Navier-Stokes equations with respect to a mean state, together with a set of equations to include such components as turbulence velocity and length scales. These models are of varying degrees of sophistication and complexity, ranging from zero-equation models ("mean-field closures") to higher order stress equation models. (For an excellent overview, see [3, 18].) From the combustion side, starting with Landau's work [11], questions of flame stability have received considerable attention over the past few decades, with much of the analysis concentrating on perturbation analysis of various models of combustion, containing such effects as mixing and flame speed dependence on curvature. An excellent, though now slightly outdated, review of such techniques may be found in [14].

Our work has been concerned with developing numerical methods to analyze the effects of such factors as viscosity, exothermicity, boundary conditions and pressure on the interaction between flame propagation and turbulent eddies. At the foundation of our investigations is the random vortex method [6], a grid-free numerical technique that is specifically designed for high Reynolds number flow, and portrays in a natural and effective manner the formation of turbulent eddies and coherent structures. Other applications of this method have included flow past an airfoil [4] and blood flow past heart values [15]. The random vortex method and the flame propagation algorithm described here were first used in a combustion setting to model turbulent combustion over a backwards facing step in [8]. In this review, we assemble the results of a series of numerical experiments we have designed to analyze some components in turbulent combustion; results described have been presented in [13, 19, 20, 21 and 22].

Statement of problem / equations of motion. We consider two-dimensional, viscous flow inside a closed square. On solid walls, we require that the normal and tangential fluid velocities be zero. Combustion is characterized by a single step irreversible chemical reaction; the fluid is a premixed fuel in which each fluid particle exists in one of two states, burnt and unburnt. When a particle burns, it undergoes an instantaneous increase in volume and becomes burnt. Thus, the flame is viewed as an infinitely thin front acting as a source of specific volume and separating the burnt regions from unburnt regions. We assume that the fluid is initially swirling in a counterclockwise direction and at $t = 0$ we ignite the fluid at a point halfway up the left side. Our goal is to analyze the interaction of the swirling fluid with the propagating flame front.

Our model is described by the *equations of zero Mach number combustion*, which hold under the assumption that the Mach number M is small, the initial pressure is spatially uniform within terms of order M^2, and the initial conditions for velocity, pressure and mass fraction are consistent within order M. Under these conditions, asymptotic limits of the full Navier-Stokes plus combustion equations can be taken to yield a set of equations that allows for large heat release, substantial temperature and density variations and interaction with the hydrodynamic flow field, but removes the detailed effects of acoustic waves and instead contains a time-dependent spatially uniform mean pressure term. This model can be viewed as existing "in between" constant density models, in which the fluid mechanics essentially decouples from the hydrodynamics, and the fully compressible combustion equations. The full derivation of this model may be found in [13]; a related model for thermally driven buoyant flows applicable to problems in fire research may be found in [17]. We summarize the equations for zero Mach number combustion below.

Let $\mathbf{v}$ be the fluid velocity vector and let $\mathbf{v} = \mathbf{w} + \nabla\varphi$ be the unique decomposition of $\mathbf{v}$ into a divergence-free component $\mathbf{w}$ and a curl-free component $\nabla\varphi$. We take the curl of the zero Mach number momentum equation to produce the *vorticity transport equation*

$$\frac{D\xi}{Dt} = \frac{1}{R}\nabla^2\xi, \tag{1}$$

where $\xi = \nabla \times \mathbf{w}$ is the vorticity and D/Dt is the total derivative $\partial_t + (\mathbf{v} \cdot \nabla)$. Here, we have ignored the term $(\nabla \times \nabla P/\rho)$ which corresponds to vorticity production across the flame front. (We hope to assess the importance of this term at a later date.) The boundary conditions are that $\mathbf{v} = 0$ on the boundary of the domain.

We view the flame front as a curve separating the burnt and unburnt regions. Let $\mathbf{r}(t)$ be a point on the front. The front burns normal to itself with speed k and is advected by the flow yielding the *Eikonal equation for the flame front*

$$d\mathbf{r}/dt = k\mathbf{n}(\mathbf{r}) + \mathbf{v}_u(\mathbf{r}), \tag{2}$$

where $\mathbf{v}_u$ is the fluid velocity on the unburnt side, $\mathbf{n}$ is the unit normal to the front at $\mathbf{r}(t)$, and the burning speed k is determined from the mass flux m across the flame front by

$$k = m(\rho_u(t), P(t))/\rho_u(t). \tag{3}$$

Here, ρ_u is the density of the unburnt fluid and P is the mean pressure. The rise in pressure, which results from fluid expansion along the front, depends on the length of the front and the volume of the vessel and is given by the *nonlinear O.D.E. for the mean pressure* $P(t)$

$$\frac{dP}{dt} = \frac{q_0 \gamma m}{\text{Vol}} L, \tag{4}$$

where q_0 is the nondimensional heat release, Vol is the volume of the vessel, and L is the length of the flame front. We assume a γ-gas law and take the mass flux to be of the form

$$m(\rho_u, P) = Q\rho_u^{1/2}P^{1/2}, \tag{5}$$

where Q is the local laminar flame velocity (see [3]); ρ_u may then be obtained from the pressure through the relation

$$\rho_u = (P(t))^{1/\gamma}\rho_u(0), \tag{6}$$

where $\rho_u(0)$ is the density of the unburnt fluid initially. Finally, the Neumann compatibility condition yields the *elliptic equation for the exothermic velocity field* $\nabla\varphi$

$$\nabla^2\varphi = \frac{1}{\gamma P}\left(-\frac{dP}{dt} + q_0 \gamma m\delta_F\right), \tag{7}$$

where δ_F is the surface Dirac measure concentrated on the flame front. Equations (1)–(7) form our equations of motion.

Numerical algorithm. Faced with the above set of equations, a standard method would be to employ finite difference techniques to produce a discrete approximation to all of the derivatives, and then solve the resulting set of algebraic equations. When applied to the turbulence part of combustion models, some of the problems inherent in such techniques are (1) the necessity of a fine grid in the boundary layer region near walls where sharp gradients exist; (2) the introduction of numerical diffusion (the error associated with the approximation equations looks like a diffusion term and hence places a computational upper bound on the size of the Reynolds number that can be effectively modelled); and (3) the intrinsic smoothing of finite difference schemes which damps out physical instabilities. The random vortex method, introduced in [6], is a grid-free approximation to the equations of viscous flow at high Reynolds number that avoids the introduction of mean states and turbulence closure relations, and concentrates on following the motion of vorticity by means of a collection of vorticity approximation elements. This technique avoids the averaging and smoothing associated with finite difference formulations and allows us to follow the development of large-scale coherent structures in the flow.

When one considers finite difference approximations to the equation for flame propagation (2), a typical method is to place marker particles along the boundary of the burnt region and formulate a set of ordinary differential equations corresponding to the motion of these marker particles. At each time step, interpolation through these markers provides an approximation to the position of the flame front. There are some problems involved with such an algorithm. It is difficult to accurately determine the normal direction (needed in (2)) from such an algorithm, and hence the numerical approximation to the propagating front usually becomes unstable and develops oscillations. Furthermore, it has been shown (see [19]) that the propagating front can develop cusps, analogous to shocks, where the front ceases to be differentiable and the normal is no longer defined. The technique of adding and subtracting marker particles as the front moves requires assumptions about differentiability and bounds on curvature. Another problem associated with marker techniques is a topological one; when two burnt regions burn into each other, such a method must "decide" how and which markers are connected and eliminate those no longer on the boundary of the flame. The numerical technique we use is based on a "volume" fraction algorithm; the technique does not require a determination of the normal direction and is not subject to the topological issue mentioned above.

Our technique will be to keep track of the vorticity as a way of computing $\mathbf{w}$ and to keep track of the flame front as a way of computing $\nabla\varphi$; combining these two at any time will yield the full velocity $\mathbf{v} = \mathbf{w} + \nabla\varphi$. We divide the square D into two regions; an interior region where we solve the full vorticity transport equation together with the boundary condition $\mathbf{v} \cdot \mathbf{n} = 0$ on ∂D (normal component vanishes), and a boundary layer region where we solve the Prandtl boundary layer equations together with the boundary condition $\mathbf{v} \equiv 0$ on ∂D (no-slip). In both regions, we use the technique of operator splitting to first update the vorticity with respect to the advection term and then with respect to the diffusion term. Similarly, we update the flame position by first allowing it to burn normal to itself and then by advecting the flame by the hydrodynamic flow field, calculating the exothermic velocity field $\nabla\varphi$ from the elliptic equation (7). Finally, we solve the nonlinear ordinary differential equation to compute the resulting rise in pressure. The separate components are as follows:

Computation of w. We update ξ (equation (1)) where $\xi = \nabla \times \mathbf{w}$ by following the motion of vorticity through the use of vortex "blobs" as introduced by Chorin in [6]. Knowledge of the position of these blobs at any time provides $\mathbf{w}$. We briefly describe the method; for details, see [6, 8, 21].

We have the vorticity advection equation $\partial_t \xi = -(\mathbf{w} \cdot \nabla)\xi$. Since $\nabla \cdot \mathbf{w} = 0$, there exists a stream function ψ such that $\mathbf{w} = (\psi_y, -\psi_x)$ and $\xi = -\nabla^2\psi$. We may write velocity as a function of vorticity through the fundamental solution to the Laplacian, namely

$$\psi(\bar{x}, t) = \int G(\bar{x} - \bar{x}')\xi(\bar{x}', t)\,d\bar{x}', \qquad (8)$$

where $G(\bar{x} - \bar{x}') = (1/2\pi)\log|\bar{x} - \bar{x}'|$ and $\bar{x} = (x, y)$. Hence

$$\mathbf{w} = \int K(\bar{x} - \bar{x}')\xi(\bar{x}', t)\,d\bar{x}', \qquad (9)$$

where $K(\bar{x} - \bar{x}') = (-y, x)/(2\pi|\bar{x} - \bar{x}'|)$. Let $\bar{x}(t)$ be the position of a particle moving in a fluid at time t. Since vorticity is advected by its own velocity field, we have that $\xi(\bar{x}(t)) = \xi(\bar{x}(0))$, that is, each particle "carries" its own vorticity with it. Our technique will be to exploit this fact; we place N of particles in the flow at $t = 0$ and follow their motion. At any later time, we have a distribution of these "delta functions" of vorticity which can be "smoothed out" to allow one to compute the resulting velocity field $\mathbf{w}$ through (9). (Alternatively, one can view this this smoothing process as something that happens to the kernel K.) There are two obvious numerical parameters involved in the above; the number N of vortex "blobs" used to describe the initial vorticity distribution and the smoothing factor σ used to compute the velocity field. We use the smoothing structure introduced in [6]; consider N vortex "blobs" placed on an initial grid in the domain, each with "smoothed" stream function

$$\psi(\bar{x}) = \begin{cases} -(k_i/2\pi)\log|\bar{x}|, & r \geqslant \sigma, \\ -(k_i/2\pi)(|\bar{x}|/\sigma + \log|\bar{x}| - 1), & r < \sigma, \end{cases}$$

where k_i is the strength (vorticity) of the ith vortex blob. To go from one time step to the next, we use the positions of the vortex blobs to determine the velocity field from (9) and numerically update these positions using Heun's method. Convergence of the vortex method was first established in [9]; for work relating to theoretical aspects of this method, see [1, 2, 9 and 10]. To satisfy the normal boundary condition $\mathbf{w} \cdot \mathbf{n} = 0$, we add a potential flow to the above motion (which, of course, adds no vorticity).

To update the vorticity with respect to the diffusion term $\frac{1}{R}\nabla^2\xi$, after the advection step we allow the vortex elements to undergo a random step, drawn from a Gaussian distribution with mean zero and variance $2\Delta t/R$. Since the random walk constitutes a solution to the diffusion equation, the combined motion of the vortex elements approximates the solution to the full vorticity transport equation (1). For details see [6, 21].

In the boundary layer, we employ similar techniques, only here our vortex elements are discrete, finite length "sheets" of vorticity. Once again, we use operator splitting to separate the Prandtl boundary layer equations into

(1) an advection equation

$$\partial_\xi = -(\mathbf{w} \cdot \nabla)\xi, \qquad \xi = -\partial_y w_x,$$

$$\nabla \cdot \mathbf{w} = 0 \quad \text{on } \partial D, \qquad \mathbf{w}(x, y = \infty) = W_\infty(x),$$

where $\mathbf{w} = (w_x, w_y)$ and W_∞ is the velocity as seen at infinity from the solid wall (the equations are written with respect to a solid wall lying on the x-axis), and

(2) a diffusion equation

$$\partial_t \xi = \tfrac{1}{R}\partial_y^2 \xi$$

(note that in this approximation diffusion only takes place in a direction normal to the solid wall). The x component of the velocity (w_x) is written as a function of vorticity and the y component (w_y) is written as a function of w_x through the incompressibility relation. As before, the positions of the vortex elements are used to approximate the vorticity distribution, allowing one to compute the advection field (w_x, w_y). The vortex sheets are advanced under this advection field and allowed to undergo a random walk in the y direction in response to the diffusion term. In addition, newly created vortex sheets are added at solid walls whenever necessary to satisfy the no-slip condition. Information is passed between the interior and the boundary layer in the following manner: the velocity from the vortex blob calculation tangential to the wall is taken as the velocity W_∞ seen at infinity from the boundary layer. Sheets diffuse away from the wall into the interior and become vortex blobs. Conservation of circulation is maintained; during this exchange, when a sheet moves too far from the wall, it becomes a blob with proper strength and vice versa. The velocity field $\mathbf{w}$ can be obtained at any time from the positions of the vortex elements; for details see [7, 21].

Computation of flame motion and $\nabla\varphi$. We keep track of the position of the flame by introducing a square grid ij on the domain and assigning each cell a number f_{ij} between 0 and 1 (a "volume fraction"; see [16]) corresponding to the amount of burnt fluid in that cell at any given time. The algorithm advances the front in a given direction by drawing in each cell for which $0 \leqslant f_{ij} \leqslant 1$ an interface which represents the boundary between the burnt and unburnt fluid and moving that interface to provide a new set of volume fractions. The orientation of the interface depends on the value of f_{ij} in both the cell and its neighbors; at any time step, the position of the flame front can be reconstructed from the field of volume fractions. The position of the flame is advanced in response to burning and advection. Burning is accomplished by allowing each cell to ignite all of its neighbors at the prescribed rate k determined from (3); this is an approximation based on Huyghens' principle, which states that the envelope of all disks centered at the front corresponds to the front displaced in a direction normal to itself; see [5]. In fact, it can be shown that this algorithm capitalizes on the geometric nature of flame propagation described in [19]. After the burning is accomplished, the exothermic velocity field $\nabla\varphi$ is determined (equation (7)) and the full velocity field $\mathbf{v} = \mathbf{w} + \nabla\varphi$ is used to advect the flame (as well as the vortex elements). Finally, the pressure is updated according to equation (5).

Results. We have performed a series of numerical experiments to analyze the various factors described in equations (1)–(7). In all of the cases described below, we shall consider a square vessel with sides of length 1m. When the initial condition is a counterclockwise swirl, this will be produced by a vortex placed in the center of the square of sufficient strength so that the velocity tangential to any wall at its midpoint is 1m/s. The initial conditions $P(0) = 1$ and $\rho_u(0) = 1$ were taken, and for viscous calculations we assumed a Reynolds number of 1000. The calculation in [8] for turbulent flow behind a step assumed a propane-air mixture with a laminar burning velocity of 12 cm/s and an inlet velocity of 6 m/s; this inlet velocity was taken as a characteristic speed scale to provide a nondimensional laminar flame speed of $Q = .02$ for a flow of unit inlet speed. Since the velocity induced by our initial vortex increases as we approach the center, it is not clear what to choose as a characteristic velocity. In [21], we took a nondimensional laminar flame speed of $Q = .14$, corresponding to a characteristic velocity 7 times that of the tangential boundary velocity; in the below results, this characteristic velocity will be used in conjunction with Q. Details about the numerical parameters used in the below calculations may be found in [20, 21 and 22].

Hydrodynamics: inviscid/viscous. In Figure 1, we show the results of a hydrodynamics calculation (no flame) comparing inviscid flow to viscous flow. Results are displayed on a 30 × 30 grid placed in the flow, where the magnitude of the vector at each point denotes the relative speed of the flow. Figure 1A, which is the initial flow, remains unchanged for all time in the inviscid case; the sole vortex remains at the center and the normal boundary condition is satisfied through the potential flow. In the viscous calculation (Figure 1A–F), small counterrotating eddies grow in each corner in response to the no-slip boundary conditions. These eddies grow, break away and diffuse downstream, and are replaced by a new set.

Flame propagation: pressure and the exothermic velocity field. In the next experiment, a motionless, inviscid fluid is ignited at the center of a closed square. We take a local laminar flame speed $Q = .2$. If the density of the burnt gas is the same as that of the unburnt gas, then the

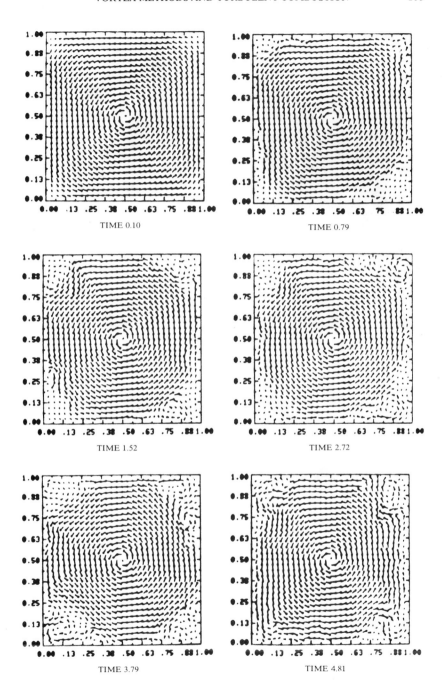

FIGURE 1.

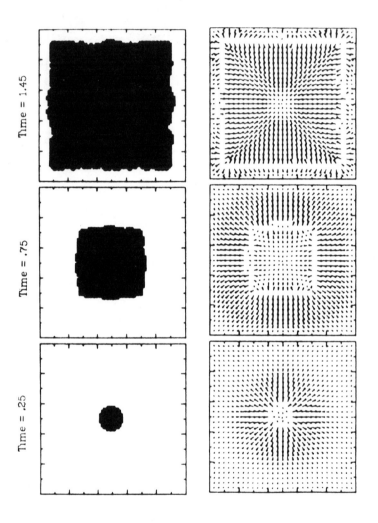

FIGURE 2. Inviscid, motionless fluid ignited in center, $g_0 = 1.333$.

nondimensional heat release is zero ($q_0 = 0$), the pressure remains constant ($dP/dt = 0$ in (4)) and the exothermic velocity field $\nabla\varphi$ is identically zero. In this case, the fluid remains still and the flame front is an expanding circle with origin at the center of the square. On the other hand, with $q_0 \neq 0$, fluid motion is induced by the propagating flame. In Figure 2, we show results in which $q_0 = 1.3333$; this corresponds to an initial ratio of burnt/unburnt of five to one. (Here, we assumed an inviscid fluid, hence the no-slip condition is violated.) The black region corresponds to the burnt region and once again the velocity is displayed

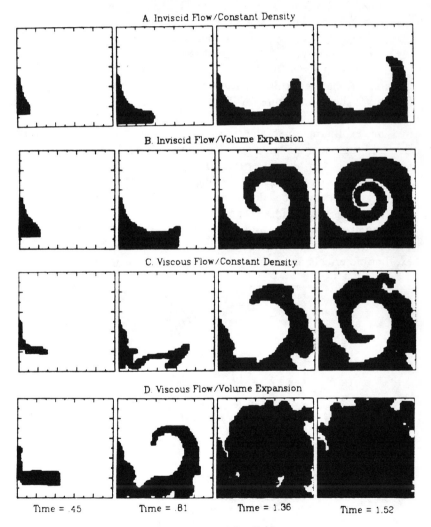

FIGURE 3. Swirling fluid.

on a 30×30 grid. One can clearly see the mechanism by which the boundary shapes the front; although the front starts off circular, it soon becomes squarelike in response to the boundary conditions on $\nabla\varphi$ and thus "burns" into the corners. When the volume was completely burnt, the pressure in the vessel is 2.93 and $k = .24$ (as compared with $k = .2$ at $t = 0$).

Hydrodynamics + combustion: exothermicity and viscosity. Next, we analyze the relative effects of viscosity and exothermicity on the flame motion. Four different experiments were performed with $Q = .14$:

A. Inviscid flow with $q_0 = 0$ (Inviscid/Constant Density),
B. Inviscid flow with $q_0 = 1.3333$ (Inviscid/Volume Expansion),
C. Viscous flow with $q_0 = 0$ (Viscous/Constant Density),
D. Viscous flow with $q_0 = 1.3333$ (Viscous/Volume Expansion).

In the two viscous runs, the flow was started two seconds before ignition so that recirculation zones would have time to develop. The results are shown in Figure 3A–D. In the inviscid, constant density case, the flame is smoothly advected by the large vortex in the center. In the inviscid, exothermic case, the velocity field produced by volume expansion and the rise in pressure and flame speed cause the flame to spiral in towards the center at a faster rate. In the viscous, constant density case, the flame front is twisted by the eddies that develop in the corners; the flame is carried over the eddies and dragged backwards into the corners. The effect of these eddies is to extend the length of the flame front, bringing it into contact with unburnt fuel and increasing the rate at which the vessel becomes fully burnt. Finally, when both viscous and exothermic effects are combined, the flame is both wrinkled due to the

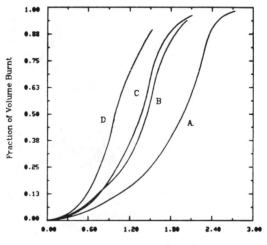

FIGURE 4.

turbulence of the flow (hence increasing the surface area of the flame) and carried by the exothermic field; in addition both the flame speed and pressure increase. The effect of these factors is to greatly decrease the amount of time required for complete conversion of reactants to products. In Figure 4 we plot the amount of volume burned versus time elapsed since ignition, illustrating the above comments.

Flame wrinkling due to viscosity as a function of laminar flame speed. Finally, to further analyze this mechanism of flame wrinkling due to hydrodynamic turbulence, with Reynolds number 1000 we repeated the viscous, constant density experiment with local laminar flame speed Q = .02, .06, .1, .14 and .2. In Figure 5, we plot the difference in volume burnt between the viscous and the inviscid case against the time elapsed since ignition for each of the above flame speeds. It is obvious that the lower the flame speed the longer the time required for the vessel to become completely burnt. However, as the flame speed decreases, viscosity plays an increasingly more important role in the combustion process, as can be seen by noting that the maximum difference between the

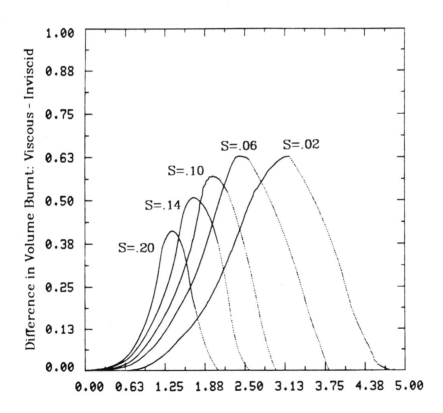

FIGURE 5.

viscous and inviscid case increases with decreasing flame speed. At low flame speeds, the burning component is overshadowed by the advection component, and it is the eddies which are responsible for the lengthening the front and bringing the flame into contact with unburnt fuel. Conversely, when the flame speed is large relative to the advection component, the faster burning rate overshadows this effect and the maximum difference is much less.

As one might expect, the above experiments merely scratch the surface of a highly complicated phenomenon. We are currently investigating such factors as flame speed dependence on curvature, the role of vorticity production along the flame front and the effect of temperature, with the hope of continuing the type of investigation discussed here.

References

1. J. T. Beale and A. Majda, *Vortex methods*. I. *Convergence in three dimensions*, Math. Comp. **39** (1982), 1–27.

2. _____, *Vortex methods*. II. *Higher order accuracy in two and three dimensions*, Math. Comp. **39** (1982), 29–52.

3. K. N. C. Bray, *Turbulent flows with premixed reactants*, Turbulent Reacting Flows (P. A. Libby and F. A. Williams, eds.), Chapter 5, Springer-Verlag, New York, 1980.

4. A. Y. Cheer, *A study of incompressible 2-D vortex flow past a circular cylinder*, SIAM J. Sci. Statist. Comput. **4** (1983), 685–705.

5. A. J. Chorin, *Flame advection and propagation algorithms*, J. Comput. Phys. **35** (1980), 1–11.

6. _____, *Numerical studies of slightly viscous flow*, J. Fluid Mech. **57** (1973), 785–796.

7. _____, *Vortex sheet approximation of boundary layers*, J. Comput. Phys. **27** (1973) 428–442.

8. A. F. Ghoniem, A. J. Chorin, and A. K. Oppenheim, *Numerical modeling of turbulent flow in a combustion tunnel*, Philos. Trans. Roy. Soc. London Ser. A **304** (1982), 303–325.

9. O. Hald, *Convergence of vortex methods*. II, SIAM J. Numer. Anal. **16** (1979), 726–755.

10. O. Hald and V. M. Del Prete, *Convergence of vortex methods for Euler's equations*, Math. Comp. **32** (1978), 791–809.

11. L. D. Landau, *On the theory of slow combustion*, J. Experimental Theoret. Phys. **14** (1944), 240–249.

12. P. A. Libby and F. A. Williams, *Fundamental aspects*, Turbulent Reacting Flows (P. A. Libby and F. A. Williams, eds.), Chapter 1, Springer-Verlag, New York, 1980.

13. A. Majda and J. A. Sethian, *The derivation and numerical solution of the equations for zero Mach number combustion*, Lawrence Berkeley Laboratory, LBL-17289, 1984 (to appear).

14. G. H. Markstein, *Nonsteady flame propagation*, Pergammon Press, MacMillan and Company, New York, 1964.

15. M. F. McCracken and C. S. Peskin, *A vortex method for blood flow through heart values*, J. Comput. Phys. **35** (1980), 183–205.

16. W. T. Noh and P. Woodward, SLIC (*Simple Line Interface Calculation*) Proc. 5th Internat. Conf. on Numerical Mathematics and Fluid Mechanics, Springer-Verlag, Berlin, 1976, pp. 330–339.

17. R. G. Rehm and H. R. Baum, *The equations of motion for thermally driven, buoyant flows*, J. Res. Nat. Bur. Standards **83** (1978), 297–308.

18. W. C. Reynolds and T. Cebeci, *Calculation of turbulent flows*, Turbulence (P. Bradshaw, ed.), Chapter 5, Springer-Verlag, Berlin, 1978.

19. J. A. Sethian, *An analysis of flame propagation*, Ph.D. Dissertation, University of California, Berkeley, June 1982.

20. _____, *Numerical simulation of flame propagation in a closed vessel*, Notes on Numerical Fluid Mechanics, **7** (Proc. Fifth Internat. GAMM Conf. on Numerical Methods and Fluid Mechanics, Rome, Italy, October 5–7, 1983, M. Pandolfi and R. Piva, eds.), Friedr. Vieweg, Weisbaden, 1984.

21. _____, *Turbulent combustion in open and closed vessels* J. Comput. Phys. (to appear).

22. _____, *The wrinkling of a flame due to viscosity*, Fire Dynamics and Heat Transfer (Proc. 21st Nat. Heat Transfer Conf., Seattle, July 24–28, 1983, J. Quintiere, ed.) ASME, New York, 1983, pp. 29–32.

DEPARTMENT OF MATHEMATICS AND LAWRENCE BERKELEY LABORATORY, UNIVERSITY OF CALIFORNIA, BERKELEY, CALIFORNIA 94720

Current address: Courant Institute of Mathematical Sciences, New York University, New York, New York 10012

Lectures in Applied Mathematics
Volume **22**, 1985

Numerical Solution of Rotating Internal Flows

Charles G. Speziale

ABSTRACT. A general numerical method is presented for the solution
of pressure-driven laminar flows in straight ducts that are subjected to
a steady spanwise rotation. The full nonlinear Navier-Stokes equations
are solved by an explicit finite difference technique where the convective terms are formulated using Arakawa's scheme and the viscous
diffusion terms are formulated using the DuFort-Frankel scheme.
Specific computer calculations are presented for rotating rectangular
ducts where the effects of secondary flows and roll-cell instablities are
explored in detail and comparisons are made with previously conducted theoretical and experimental investigations. The advantages
and disadvantages of this approach are discussed along with the
extension of this method to include dilute viscoelastic fluid behavior.

1. Introduction. The pressure driven flow of a viscous fluid through
straight ducts that are subjected to a spanwise rotation constitutes a
problem with a wealth of interesting nonlinear mathematical behavior
and important technological applications in the design of various types of
rotating machinery. Since the fully developed laminar flow where the
velocity field is independent of the axial coordinate is to be considered,
an axial rotation is of little consequence (i.e., relative to an observer
moving with the duct, an axial rotation leaves the velocity field unaf-
fected). Previously conducted theoretical and experimental investigations
(see Hart [1] and Lezius and Johnston [2]) on rotating channel flow have
demonstrated the existence of three flow regimes. At weak rotation rates,
there is a double vortex secondary flow; at intermediate rotation rates,
there is an instability in the form of longitudinal roll-cells; and at more
rapid rotation rates, there is a restabilization of the flow to a Taylor-
Proudman regime where the axial velocity profiles, in the interior if the

---

1980 *Mathematics Subject Classification.* Primary 76U05, 76A60, 35Q10, 39A10.

channel, do not vary along the direction of the axis of rotation. It is interesting to note that qualitatively similar regimes occur in stationary curved ducts (c.f., Cheng, Lin and Ou [3] and Dennis and Ng [4]) and, hence, the results to be obtained herein can have even wider applications than those mentioned above.

The purpose of the present paper is to examine in more detail a finite difference approach that was utilized in Speziale [5] and Speziale and Thangam [6] to study pressure-driven viscous flows in rectangular ducts that are subjected to a spanwise rotation. In this approach, the complete nonlinear and time-dependent Navier-Stokes equations in a rotating framework are solved by an explicit finite difference technique where the convective terms are formulated using Arakawa's scheme and the viscous diffusion terms are formulated using the DuFort-Frankel scheme. Consequently, it becomes possible to examine in detail the way that the axial flow interacts with the resulting secondary flows and roll-cell instabilities —an important, but highly nonlinear effect. Furthermore, it becomes possible to examine the time evolution of these flow structures and to determine the effect of the boundaries on the roll-cell instabilities in rotating channel flow. In all of the previous studies where the full nonlinear equations of motion are not solved, the results obtained are limited to either the determination of the secondary flow structure in the limiting cases of very weak or very rapid rotations, or to the determination of the stability boundaries for rotating plane Poiseuille flow. It will thus be demonstrated that the numerical approach to be presented here gives a much more complete picture of the various physical mechanisms that are manifested in rotating internal flows.

Extensive comparisons will be made between the results obtained by this method and those obtained in previous theoretical and experimental investigations. The advantages and disadvantages of this approach will be examined in detail, along with the extension of this method to deal with dilute viscoelastic fluids. The prospects for future research will be discussed briefly in the last section.

2. Numerical formulation of the physical problem. The problem to be considered is that of the pressure-driven laminar flow of an incompressible viscous fluid through a straight duct that is subjected to a steady spanwise rotation Ω (see Figure 1). Here, the governing equations are the Navier-Stokes equations and continuity equation which take the form

$$\frac{\partial \mathbf{v}}{\partial t} + \mathbf{v} \cdot \nabla \mathbf{v} = -\frac{1}{\rho} \nabla P + \nu \nabla^2 \mathbf{v} - 2\Omega \times \mathbf{v}, \qquad (1)$$

$$\nabla \cdot \mathbf{v} = 0 \qquad (2)$$

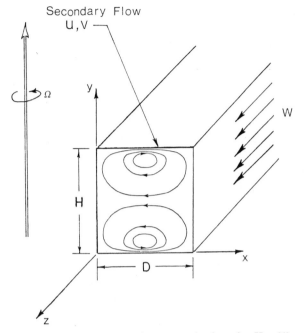

FIGURE 1. Flow in a rotating rectangular duct after Hart [1].

(Reprinted by permission of the Cambridge University Press from C. Speziale, J. Fluid Mechanics **122** (1982), 251–271.)

in a rotating framework (c.f. Batchelor [7]). In (1) and (2), $\mathbf{v}$ is the velocity vector, P is the modified pressure which includes the gravitational and centrifugal force potentials, ρ is the density of the fluid, and ν is the kinematic viscosity of the fluid. The axial pressure gradient $\partial P/\partial z = -G$ is constant and is maintained by external means. The duct is sufficiently long so that there exists an interior portion where end effects can be suppressed and the flow properties are independent of the axial coordinate z. In the absence of rotations (i.e., for $\Omega = 0$), the fully developed velocity field is of the unidirectional form

$$\mathbf{v} = w(x, y)\mathbf{k}, \tag{3}$$

where w is a solution of the Poisson equation

$$\nabla^2 w = -G/\rho\nu \tag{4}$$

which the Navier-Stokes equations reduce to (c.f., Batchelor [7]). However, for nonzero rotation rates, the fully developed velocity field is three-dimensional relative to an observer that rotates with the duct (see Hart [1]). To be specific, $\mathbf{v}$ is of the fully developed form

$$\mathbf{v} = u(x, y)\mathbf{i} + v(x, y)\mathbf{j} + w(x, y)\mathbf{k}, \tag{5}$$

where w represents the axial velocity, and u and v constitute the secondary flow. As a result of (5), fluid particles undergo a spiraling motion down the duct in contrast to the nonrotating case where the trajectories are straight lines. By making use of the fact that $\Omega = \Omega \mathbf{j}$, the equations of motion (1) and (2), in component form, are as follows:

$$\frac{\partial u}{\partial t} + u \frac{\partial u}{\partial x} + v \frac{\partial u}{\partial y} = -\frac{1}{\rho} \frac{\partial P}{\partial x} + \nu \nabla^2 u - 2\Omega w, \tag{6}$$

$$\frac{\partial v}{\partial t} + u \frac{\partial v}{\partial x} + v \frac{\partial v}{\partial y} = -\frac{1}{\rho} \frac{\partial P}{\partial y} + \nu \nabla^2 v, \tag{7}$$

$$\frac{\partial w}{\partial t} + u \frac{\partial w}{\partial x} + v \frac{\partial w}{\partial y} = \frac{G}{\rho} + \nu \nabla^2 w + 2\Omega u, \tag{8}$$

$$\frac{\partial u}{\partial x} + \frac{\partial v}{\partial y} = 0, \tag{9}$$

where we have made use of the fact that the flow properties are independent of the axial coordinate z. As a direct consequence of the simplified form of the continuity equation (9), there exists a secondary flow stream function ψ such that

$$u = -\partial \psi / \partial y, \qquad v = \partial \psi / \partial x. \tag{10}$$

Here, ψ is a solution of the Poisson equation

$$\nabla^2 \psi = \zeta, \tag{11}$$

where

$$\zeta = \frac{\partial v}{\partial x} - \frac{\partial u}{\partial y} \tag{12}$$

is the axial component of the vorticity vector. The axial vorticity ζ is determined from the z-component of the vorticity transport equation which is given by

$$\frac{\partial \zeta}{\partial t} + u \frac{\partial \zeta}{\partial x} + v \frac{\partial \zeta}{\partial y} = \nu \nabla^2 \zeta + 2\Omega \frac{\partial w}{\partial y}. \tag{13}$$

This equation is obtained by differencing the derivative of (7) with respect to x and the derivative of (6) with respect to y.

The equations of motion for rotating duct flow will thus be solved numerically in a modified vorticity-stream function form which is as follows:

$$\frac{\partial w}{\partial t} + u\frac{\partial w}{\partial x} + v\frac{\partial w}{\partial y} = \frac{G}{\rho} + \nu\nabla^2 w + 2\Omega u, \tag{14}$$

$$\frac{\partial \zeta}{\partial t} + u\frac{\partial \zeta}{\partial x} + v\frac{\partial \zeta}{\partial y} = \nu\nabla^2\zeta + 2\Omega\frac{\partial w}{\partial y}, \tag{15}$$

$$\nabla^2\psi = \zeta, \tag{16}$$

$$u = -\frac{\partial\psi}{\partial y}, \qquad v = \frac{\partial\psi}{\partial x}. \tag{17}$$

As a result of (16) and (17), it is quite clear that secondary flows arise from a *nonzero* axial vorticity ζ. Hence, the Coriolis term $2\Omega\,\partial w/\partial y$, which serves as an axial vorticity source term in (15), is the driving mechanism for the creation of secondary flows in a rotating duct. This coupled system of nonlinear partial differential equations (14)–(17) is solved subject to the boundary conditions

$$u = 0, \qquad v = 0, \qquad w = 0, \qquad \psi = 0, \tag{18}$$

on the walls of the duct. The boundary conditions on the axial vorticity ζ are derived by a Taylor expansion of (16). These will be presented later for the case of a rectangular duct. The initial conditions that will be utilized correspond to those associated with the spin-up of a fully developed duct flow. Hence, the angular velocity Ω is impulsively applied at time $t = 0$ with

$$u = 0, \quad v = 0, \quad w = w_i, \quad \zeta = 0, \quad \psi = 0, \tag{19}$$

where w_i is the unidirectional velocity profile for the nonrotating case which is obtained from (4). This actually constitutes the way in which the experiments on rotating duct flows are usually conducted (c.f. Hart [1]).

The finite difference technique to be used will now be presented for a rectangular duct that is discretized into an $M \times N$ mesh. Both the axial momentum and vorticity transport equations will be solved by a modified form of Arakawa's method that utilizes the DuFort-Frankel scheme for the viscous diffusion terms. Arakawa's method is advantageous since it is an explicit and conservative finite difference scheme which has no boundary condition problems and is, furthermore, not subject to nonlinear instabilities which arise from aliasing errors (all aliasing errors are bounded). In Arakawa's method, the convective derivative of any flow variable f is represented as follows (c.f., Roache [8]):

$$
\begin{aligned}
\left(\frac{Df}{Dt}\right)_{i,j}^{n} &= \left(\frac{\partial f}{\partial t} + u\frac{\partial f}{\partial x} + v\frac{\partial f}{\partial y}\right)_{i,j}^{n} \\
&= \frac{f_{i,j}^{n+1} - f_{i,j}^{n-1}}{2\Delta t} \\
&\quad + \frac{1}{12\Delta x \Delta y}\Big[\left(\psi_{i+1,j}^{n} - \psi_{i-1,j}^{n}\right)\left(f_{i,j+1}^{n} - f_{i,j-1}^{n}\right) \\
&\quad - \left(\psi_{i,j+1}^{n} - \psi_{i,j-1}^{n}\right)\left(f_{i+1,j}^{n} - f_{i-1,j}^{n}\right) \\
&\quad + \psi_{i+1,j}^{n}\left(f_{i+1,j+1}^{n} - f_{i+1,j-1}^{n}\right) - \psi_{i-1,j}^{n}\left(f_{i-1,j+1}^{n} - f_{i-1,j-1}^{n}\right) \\
&\quad - \psi_{i,j+1}^{n}\left(f_{i+1,j+1}^{n} - f_{i-1,j+1}^{n}\right) \\
&\quad + \psi_{i,j-1}^{n}\left(f_{i+1,j-1}^{n} - f_{i-1,j-1}^{n}\right) + f_{i,j+1}^{n}\left(\psi_{i+1,j+1}^{n} - \psi_{i-1,j+1}^{n}\right) \\
&\quad - f_{i,j-1}^{n}\left(\psi_{i+1,j-1}^{n} - \psi_{i-1,j-1}^{n}\right) \\
&\quad - f_{i+1,j}^{n}\left(\psi_{i+1,j+1}^{n} - \psi_{i+1,j-1}^{n}\right) + f_{i-1,j}^{n}\left(\psi_{i-1,j+1}^{n} - \psi_{i-1,j-1}^{n}\right)\Big],
\end{aligned}
\tag{20}
$$

where

$$
f_{i,j}^{n} = f(i\Delta x, j\Delta y, n\Delta t), \qquad \psi_{i,j}^{n} = \psi(i\Delta x, j\Delta y, n\Delta t), \tag{21}
$$

$$
i = 0,1,\ldots,M, \qquad j = 0,1,\ldots,N, \qquad n = 0,1,\ldots,
$$

and Δx, Δy and Δt are, respectively, the mesh sizes in the x and y directions and the time interval. The axial momentum equation (14) and the axial vorticity transport equation (15) are solved in the finite difference form

$$
\begin{aligned}
\left(\frac{Dw}{Dt}\right)_{i,j}^{n} &= \frac{G}{\rho} + \nu\left[\frac{w_{i+1,j}^{n} - w_{i,j}^{n-1} - w_{i,j}^{n+1} + w_{i-1,j}^{n}}{(\Delta x)^2}\right. \\
&\quad \left. + \frac{w_{i,j+1}^{n} - w_{i,j}^{n-1} - w_{i,j}^{n+1} + w_{i,j-1}^{n}}{(\Delta y)^2}\right] + 2\Omega u_{i,j}^{n}, \tag{22}
\end{aligned}
$$

$$
\begin{aligned}
\left(\frac{D\zeta}{Dt}\right)_{i,j}^{n} &= \nu\left[\frac{\zeta_{i+1,j}^{n} - \zeta_{i,j}^{n-1} - \zeta_{i,j}^{n+1} + \zeta_{i-1,j}^{n}}{(\Delta x)^2}\right. \\
&\quad \left. + \frac{\zeta_{i,j+1}^{n} - \zeta_{i,j}^{n-1} - \zeta_{i,j}^{n+1} + \zeta_{i,j-1}^{n}}{(\Delta y)^2}\right] + 2\Omega\left(\frac{w_{i,j+1}^{n} - w_{i,j-1}^{n}}{2\Delta y}\right), \\
&\tag{23}
\end{aligned}
$$

where the convective derivatives are formulated by Arakawa's method given by (20). The viscous diffusion terms in (22) and (23) are formulated with the DuFort-Frankel scheme (c.f. Richtmyer and Morton [9]) which has the advantage of being unconditionally stable. It does, however, require that $\Delta t / \Delta x \to 0$ as Δt and $\Delta x \to 0$ in order to be consistent with the diffusion equation (otherwise it is consistent with a more complicated hyperbolic equation). The Coriolis terms in (22) and (23) are centered in time for the purposes of stability (the latter term is central differenced for second-order accuracy).

The Poisson equation is solved in the standard central difference form

$$\frac{\psi_{i+1,j}^n - 2\psi_{i,j}^n + \psi_{i-1,j}^n}{(\Delta x)^2} + \frac{\psi_{i,j+1}^n - 2\psi_{i,j}^n + \psi_{i,j-1}^n}{(\Delta y)^2} = \zeta_{i,j}^n, \quad (24)$$

by making use of a compact noniterative Poisson solver due to Buneman [10] which employs cyclic reduction. This Poisson solver is extremely efficient since it was constructed especially for use in rectangular domains. It suffers only from the minor inconvenience of requiring that M and N be powers of two. The secondary flow velocities which are determined from (17), are solved in the central difference form

$$u_{i,j}^n = -\left(\frac{\psi_{i,j+1}^n - \psi_{i,j-1}^n}{2\Delta y}\right), \qquad v_{i,j}^n = \frac{\psi_{i+1,j}^n - \psi_{i-1,j}^n}{2\Delta x}. \quad (25)$$

Equations (22)–(25) constitute the complete finite difference formulation of the equations of motion. Of course, these equations must be solved in conjunction with the initial and boundary conditions given in (18) and (19) which can be easily implemented. However, as alluded to earlier, the boundary conditions on the axial vorticity must be derived. This is accomplished by making a Taylor expansion of (16) in the vicinity of the walls of the duct. To second-order accuracy, the boundary conditions obtained are of the form

$$\zeta_{0,j}^n = \frac{8\psi_{1,j}^n - \psi_{2,j}^n}{2(\Delta x)^2}, \qquad \zeta_{M-1,j}^n = \frac{8\psi_{M-1,j}^n - \psi_{M-2,j}^n}{2(\Delta x)^2}, \quad (26)$$

$$\zeta_{i,0}^n = \frac{8\psi_{i,1}^n - \psi_{i,2}^n}{2(\Delta y)^2}, \qquad \zeta_{i,N}^n = \frac{8\psi_{i,N-1}^n - \psi_{i,N-2}^n}{2(\Delta y)^2}, \quad (27)$$

where we have made use of the fact that ψ vanishes on the walls of the duct. This completes the finite difference formulation of the problem. The specific calculation sequence is as follows: the flow variables are initialized using (19); the values of the axial velocity w and axial vorticity ζ are updated by using (22) and (23); the secondary flow stream function ψ and secondary flow velocity fields u and v are obtained from (24) and

(25); and, finally, the boundary conditions on the axial vorticity ζ are updated by using (26) and (27). This calculation procedure is carried out until convergence is obtained to three significant figures.

At this point, some comments should be made concerning the accuracy and stability of these computations. The finite difference formulation of this problem, in its entirety, is second-order accurate, i.e., the truncation error τ is such that

$$\|\tau\| = O(\Delta x^2, \Delta y^2, \Delta t^2), \tag{28}$$

where $\| \cdot \|$ denotes a suitable norm. Unfortunately, as a result of the complex nonlinear nature of the governing equations of motion, no formal error estimates can be derived at this time. As far as stability is concerned, a stability criterion obtained from the simple superposition of the generalized CFL condition and Neumann condition was utilized (see Roache [8]). This requires that Δt satisfy the constraint

$$\Delta t \leqslant \left[2\nu \left(\frac{1}{\Delta x^2} + \frac{1}{\Delta y^2} \right) + \frac{|u|_{\max}}{\Delta x} + \frac{|v|_{\max}}{\Delta y} \right]^{-1}, \tag{29}$$

which yielded stable calculations for all of the cases that were computed. It should be noted that while the von Neumann condition is not a necessary condition for the linear stability of the DuFort-Frankel scheme, it was nevertheless implemented since it guarantees consistency with the diffusion equation (i.e., it guarantees that $\Delta t/\Delta x \to 0$ as $\Delta t, \Delta x \to 0$). Of course, the generalized CFL condition is usually a necessary condition for the linear stability of incompressible Euler equations. As mentioned earlier, Arakawa's method has no problems with nonlinear instabilities that arise from aliasing errors or from the implementation of boundary conditions. On the other hand, this method has been known to occasionally give rise to nonlinear time-splitting instabilities (c.f. Roache [8]). However, these can usually be eliminated quite easily (e.g., the flow variables at two successive iterations can be periodically equated). In all of the calculations to be presented in this paper, no problems with time-splitting instabilities were encountered.

3. Numerical results. Calculations have been conducted for 2×1 and 8×1 ducts (i.e., ducts with an aspect ratio H/D of 2 and 8, respectively). An 8×1 duct is in the range of aspect ratios used in the experiments of Hart [1] and Lezius and Johnston [2] to simulate flow in a rotating channel (i.e., a duct with an infinite aspect ratio). The level of discretization chosen for these computations in a 2×1 and 8×1 duct was $M = 32$, $N = 64$ and $M = 16$, $N = 128$, respectively. Hence, for both cases, the cross section of the duct was discretized into more than 2000 mesh points.

In the previous experiments by Hart [1] and Lezius and Johnston [2], the results obtained were nondimensionalized with a velocity scale based on the integrated average axial velocity in the *rotating* duct which is part of the solution to be obtained (as we shall soon see, a spanwise rotation leads to a reduction in the flowrate for a given physical pressure gradient). Consequently, there was no advantage to be gained by nondimensionalizing the equations of motion a priori. Calculations were carried out for water (at room temperature) with the physical properties of $\rho = 1.936$ slugs/ft^3 and $\nu = 1.1 \times 10^{-5}$ ft^2/sec. For the 2×1 duct, uniform mesh sizes of $\Delta x = \Delta y = 0.005$ ft were chosen which yield a duct with the dimensions of $D = 1.92$ in and $H = 3.84$ in. For the 8×1 duct, uniform mesh sizes of $\Delta x = \Delta y = 0.01$ ft were chosen which yield a duct with the dimensions of $D = 1.92$ in and $H = 15.36$ in. Hence, both ducts have the same characteristic dimension D of approximately 2 in. The flow variables will be nondimensionalized by the introduction of a velocity scale W_0 and length scale D which are, respectively, the fully-developed integrated average axial velocity and the width of the duct. With the use of these scales, the equations of motion (14)–(17) take the dimensionless form

$$\frac{\partial w^*}{\partial t^*} + u^* \frac{\partial w^*}{\partial x^*} + v^* \frac{\partial w^*}{\partial y^*} = C + \frac{1}{\mathrm{Re}} \nabla^{*2} w^* + 2\,\mathrm{Ro}\, u^*, \quad (30)$$

$$\frac{\partial \zeta^*}{\partial t^*} + u^* \frac{\partial \zeta^*}{\partial x^*} + v^* \frac{\partial \zeta^*}{\partial y^*} = \frac{1}{\mathrm{Re}} \nabla^{*2} \zeta^* + 2\,\mathrm{Ro}\, \frac{\partial w^*}{\partial y^*}, \quad (31)$$

$$\nabla^{*2} \psi^* = \zeta^*, \quad (32)$$

$$u^* = -\frac{\partial \psi^*}{\partial y^*}, \qquad v^* = \frac{\partial \psi^*}{\partial x^*}, \quad (33)$$

where an asterisk represents a dimensionless flow variable and

$$\mathrm{Re} = W_0 D / \nu, \qquad \mathrm{Ro} = \Omega D / W_0, \qquad C = GD / \rho W_0^2 \quad (34)$$

are, respectively, the Reynolds number, rotation number, and dimensionless pressure gradient. It is thus clear that flow similitude is achieved for any Newtonian fluid or duct size provided that the values of the parameters Re and Ro do not change. At this point, the Ekman number

$$E = \nu / 2\Omega D^2 = (2\,\mathrm{Re}\,\mathrm{Ro})^{-1} \quad (35)$$

will be introduced which will be utilized later in the analysis of some of the numerical results. The calculations that were conducted were made dimensionless, a posteriori, by utilizing (34) and (35). All of these calculations required from 1500 and 2500 iterations for convergence, where the later number corresponds to the case when the secondary flow was strong so that (29) necessitates that Δt be smaller. Approximately 40 min. to 1 hr. of CPU time was required on a DEC system-10 computer to achieve a steady state.

The computations conducted in a 2×1 duct will be presented first. The fully developed secondary flow streamlines that were computed for Re = 86 and Ro = 0.27 are shown in Figure 2. This double vortex secondary flow pattern, in which the two counterrotating vortices are slightly compressed against the upper and lower walls of the duct to which they are adjacent, is in excellent agreement with the experimental observations of Hart [1] shown in Figure 1. For this as well as all subsequent computer contour maps of the secondary flow streamlines, the upper vortex rotates in the clockwise direction while the vortices below counterrotate. The rotation number is large enough here so that the resulting secondary flow has a profound distortional effect on the axial velocity profiles as shown in Figures 3(a) and 3(b). To be more specific, the axial velocity profile along the horizontal centerline of the duct shown in Figure 3(a) is asymmetric with its maximum value shifted toward the low pressure side of the duct. On the other hand, the axial velocity profile along the vertical centerline of the duct becomes flat in the interior region as would be expected from the Taylor-Proudman Theorem (see Hart [1]). It is clear from Figures 3(a) and 3(b) that there is a substantial reduction in the flowrate (approximately 10%) as a result of the presence of this Coriolis driven secondary flow. Such friction losses would be expected on physical grounds since the secondary flow leads to the dissipation of energy in a transverse motion which does not contribute to the flowrate.

At intermediate rotation rates and somewhat higher Reynolds numbers, the fully-developed double vortex secondary flow becomes unstable

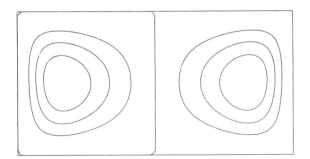

FIGURE 2. Computer generated contour maps of the fully developed secondary flow streamlines in a 2×1 duct. Re = 86, Ro = 0.27 ($\Omega = 0.01$ rad/sec, $G = 1 \times 10^{-4}$ lb/ft^3).

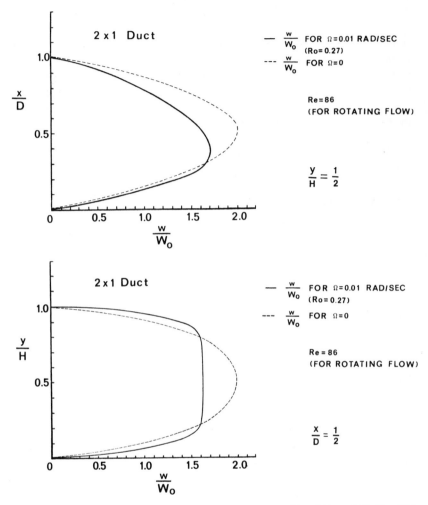

FIGURE 3. Axial velocity profiles in a 2 × 1 duct for Re = 86 and Ro = 0.27 (Ω = 0.01 rad/sec, $G = 1 \times 10^{-4}$ lb/ft^3).

(a) Along the horizontal centerline of the duct.

(b) Along the vertical centerline of the duct.

(Reprinted by permission of the Cambridge University Press from C. Speziale, J. Fluid Mechanics **122** (1982), 251–271.)

in the presence of small disturbances. A pair of counterrotating rolls appear on the low pressure side of the duct, thus leading to a four vortex secondary flow. This effect is illustrated in the computer contour map of the fully developed secondary flow shown in Figure 4 for Re = 279 and Ro = 0.833. Such secondary flows lead to a more substantial distortion of the axial velocity profiles as shown in Figures 5(a) and 5(b). In Figure 5(a) the axial velocity profile along the horizontal centerline of the duct is shown. This profile is asymmetric with the maximum velocity shifted toward the low pressure side of the duct. A point of inflection occurs near the center of the duct which is suggestive of the presence of an instability. The axial velocity profile along the vertical centerline of the duct shown in Figure 5(b) is even more profoundly distorted. The peaks in the axial velocity near the upper and lower walls of the duct arise from Ekman suction, whereas the peak near the center of the duct results from the presence of the rolls. As expected on physical grounds, the flow assumes a Taylor-Proudman configuration elsewhere. Again, there is a substantial reduction in the flowrate as a result of the presence of the secondary flow.

At more rapid rotation rates, the flow restabilizes to a Taylor-Proudman regime where the secondary flow returns to a double vortex configuration and the axial velocity profiles in the interior of the duct do not vary along the direction of the axis of rotation.

A computer contour map of the fully developed secondary flow streamlines is shown in Figure 6 for Re = 220 and Ro = 2.08. This figure clearly shows the existence of a slightly asymmetric double vortex secondary flow in which the vortices are strongly compressed against the upper and lower walls of the duct. The axial velocity profile along the horizontal centerline of the duct for this flow is shown in Figure 7(a). As before, this profile is asymmetric with the maximum velocity shifted toward the low pressure side of the duct. The axial velocity profile along the vertical centerline of the duct shown in Figure 7(b) assumes a Taylor-Proudman configuration in the interior of the duct (i.e., the axial

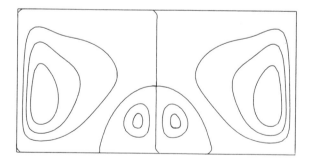

FIGURE 4. Computer generated contour maps of the fully-developed secondary flow streamlines in a 2×1 duct. Re = 279, Ro = 0.833 ($\Omega = 0.1$ rad/sec, $G = 6 \times 10^{-4}$ lb/ft^3).

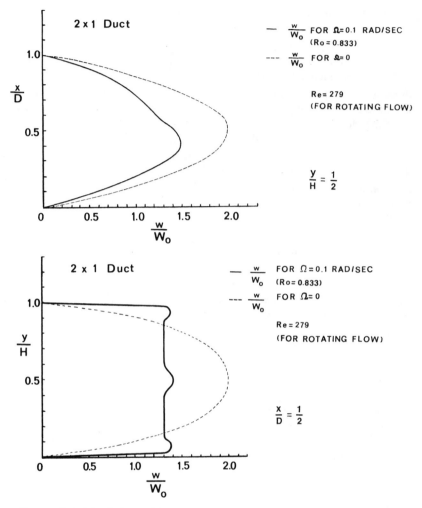

FIGURE 5. Axial velocity profiles in a 2×1 duct for Re = 279 and Ro = 0.83 ($\Omega = 0.1$ rad/sec, $G = 6 \times 10^{-4}$ lb/ft^3).

(a) Along the horizontal centerline of the duct.

(b) Along the vertical centerline of the duct.

(Reprinted by permission of the Cambridge University Press from C. Speziale, J. Fluid Mechanics **122** (1982), 251–271.)

velocity does not vary along the axis of rotation in the interior of the duct). Here, the two peaks in the axial velocity profile near the walls of the duct have their maximum values at a distance approximately equal to $2E^{1/2}D$ from each wall. This is consistent with the position of the overshoot obtained from the linear theory of the Ekman layer and, hence, these peaks can be thought of as arising from Ekman suction (see Hart [1]). The secondary flow in this case leads to an even more substantial reduction in the flowrate (of the order of 30%).

Now, the calculations that were conducted in an 8×1 duct will be discussed. As mentioned earlier, this aspect ratio is in the same range as those used in the experiments of Hart [1] and Lezius and Johnston [2] to simulate rotating channel flow. In Figure 8, a computer contour map of the fully developed secondary flow streamlines is shown for $Re = 80.4$ and $Ro = 2.9 \times 10^{-4}$. For this rather weak rotation, the secondary flow assumes a double vortex configuration where each vortex, whose size is of the order of the width of the channel, is slightly compressed against the horizontal wall of the channel to which it is adjacent. Since this secondary flow is weak, it has almost no effect on the axial velocity profiles. As the rotation rate is increased, this double vortex secondary flow starts to extend far into the interior of the channel. The computer generated secondary flow streamlines, at various times, for $Re = 107$ and $Ro = 0.5$, shown in Figure 9, illustrate this phenomenon.

At intermediate rotation rates, the stretched double vortex secondary flow shown in Figure 9 becomes unstable. This simple secondary flow is replaced with a double vortex secondary flow combined with roll-cells in the interior of the channel. This roll-cell instability is illustrated in Figure 10 where the computed secondary flow streamlines, at various times, are shown for $Re = 248$ and $Ro = 0.047$ which is in the roll-cell regime. The corresponding axial velocity profile along the horizontal centerline of the channel is shown in Figure 11(a). This profile is asymmetric with the maximum velocity shifted toward the high pressure side of the channel. The axial velocity profile along the vertical centerline of the channel

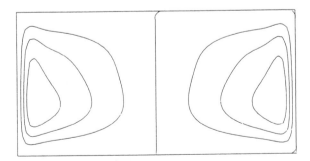

FIGURE 6. Computer generated contour map of the fully-developed secondary flow streamlines in a 2×1 duct. $Re = 220$, $Ro = 2.08$ ($\Omega = 0.2$ rad/sec, $G = 6 \times 10^{-4}$ lb/ft^3).

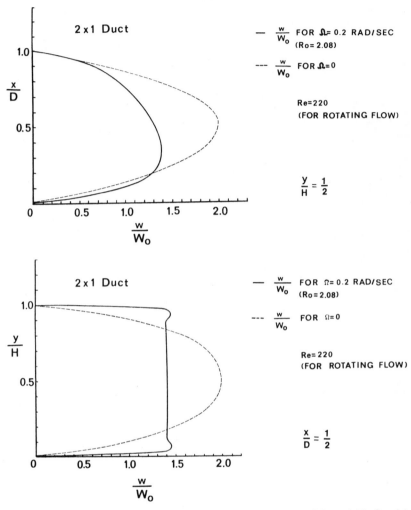

FIGURE 7. Axial velocity profiles in a 2×1 duct for Re = 220 and Ro = 2.08 ($\Omega = 0.2$ rad/sec, $G = 6 \times 10^{-4}$ lb/ft^3).

(a) Along the horizontal centerline of the duct.

(b) Along the vertical centerline of the duct.

(Reprinted by permission of the Cambridge University Press from C. Speziale, J. Fluid Mechanics **122** (1982), 251–271.)

shown in Figure 11(b) has a wavy structure which is in excellent qualitative agreement with the experimental observations of Hart [1] shown in Figure 12(c), (d). For this Reynolds number and rotation number the secondary flow gives rise to a 7.6% reduction in the flowrate.[1]

---

[1] $Q_r/Q \equiv$ ratio of the flowrate in a rotating duct to that in a stationary duct for a given physical pressure gradient (Figure 11).

fully-developed

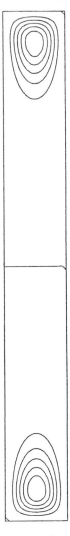

$$\Omega = 1 \times 10^{-5} \, \text{rad/s}$$
$$\text{Re} = 80.4$$
$$\text{Ro} = 2.9 \times 10^{-4}$$

FIGURE 8. Computer generated fully-developed secondary flow streamlines in an 8×1 duct. Re = 80.4, Ro = 2.9×10^{-4} ($\Omega = 10^{-5}$ rad/sec, $G = 6 \times 10^{-5}$ lb/ft^3).
(Reprinted by permission of the Cambridge University Press from C. Speziale and S. Thangam, J. Fluid Mechanics **130** (1983), 377–395.)

(a)   (b)   (c)

$t = 1\,s$ $t = 75\,s$ fully-developed

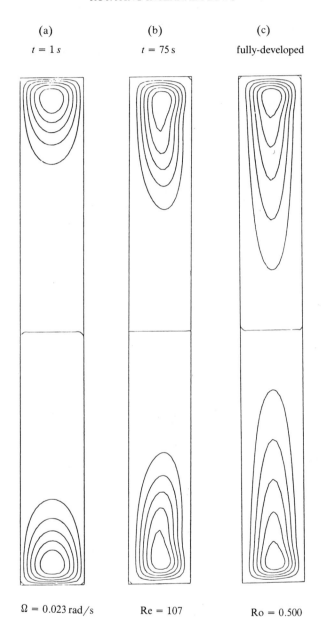

$\Omega = 0.023\,\mathrm{rad/s}$ $\mathrm{Re} = 107$ $\mathrm{Ro} = 0.500$

FIGURE 9. Computer generated secondary flow streamlines in an 8×1 duct. $\mathrm{Re} = 107$, $\mathrm{Ro} = 0.5$ ($\Omega = 0.023$ rad/sec, $G = 8.6 \times 10^{-5}$ lb/ft^3). (a) $t = 1$ s. (b) $t = 75$ s. (c) fully-developed.
(Reprinted by permission of the Cambridge University Press from C. Speziale and S. Thangam, J. Fluid Mechanics **130** (1983), 377–395.)

(a)   (b)   (c)   (d)
$t = 10\,s$ $t = 500\,s$ $t = 1600\,s$ fully-developed

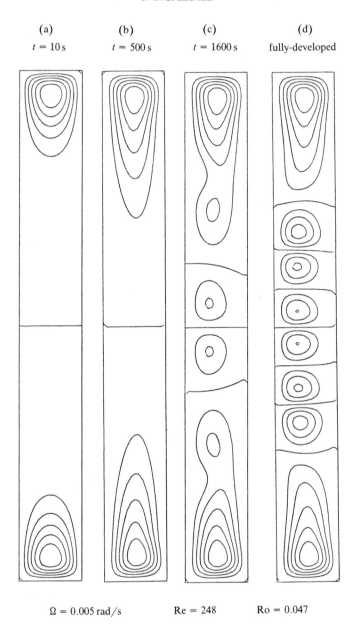

$\Omega = 0.005\,\text{rad}/s$ Re $= 248$ Ro $= 0.047$

FIGURE 10. Computer generated secondary flow streamlines in an 8×1 duct. Re $= 248$, Ro $= 0.047$ ($\Omega = 0.005$ rad/sec, $G = 2 \times 10^{-4}$ lb/ft^3). (a) $t = 10\ s$. (b) $t = 500\ s$. (c) $t = 1600\ s$. (d) fully-developed.

(Reprinted by permission of the Cambridge University Press from C. Speziale and S. Thangam, J. Fluid Mechanics **130** (1983), 377–395.)

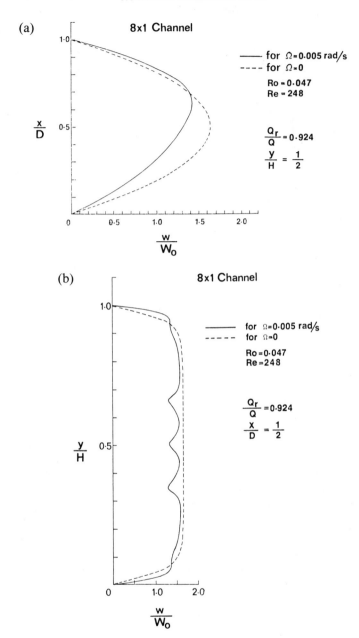

FIGURE 11. Axial velocity profiles in an 8×1 duct. Re = 248, Ro = 0.047 ($\Omega = 0.005$ rad/sec, $G = 2 \times 10^{-4}$ lb/ft^3).

(a) Along the horizontal centerline of the channel.

(b) Along the vertical centerline of the channel.

(Reprinted by permission of the Cambridge University Press from C. Speziale and S. Thangam, J. Fluid Mechanics **130** (1983), 377–395.)

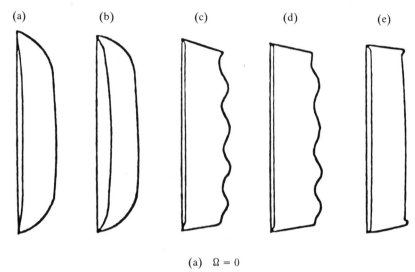

(a) $\Omega = 0$

(b) $\Omega > 0$, with double-vortex secondary flow

(c, d) $\Omega > 0$, with roll cells

(e) $\Omega > 0$, Taylor-Proudman regime

FIGURE 12. Experimental axial velocity profiles along the vertical centerline of a rotating channel after Hart [1]. (a) $\Omega = 0$, (b) $\Omega > 0$, with double-vortex secondary flow. (c), (d) $\Omega > 0$, with roll-cells. (e) $\Omega > 0$, Taylor-Proudman regime.

(Reprinted by permission of the Cambridge University Press from C. Speziale and S. Thangam, J. Fluid Mechanics 130 (1983), 377–395.)

At more rapid rotation rates, the secondary flow restabilizes to a stretched double vortex configuration where the axial velocity is in a Taylor-Proudman regime. The computed secondary flow streamlines, at various times, shown in Figure 13 for Re = 171 and Ro = 2.73, illustrate this effect. In Figure 14(a), the corresponding axial velocity profile along the horizontal centerline of the channel is shown. This velocity profile is only slightly asymmetric for this restabilized flow. In Figure 14(b), the corresponding axial velocity profile along the vertical centerline of the channel is shown. As in the case of a rapidly rotating 2×1 duct, this velocity profile assumes a Taylor-Proudman configuration in the interior of the channel with Ekman suction peaks near the upper and lower walls of the channel. This axial velocity profile is an excellent qualitative agreement with the experimental observations of Hart [1] shown in Figure 12(e). For this rapidly rotating channel, there is an enormous reduction in the flowrate of approximately 36.4%.

(a)   (b)   (c)   (d)

$t = 1\,s$ $t = 25\,s$ $t = 50\,s$ fully-developed

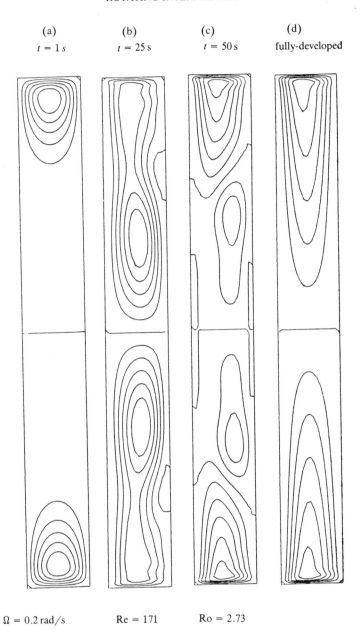

$\Omega = 0.2\,\text{rad}/s$ Re $= 171$ Ro $= 2.73$

FIGURE 13. Computer generated secondary flow streamlines in an 8×1 duct. Re $= 171$, Ro $= 2.73$ ($\Omega = 0.2$ rad/sec, $G = 2 \times 10^{-4}$ lb/ft^3). (a) $t = 1$ s. (b) $t = 25$ s. (c) $t = 50$ s. (d) fully-developed.

(Reprinted by permission of the Cambridge University Press from C. Speziale and S. Thangam, J. Fluid Mechanics **130** (1983), 377–395.)

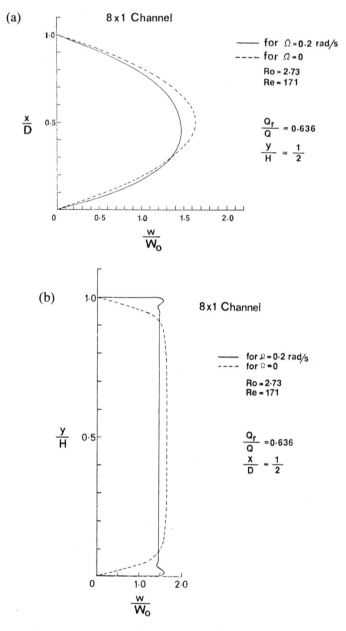

FIGURE 14. Axial velocity profiles in an 8×1 duct. Re = 171, Ro = 2.73 ($\Omega = 0.2$ rad/sec, $G = 2 \times 10^{-4}$ lb/ft^3).

(a) Along the horizontal centerline of the channel.

(b) Along the vertical centerline of the channel.

(Reprinted by permission of the Cambridge University Press from C. Speziale and S. Thangam, J. Fluid Mechanics **130** (1983), 377–395.)

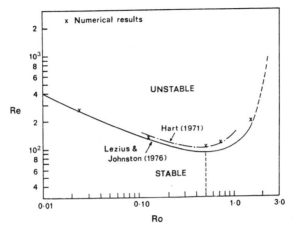

FIGURE 15. Numerically obtained stability boundary points for the onset of roll-cell instabilities in rotating channel flow.

(Reprinted by permission of the Cambridge University Press from C. Speziale and S. Thangam, J. Fluid Mechanics **130** (1983), 377–395.)

The numerically obtained stability boundary points are shown in Figure 15 alongside the theoretical results of Hart [1] and Lezius and Johnston [2] which were obtained from a linear stability analysis on rotating plane Poiseuille flow (i.e., flow in a rotating channel with an infinite aspect ratio). A critical point with the values of $Re_c = 110$ and $Ro_c = 0.5$ was obtained which are somewhat higher than the values of $Re_c = 88.53$ and $Ro_c = 0.5$ obtained by Lezius and Johnston [2]. However, it should be noted that since the calculations presented herein are for a finite aspect ratio channel, somewhat higher stability limits would be expected on physical grounds (i.e., the presence of the upper and lower walls of the channel tend to have a stabilizing effect on the flow). Thus, it is clear that the numerically obtained stability boundary points are in the correct range of the previously conducted theoretical analyses. For more detailed numerical results on flow in rotating rectangular ducts, the reader is referred to Speziale [5] and Speziale and Thangam [6].

4. Extension of the method to dilute viscoelastic fluids. It has been known for quite some time that the introduction of a minute amount of a long chain polymer (e.g., 50–100 parts/million by weight) to a Newtonian liquid produces a highly dilute viscoelastic liquid which can be considerably more stable in the presence of rotations. Furthermore, the rotational secondary flows associated with such dilute viscoelastic fluids give rise to friction losses that are considerably lower than for the Newtonian case (c.f. Ginn and Denn [11] and Denn and Roisman [12]). However, to date, there have been no nonlinear computations that illustrate this phenomenon. It will now be shown how the finite difference technique presented in this paper can be extended to describe such effects in rotating channel flow.

One of the more common models describing the flow of a dilute viscoelastic liquid is the Maxwell fluid for which the dissipative stress τ is of the form

$$\tau_{kl} + \lambda \overset{\circ}{\tau}_{kl} = 2\mu d_{kl} \tag{36}$$

where $d_{kl} = \frac{1}{2}(\partial v_k/\partial x_l + \partial v_l/\partial x_k)$ is the rate of deformation tensor, $\mu = \rho \nu$ is the dynamic viscosity, λ is the relaxation time, and

$$\overset{\circ}{\tau}_{kl} = \frac{\partial \tau_{kl}}{\partial t} + \mathbf{v} \cdot \nabla \tau_{kl} + \tau_{km} \frac{\partial v_m}{\partial x_l} + \tau_{lm} \frac{\partial v_m}{\partial x_k} \tag{37}$$

is the frame-indifferent convected time rate or Oldroyd derivative (c.f. Schowalter [13]). For the rotating channel flow of a dilute viscoelastic fluid, the relaxation time is small and it will be assumed that the secondary flow is relatively weak, i.e.,

$$\frac{\lambda \nu}{D^2} \ll 1, \qquad \frac{\|u\|}{\|w\|} \ll 1, \qquad \frac{\|v\|}{\|w\|} \ll 1, \tag{38}$$

where $\| \cdot \|$ is the maximum norm. Here, it should be noted that since the viscoelastic fluid is highly dilute, the fluid density ρ and kinematic viscosity ν are virtually the same as for the Newtonian fluid from which it is constructed. Provided that (38) is satisfied, the components of the dissipative stress tensor (36) can be approximated as follows:

$$\tau_{xx} = 2\mu \frac{\partial u}{\partial x} - 2\lambda u \left(\frac{\partial w}{\partial x} \right)^2, \tag{39}$$

$$\tau_{yy} = 2\mu \frac{\partial v}{\partial y} - 2\lambda \mu \left(\frac{\partial w}{\partial y} \right)^2, \tag{40}$$

$$\tau_{xy} = \mu \left(\frac{\partial u}{\partial y} + \frac{\partial v}{\partial x} \right) - 2\lambda \mu \frac{\partial w}{\partial x} \frac{\partial w}{\partial y}, \tag{41}$$

$$\tau_{xz} = \mu \frac{\partial w}{\partial x}, \tag{42}$$

$$\tau_{yz} = \mu \frac{\partial w}{\partial y} \tag{43}$$

in fully developed rotating channel flow (see Speziale [14]). These stresses give rise to the following equations of motion for the rotating channel flow of a dilute viscoelastic fluid (see Speziale [14]):

$$\frac{\partial w}{\partial t} + u \frac{\partial w}{\partial x} + v \frac{\partial w}{\partial y} = \frac{G}{\rho} + \nu \nabla^2 w + 2\Omega u, \tag{44}$$

$$\frac{\partial \zeta}{\partial t} + u \frac{\partial \zeta}{\partial x} + v \frac{\partial \zeta}{\partial y} = \nu \nabla^2 \zeta + 2\lambda \nu \left(\frac{\partial w}{\partial x} \frac{\partial}{\partial y} \nabla^2 w - \frac{\partial w}{\partial y} \frac{\partial}{\partial x} \nabla^2 w \right)$$
$$+ 2\Omega \frac{\partial w}{\partial y},$$
(45)

$$\nabla^2 \psi = \zeta,$$
(46)

$$u = -\frac{\partial \psi}{\partial y}, \qquad v = \frac{\partial \psi}{\partial x}.$$
(47)

These equations are the same as those for the corresponding Newtonian case with one exception—the axial vorticity transport equation (45) has the additional source term

$$2\lambda \nu \left(\frac{\partial w}{\partial x} \frac{\partial}{\partial y} \nabla^2 w - \frac{\partial w}{\partial y} \frac{\partial}{\partial x} \nabla^2 w \right)$$
(48)

which arises from relaxation effects. It can be shown that this term reduces the dissipation associated with the secondary flow and thus gives rise to a drag reduction.

The finite difference technique presented in §2 can be easily modified to solve the equations of motion (44)–(47) since only the axial vorticity transport equation is altered. To be specific, the same finite difference formulation is utilized where equation (23) is simply replaced with the equation

$$\left(\frac{D\zeta}{Dt} \right)_{i,j}^n = \nu \left[\frac{\zeta_{i+1,j}^n - \zeta_{i,j}^{n-1} - \zeta_{i,j}^{n+1} + \zeta_{i-1,j}^n}{(\Delta x)^2} \right. $$
$$\left. + \frac{\zeta_{i,j+1}^n - \zeta_{i,j}^{n-1} - \zeta_{i,j}^{n+1} + \zeta_{i,j-1}^n}{(\Delta y)^2} \right] + 2\Omega \left(\frac{w_{i,j+1}^n - w_{i,j-1}^n}{2\Delta y} \right)$$
$$+ 2\lambda \nu \left[\left(\frac{w_{i+1,j}^n - w_{i-1,j}^n}{2\Delta x} \right) \left(\frac{A_{i,j+1}^n - A_{i,j-1}^n}{2\Delta y} \right) \right.$$
$$\left. - \left(\frac{w_{i,j+1}^n - w_{i,j-1}^n}{2\Delta y} \right) \left(\frac{A_{i+1,j}^n - A_{i-1,j}^n}{2\Delta x} \right) \right],$$
(49)

where

$$A_{i,j}^n = (\nabla^2 w)_{i,j}^n = \frac{w_{i+1,j}^n - 2w_{i,j}^n + w_{i-1,j}^n}{(\Delta x)^2} + \frac{w_{i,j+1}^n - 2w_{i,j}^n + w_{i,j-1}^n}{(\Delta y)^2}$$
(50)

is the finite difference representation of $\nabla^2 w$. The viscoelastic vorticity source term (48) is central differenced spatially for second-order accuracy and centered in time for numerical stability.

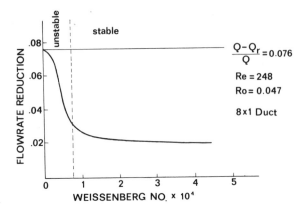

FIGURE 16. Flowrate reduction as a function of the Weissenberg number in the rotating channel flow of a dilute viscoelastic fluid.

Some preliminary calculations have been conducted utilizing this model. The addition of a minute amount of a long chain polymer to the rotationally unstable Newtonian flow shown in Figures 10 and 11 where Re = 248 and Ro = 0.047 was considered. It was found that when the Weissenberg number

$$\lambda \nu / D^2 > 7.4 \times 10^{-5}, \qquad (51)$$

the secondary flow restabilized to a stretched double vortex configuration which was qualitatively quite similar to that shown in Figure 9(c). Furthermore, immediately before as well as after the restabilization occurs, the secondary flows associated with this dilute viscoelastic fluid give rise to friction losses that are substantially lower than those for the Newtonian case. For instance, for Weissenberg numbers greater than 7×10^{-5}, the dilute viscoelastic secondary flows give rise to a reduction in flowrate in the range of 2–3%, as compared to the value of 7.6% in the Newtonian case (see Figure 16). Hence, it is clear that the addition of a minute amount of a long chain polymer to a Newtonian fluid in rotating channel flow yields secondary flows with a substantially lower frictional drag—a striking nonlinear effect which can have important technological applications. To the best knowledge of the author, these are the first computations to illustrate this phenomenon in rotating channel flow.

5. Conclusion. An explicit finite difference technique has been presented for the solution of laminar internal flows which are subjected to a steady spanwise rotation. The complete time-dependent Navier-Stokes equations are solved by this method where the convective terms are formulated using Arakawa's scheme, and the viscous diffusion terms are formulated using the DuFort-Frankel scheme. Beyond being conservative and linearly stable (provided that Δt is sufficiently small), this technique is not subject to nonlinear instabilities that can arise from boundary conditions or aliasing errors (all aliasing are bounded). Furthermore, time-splitting instabilities can be eliminated expeditiously. The only significant disadvantage of this method, which is intrinsic to any finite difference approach, lies in the difficulty of extending the calculations to irregularly-shaped domains. However, this is much less of a problem with Arakawa's method (or any other comparable nine-point second-order accurate scheme which is conservative), since this method can be shown to come from a linear combination of two finite element schemes—bilinear elements in rectangles and linear elements in triangles (see Jespersen [15]). Hence, it is practical to generalize the method presented herein to deal with ducts of arbitrary cross section.

Detailed numerical calculations were presented for pressure-driven viscous flows in rotating rectangular ducts. These calculations demonstrated the existence of three flow regimes: at weak rotation rates there is a double vortex secondary flow; at intermediate rotation rates, there is an instability in the form of longitudinal rolls superimposed on a double vortex secondary flow; and at rapid rotation rates, there is a restabilization of the flow to a Taylor-Proudman regime. The specific numerical results obtained were found to be in excellent agreement with previously conducted theoretical and experimental studies. An extension of this numerical method to deal with dilute viscoelastic fluids was also presented. Specific calculations were presented which demonstrate that the addition of a minute amount of a long chain polymer to a Newtonian liquid can have a stabilizing effect on rotating channel flow and give rise to secondary flows with a substantially reduced frictional drag. This is a striking nonlinear effect which, although observed experimentally in analogous flow configurations, has never before been calculated in rotating channel flow.

Future research is needed on turbulent rotating internal flows. Furthermore, much more work needs to be done on the rotating internal flows of more concentrated viscoelastic fluids for which there is a problem with numerical instabilities. Nevertheless, as a result of the research effort of the past decade, a much better understanding of the various types of intriguing nonlinear phenomena that occur in rotating internal flows has been achieved.

Acknowledgements. The author would like to thank Dr. Sivagnanam Thangam for his assistance in the calculation of roll-cell instabilities in rotating channel flow. Thanks are also due to Dr. Gareth Williams for some valuable comments concerning the numerical method used and Miss Joanne Fendell for running some of the computer programs. This work was supported in part by the Exxon Education Foundation and the National Science Foundation under Grant No. ENG 79-08180.

288   C. G. SPEZIALE

REFERENCES

1. J. Hart, *Instability and secondary motion in a rotating channel flow*, J. Fluid Mech. **45** (1971), 341–351.

2. D. K. Lezius and J. P. Johnston, *Roll-cell instabilities in rotating laminar and turbulent channel flow*, J. Fluid Mech. **77** (1976), 153–175.

3. K. C. Cheng, R. C. Lin and J. W. Ou, *Fully-developed laminar flow in curved rectangular channels*, ASME J. Fluids Engrg. **98** (1976), 41–48.

4. S. C. Dennis and M. Ng, *Dual solutions for steady laminar flow through a curved tube*, Quart J. Mech. Appl. Math. **35** (1982), 305–324.

5. C. G. Speziale, *Numerical study of viscous flow in rotating rectangular ducts*, J. Fluid Mech. **122** (1982), 251–271.

6. C. G. Speziale and S. Thangam, *Numerical study of secondary flows and roll-cell instabilities in rotating channel flow*, J. Fluid Mech. 130 (1983), 377–395.

7. G. K. Batchelor, *Introduction to fluid dynamics*, Cambridge Univ. Press, London, 1967.

8. P. J. Roache, *Computational fluid dynamics*, Hermosa, Albuquerque, N.M., 1972.

9. R. D. Richtmyer and K. W. Morton, *Difference methods for initial value problems*, Interscience, New York, 1967.

10. O. Buneman, *A compact noniterative Poisson solver*, Stanford Univ. Inst. for Plasma Research Report SUIPR 294, 1969.

11. R. F. Ginn and M. M. Denn, *Rotational stability in viscoelastic liquids: Theory*, AIChE J. **15** (1969), 450–454.

12. M. M. Denn and J. J. Roisman, *Rotational stability and measurement of normal stress functions in dilute polymer solutions*, AIChE J. **15** (1969), 454–459.

13. W. R. Schowalter, *Mechanics of non-Newtonian fluids*, Pergamon, New York, 1978.

14. C. G. Speziale, *Instabilities of the Taylor type and the Taylor-Proudman theorem for rotating non-Newtonian fluids*, Proc. 12th Southeastern Conf. on Theoretical and Appl. Mechanics **II** (1984), 3–9.

15. D. C. Jespersen, *Arakawa's method is a finite element method*, J. Comput. Phys. **16** (1974), 383–390.

MECHANICAL ENGINEERING DEPARTMENT, STEVENS INSTITUTE OF TECHNOLOGY, HOBOKEN, NEW JERSEY 07030

Lectures in Applied Mathematics
Volume **22**, 1985

High Resolution TVD Schemes Using Flux Limiters

Peter K. Sweby[1]

ABSTRACT. Recently much effort has been placed into designing numerical schemes for conservation laws giving high resolution to discontinuities of the solution, whilst being devoid of the spurious oscillations plaguing the more classical second and higher order accurate schemes. (Absence of such oscillations is a characteristic of total variation diminishing (TVD) schemes.) One such method of achieving this goal is the utilisation of flux limiters to add only enough antidiffusive flux to gain resolution whilst still ensuring the absence of oscillations. A systematic approach to the notion of flux limiters is given and numerical comparison of various limiters on simple test problems is presented.

1. Introduction. In recent years much effort has been placed into obtaining second order accurate schemes for conservation laws which do not exhibit the spurious oscillations, near discontinuities of the solution, generated by the more classical second order schemes such as Lax-Wendroff [7]. The reason for this effort is because, although oscillation-free, first order accurate schemes, in general, give poor resolution to discontinuities. One approach used has been that of flux limiters, where the numerical flux is limited so as to prevent these oscillations from occuring. This method is similar to the flux corrected transport (FCT) of

1980 *Mathematics Subject Classification*. Primary 65M05; Secondary 65M10.
Key words and phrases. Hyperbolic conservation laws. finite difference approximations, flux limiters.
[1] Research supported by NASA grant number NAG1-273.

Boris and Book [1] and Zalesak [27] although theirs was an essentially two-step approach.

Some years ago Van Leer [11] combined limited nonconservative forms of the second order Lax-Wendroff and Warming-Beam [25] schemes to obtain a monotonicity preserving conservative scheme. More recently, Roe [16, 17, 22, 23, 18] has utilised flux limiters to obtain second order monotonicity preserving schemes. Also in the last couple of years, others such as Chakravarthy and Osher [2], have adopted this approach.

In proving convergence for nonlinear schemes [19, 22, 9], a crucial estimate needed is a uniform bound on the variation of the solution, which is certainly obtained if that variation decreases in time, and Harten [5] has named such schemes total variation diminishing (TVD). Apart from the implied bound for convergence proofs, a more immediate property of such schemes are that they are monotonicity preserving, i.e. do not suffer from spurious oscillations.

In §2 we lay the foundation for flux limiters in the formulation of first order 3-point schemes; then, in §3 we systematically outline the method of using flux limiters to obtain second order accurate TVD schemes.

In §4 we present brief numerical comparison of specific limiters and, finally, in §5 we mention some extensions of the technique as well as other comments.

A slightly more detailed study may be found in [21].

2. Preliminaries. We consider fully explicit approximations to the scalar nonlinear conservation law

$$u_t + f(u)_x = 0, \qquad u(x,0) = u_0(x). \qquad (2.1)$$

In particular, we consider 3-point schemes written in conservation form

$$u^k = u_k - \lambda(h_{k+1/2} - h_{k-1/2}), \qquad (2.2)$$

where

$$h_{k+1/2} = h(u_{k+1}, u_k), \qquad h(u, u) = f(u), \qquad (2.3)$$

is a consistent numerical flux function, λ is the mesh ratio

$$\lambda = \Delta t / \Delta x, \qquad (2.4)$$

and u_k are the nodal values of the piecewise constant mesh function $u_{\Delta x}(x, t)$ approximating $u(x, t)$. (We use the notation

$$u^k \equiv u_k^{n+1}, \qquad u_k \equiv u_k^n, \qquad (2.5)$$

where k and n are the spatial and temporal indices, respectively.)

For clarity we restrict our attention to regular girds, Δx constant, although results for irregular grids follow in a similar manner. (See §5.)

If the numerical flux, $h_{k+1/2}$ is that of an E-scheme [13], then it satisfies

$$\text{sgn}(u_{k+1} - u_k)\left[h_{k+1/2} - f(u)\right] \leqslant 0 \qquad (2.6)$$

for all u between u_k and u_{k+1}. E-schemes are entropy satisfying [8, 24] but at most first order accurate. (It should be noted that monotone schemes [6] belong to the class of E-schemes.)

For each computational cell (x_k, x_{k+1}), we define flux differences

$$\begin{aligned}
\left(\Delta f_{k+1/2}\right)^+ &= -\left(h_{k+1/2} - f(u_{k+1})\right), \\
\left(\Delta f_{k+1/2}\right)^- &= \left(h_{k+1/2} - f(u_k)\right),
\end{aligned} \qquad (2.7)$$

and note that

$$\left(\Delta f_{k+1/2}\right)^+ + \left(\Delta f_{k+1/2}\right)^- = \Delta f_{k+1/2} \qquad (2.8)$$

(using the notation $\Delta f_{k+1/2} \equiv f_{k+1} - f_k$).

We now use these flux differences to define local CFL numbers

$$\nu^+_{k+1/2} = \lambda \frac{\left(\Delta f_{k+1/2}\right)^+}{\Delta u_{k+1/2}}, \qquad (2.9a)$$

$$\nu^-_{k+1/2} = \lambda \frac{\left(\Delta f_{k+1/2}\right)^-}{\Delta u_{k+1/2}}, \qquad (2.9b)$$

and

$$\nu_{k+1/2} = \nu^+_{k+1/2} + \nu^-_{k+1/2} = \lambda \frac{\Delta f_{k+1/2}}{\Delta u_{k+1/2}}. \qquad (2.9c)$$

Note that, for E-schemes, (2.6) ensures that

$$\nu^+_{k+1/2} \geqslant 0, \qquad \nu^-_{k+1/2} \leqslant 0, \qquad (2.10)$$

justifying the superscripts.

The total variation of the solution at time $(n + 1)\Delta t$ is defined to be

$$\text{TV}(u^{n+1}) = \sum_k \left|u^{n+1}_{k+1} - u^{n+1}_k\right|, \qquad (2.11)$$

and so a scheme which is total variation diminishing (TVD) is one where

$$\text{TV}(u^{n+1}) \leqslant \text{TV}(u^n). \qquad (2.12)$$

Total variation diminishing is very useful for convergence proofs, where a uniform bound on the variation is required (see, for example, [5, 22, 19, 23]), but as mentioned previously a more immediate consequence of TVD is the lack of spurious oscillations in the solution.

For a general scheme written in the form

$$u^k = u_k - C_{k-1/2}\Delta u_{k-1/2} + D_{k+1/2}\Delta u_{k+1/2}, \qquad (2.13)$$

where $C_{k-1/2}$ and $D_{k+1/2}$ are data dependent (i.e. functions of the set $\{u_k\}$), it is easily shown [5, 23, 22] that *sufficient* conditions for it to be TVD are the inequalities

$$0 \leqslant C_{k+1/2}, \quad 0 \leqslant D_{k+1/2}, \quad 0 \leqslant C_{k+1/2} + D_{k+1/2} \leqslant 1. \quad (2.14)$$

Since from (2.7)

$$h_{k+1/2} - h_{k-1/2} = \left(\Delta f_{k+1/2}\right)^- + \left(\Delta f_{k-1/2}\right)^+, \qquad (2.15)$$

we may write the scheme (2.2) as

$$u^k = u_k - v^+_{k-1/2}\Delta u_{k-1/2} - v^-_{k+1/2}\Delta u_{k+1/2}, \qquad (2.16)$$

i.e. taking

$$C_{k+1/2} = v^+_{k+1/2}, \qquad D_{k+1/2} = -v^-_{k+1/2}, \qquad (2.17)$$

and therefore (cf. (2.10)) an E-scheme is TVD under the CFL-like condition

$$v^+_{k+1/2} - v^-_{k+1/2} \leqslant 1. \qquad (2.18)$$

In general we shall assume that a given scheme is TVD under the CFL condition

$$\sup_{\xi} \left(\lambda |f'(\xi)|\right) \leqslant \mu \leqslant 1. \qquad (2.19)$$

EXAMPLE. One example of an E-scheme is the Engquist-Osher scheme [3], which is also a monotone scheme and has numerical flux

$$h^{\mathrm{EO}}_{k+1/2} = f^+_k + f^-_{k+1} + f(\bar{u}), \qquad (2.20)$$

where

$$f^+_k = \int_{\bar{u}}^{u_k} \chi(s)f'(s)\,ds, \qquad f^-_k = \int_{\bar{u}}^{u_k} (1 - \chi(s))f'(s)\,ds,$$

$$(2.21)$$

and

$$\chi(s) = \begin{cases} 1, & f'(s) > 0, \\ 0, & f'(s) \leqslant 0 \end{cases}$$

($\bar{u}$ is the sonic point of $f(u)$, $f'(\bar{u}) = 0$).

Using the definitions (2.7) and (2.9a)–(2.9c) we get

$$\left(\Delta f_{k+1/2}\right)^+ = f_{k+1}^+ - f_k^+ = \int_{u_k}^{u_{k+1}} \chi(s) f'(s)\, ds, \qquad (2.22a)$$

and similarly

$$\left(\Delta f_{k+1/2}\right)^- = \int_{u_k}^{u_{k+1}} \left(1 - \chi(s)\right) f'(s)\, ds, \qquad (2.22b)$$

giving

$$\nu_{k+1/2}^+ - \nu_{k+1/2}^- = \frac{\lambda}{\Delta u_{k+1/2}} \int_{u_k}^{u_{k+1}} |f'(s)|\, ds \leqslant \lambda |f'|_{\max}. \qquad (2.23)$$

Therefore the Engquist-Osher scheme is TVD subject to the CFL condition

$$\sup_{\xi}\left(\lambda |f'(\xi)|\right) \leqslant 1. \qquad (2.24)$$

As mentioned in the Introduction, first order accurate schemes suffer from numerical diffusion; yet the more classical second order accurate schemes, although giving better resolution, are prone to spurious oscillations near discontinuities of the solution. In the next section we show how to overcome this problem with the use of flux limiters to construct second order accurate TVD schemes.

3. High resolution schemes. For clarity we first consider the scalar linear equation

$$u_t + a u_x = 0, \qquad a = \text{const} > 0. \qquad (3.1)$$

The second order Lax-Wendroff scheme [7] may be written, in this case, as $(\nu = \lambda a)$

$$u^k = u_k - \nu \Delta u_{k-1/2} - \Delta_-\left\{\tfrac{1}{2}(1 - \nu)\nu \Delta u_{k+1/2}\right\}, \qquad (3.2)$$

i.e. as a first order scheme

$$u^k = u_k - \nu \Delta u_{k-1/2}, \qquad (3.3)$$

plus an "antidiffusive flux"

$$-\Delta_-\left\{\tfrac{1}{2}(1 - \nu)\nu \Delta u_{k+1/2}\right\}. \qquad (3.4)$$

It is well known that the Lax-Wendroff scheme produces spurious oscillations at discontinuities of the solution, yet the first order scheme does not. We therefore limit the amount of antidiffusive flux that we add to the first order scheme, in much the same way as the flux corrected transport (FCT) of Boris and Book [1] and Zalesak [27], except that here we do so as a one-step scheme and also do not impose an upper limit of unity on the limiter, i.e. we add

$$-\Delta_-\left\{\phi_k\tfrac{1}{2}(1-\nu)\nu\Delta u_{k+1/2}\right\}, \qquad (3.5)$$

to the first order scheme (3.3), where ϕ_k is the limiter.

Following Roe [16], Van Leer [11] and others, we take the limiter to be a function of the ratio of consecutive gradients, i.e.

$$\phi_k = \phi(r_k), \quad \text{where} \quad r_k = \frac{\Delta u_{k-1/2}}{\Delta u_{k+1/2}}, \qquad (3.6)$$

and we choose the function $\phi(r)$ such that the resulting scheme is TVD. To write the limiter scheme in the form of (2.13) we may take

$$C_{k-1/2} = \nu\left\{1 + \tfrac{1}{2}(1-\nu)\left[\phi(r_k)/r_k - \phi(r_{k-1})\right]\right\}, \quad D_{k+1/2} = 0, \qquad (3.7)$$

and so a sufficient condition for the scheme to be TVD is

$$\left|\phi(r_k)/r_k - \phi(r_{k-1})\right| \leqslant \Phi \leqslant 2 \qquad (3.8)$$

(i.e. $0 \leqslant C_{k-1/2} \leqslant 1$).

We now stipulate that $\phi(r) \geqslant 0$ so as to maintain the sign of the antidiffusive flux and further that $\phi(r) = 0$ for $r < 0$ (i.e. "turn off" the antidiffusive flux at extrema) and the bound (3.8) now becomes

$$0 \leqslant (\phi(r)/r, \phi(r)) \leqslant \Phi \leqslant 2, \qquad (3.9)$$

and is shown pictorially in Figure 1, where the limiters corresponding to the Lax-Wendroff and Warming-Beam [25] schemes are also indicated. Note that for these schemes the limiters do not lie entirely in the TVD region.

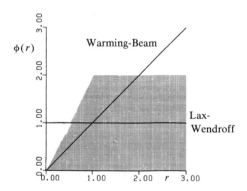

FIGURE 1. TVD region.

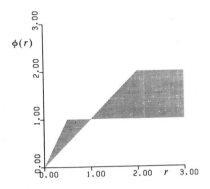

FIGURE 2. Second order TVD region.

For the scheme to be second order accurate, it must be an average of the second order centered Lax-Wendroff scheme and the second order upwind Warming-Beam scheme, i.e., in terms of limiters,

$$\phi(r) = (1 - \theta(r))\phi^{\mathrm{LW}}(r) + \theta(r)\phi^{\mathrm{WB}}(r), \qquad (3.10)$$

where $0 \leqslant \theta(r) \leqslant 1$. Since

$$\phi^{\mathrm{LW}}(r) \equiv 1, \qquad \phi^{\mathrm{WB}}(r) \equiv r, \qquad (3.11)$$

we have

$$\phi(r) = 1 + \theta(r)(r - 1), \qquad (3.12)$$

and the corresponding region is shown in Figure 2. (Numerical experiments reveal that for $\theta(r)$ outside the prescribed range, i.e. a nonconvex average, sine wave type data becomes squared after a few time steps.)

We now extend this idea to the nonlinear equation (2.1) replacing the constant ν by the local CFL numbers (2.9a)–(2.9c), using a general 3-point first order scheme (e.g. an E-scheme) as a base scheme and then adding a limited antidiffusive flux, which will now be influenced by the local direction of flow. In full,

$$u^k = u_k - \lambda \Delta_- h^1_{k+1/2} \qquad (3.13a)$$

$$- \lambda \Delta_- \left\{ \phi(r_k^+) \alpha_{k+1/2}^+ (\Delta f_{k+1/2})^+ - \phi(r_{k+1}^-) \alpha_{k+1/2}^- (\Delta f_{k+1/2})^- \right\},$$

$$(3.13b)$$

where

$$\alpha_{k+1/2}^+ = \tfrac{1}{2}\big(1 - \nu_{k+1/2}^+\big), \qquad \alpha_{k+1/2}^- = \tfrac{1}{2}\big(1 + \nu_{k+1/2}^-\big), \qquad (3.14)$$

and

$$r_k^+ = \frac{\alpha_{k-1/2}^+\big(\Delta f_{k-1/2}\big)^+}{\alpha_{k+1/2}^+\big(\Delta f_{k+1/2}\big)^+}, \qquad r_k^- = \frac{\alpha_{k+1/2}^-\big(\Delta f_{k+1/2}\big)^-}{\alpha_{k-1/2}^-\big(\Delta f_{k-1/2}\big)^-}. \qquad (3.15)$$

Notice the revised definition of the ratio $r_k^\pm$ which ensures that the scheme is still an average of the Lax-Wendroff and Warming-Beam schemes in the nonlinear case.

If we now write the scheme in the form (2.13), in a similar manner as in (3.7) except that $D_{k+1/2}$ is no longer zero, we get the CFL restriction (full details are given in [21])

$$\sup_{\xi}\big(\lambda|f'(\xi)|\big) \leqslant \left(\frac{2}{2 + \Phi}\right)\mu \leqslant \frac{2}{3} \qquad (3.16)$$

to guarantee that the scheme is TVD.

In the next section we look at various specific limiters which lie in the region of Figure 2 and present brief numerical comparisons.

4. Various limiters. We now present various specific second order TVD limiters together with a brief numerical comparison.

(a) Φ *limiters*. These are a class of limiters [21] which, together, cover the entire second order TVD region of Figure 2. For any Φ between 1 and 2 they are defined as

$$\phi_\Phi(r) = \mathrm{Max}(0, \mathrm{Min}(\Phi r, 1), \mathrm{Min}(r, \Phi)) \qquad (4.1)$$

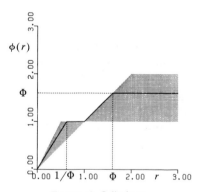

FIGURE 3. Φ limiters.

(see Figure 3) and contain two special cases:

ϕ_1: the "minmod" limiter [23, 22, 17],

$$\phi_1(r) = \text{Max}(0, \text{Min}(r, 1)) \qquad (4.2)$$

which was Roe's original limiter and is also a special case of the Chakravarthy-Osher limiter (see below) as well as the limiter used by various other authors (e.g. Harten [5] and LeVeque and Goodman [10]) to achieve TVD schemes via a different approach. (Note that these other approaches all reduce to the same scheme as that of the ϕ_1 limiter in the case of the linear equation (3.1).) This limiter corresponds to the bottom boundary of the region of Figure 2.

ϕ_2: Roe's "superbee" limiter [18]

$$\phi_2(r) = \text{Max}(0, \text{Min}(2r, 1), \text{Min}(r, 2)) \qquad (4.3)$$

which gives very sharp profiles and corresponds to the upper boundary of the region of Figure 2. These limiters are shown in Figure 4.

(b) *Van Leer limiter.* This limiter is proposed in [11] and, when reformulated to fit into the framework of this paper, is defined by

$$\phi_{\text{VL}}(r) = \frac{|r| + r}{1 + |r|}, \qquad (4.4)$$

and is a smooth curve in the TVD region as shown in Figure 5.

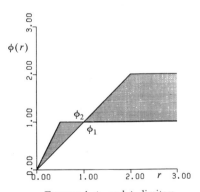

FIGURE 4. ϕ_1 and ϕ_2 limiters.

(c) *Chakravarthy-Osher limiter* [2].

$$\phi_{CO} = Max(0, Min(r, \psi)), \qquad 1 \leqslant \psi \leqslant 2. \qquad (4.5)$$

This limiter is shown in Figure 6 and is the only limiter presented here which lacks the symmetry property

$$\phi(r)/r \equiv \phi(1/r), \qquad (4.6)$$

except in the special case $\psi = 1$ when it is equivalent to the ϕ_1 limiter. We shall later see the result of this lack of symmetry.

Figures 7–16 show the results of the limiters ϕ_1, ϕ_2, ϕ_{VL} and ϕ_{CO} (with $\psi = 2$), as well as the first order scheme without any antidiffusive flux added, when applied to simple test problems.

For Figures 7 and 8 the test problem was the linear advection equation with both square wave and $\sin^2$ wave initial data. (The exact solution is depicted by the solid line.) Although the results are shown after only 25 time steps, the difference between the schemes is already apparent. The ϕ_2 limiter displays the sharpest discontinuity (which Roe [18] notes is maintained even after many hundreds of time steps) but does exhibit a slight squaring of the smooth data. The Van Leer limiter ϕ_{VL} and the ϕ_1 limiter both give comparable results, ϕ_{VL} being the slightly better of the two, whilst the lack of symmetry of the ϕ_{CO} limiter is clearly noticeable even after so few time steps.

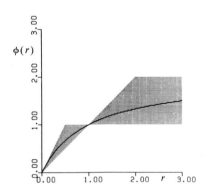

FIGURE 5. Van Leer limiter ϕ_{VL}.

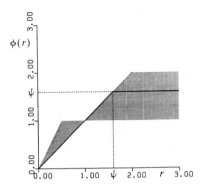

FIGURE 6. Chakravarthy-Osher limiter ϕ_{CO}.

Figure 9 shows the results when applied to the inviscid Burger's equation ($f \equiv \frac{1}{2}u^2$) with square wave initial data (and periodic boundary conditions). Here the difference between the limiters is less pronounced, although all give an improvement over the first order scheme. Note how the limiter schemes reduce the "dogleg" (or "entropy glitch") in the expansion fan.

Figures 10–16 show the results of a slightly more ambitious (yet still simple) test problem, that of Sod's shock tube problem [20], using Roe's approximate Riemann solver [15] to solve the system of Euler equations. Here the results are best judged by looking at the energy plot which contains (left to right) an expansion fan, a contact discontinuity and a shock. Figures 10–14 show the first order scheme and the limiter schemes applied in turn. Note that in Figure 12 the slight overshoot in the shock given by the ϕ_2 limiter due to its more restrictive CFL condition. (All tests were run with the same mesh ratio.) Despite this overshoot at the shock, ϕ_2 clearly gives the sharpest contact whilst ϕ_{VL} gives the cleanest shock (without overshoot).

Roe has often noted that with his decomposition the same scheme need not be applied to all characteristic fields, and Figures 15 and 16 illustrate this fact and the improvement that can be gained. In Figure 15, ϕ_1 has been applied to the two genuinely nonlinear fields whilst ϕ_2 has been used for the linearly degenerate field. Note how both a sharp contact and a sharp shock are obtained without any overshoot and, in Figure 16, it is seen how the shock is improved still further by use of the Van Leer limiter on the nonlinear fields instead of ϕ_1.

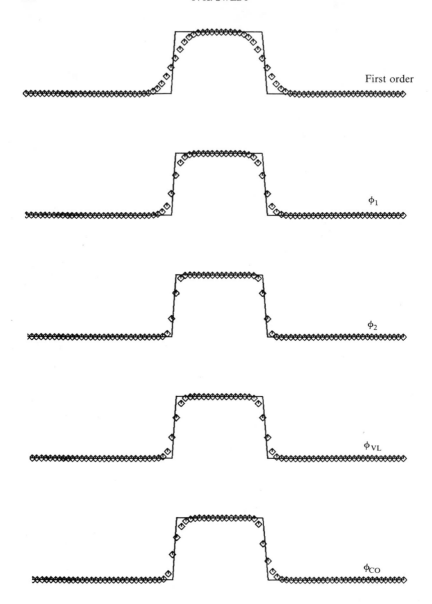

First order

ϕ_1

ϕ_2

ϕ_{VL}

ϕ_{CO}

FIGURE 7. Linear advection equation, square wave data.

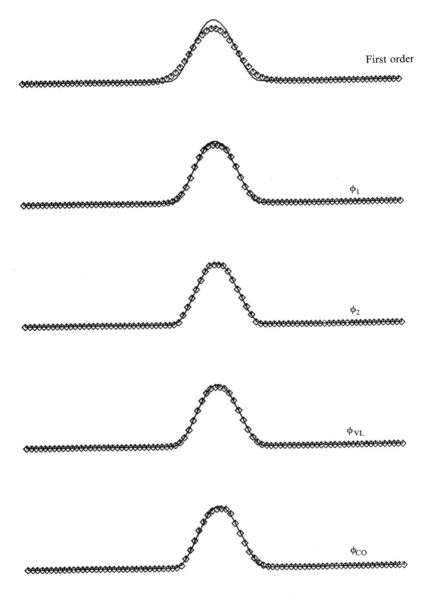

FIGURE 8. Linear advection equation, $\sin^2$ wave data.

P. K. SWEBY

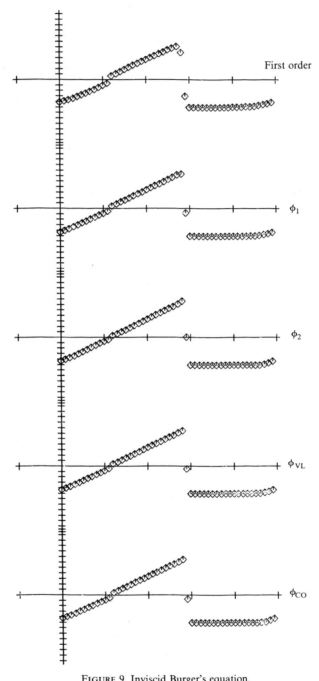

First order

ϕ_1

ϕ_2

ϕ_{VL}

ϕ_{CO}

FIGURE 9. Inviscid Burger's equation.

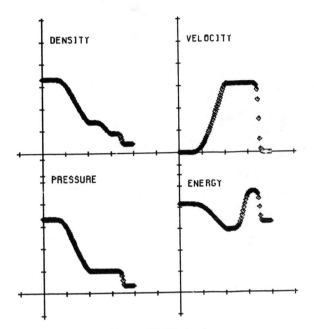

FIGURE 10. First order.

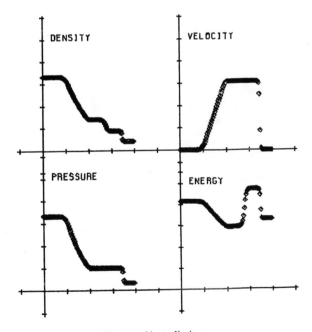

FIGURE 11. ϕ_1 limiter.

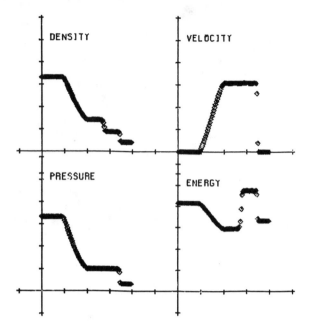

FIGURE 12. ϕ_2 limiter.

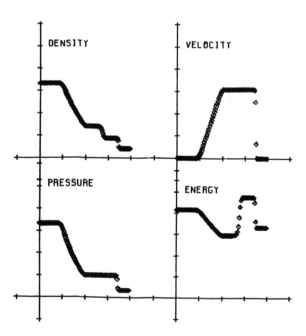

FIGURE 13. ϕ_{VL} limiter.

5. Extensions and comments. In this section we briefly mention the extension of flux limiters to irregular grids, systems of equations (cf. Sod's shock tube problem in §4) and implicit calculations. We then conclude with comments including ones on entropy satisfaction and a contemporary technique for obtaining high resolution TVD schemes.

When we are dealing with irregular grids, the mesh ratio λ is no longer a constant since the computational cell size Δx_k varies. This must also be taken into account when computing the ratios $r^{\pm}$ which are now defined as

$$r_k^+ = \frac{\lambda_{k-1/2}\alpha_{k-1/2}^+\left(\Delta f_{k-1/2}\right)^+}{\lambda_{k+1/2}\alpha_{k+1/2}^+\left(\Delta f_{k+1/2}\right)^+} = \frac{\alpha_{k-1/2}^+\left(\Delta f_{k-1/2}\right)^+/\Delta x_{k-1/2}}{\alpha_{k+1/2}^+\left(\Delta f_{k+1/2}\right)^+/\Delta x_{k+1/2}},$$

and similarly for r_k^- where $\Delta x_{k+1/2}$ is the appropriate mesh length. For a more detailed discussion of irregular grid schemes, see [23].

For systems of conservation laws, the flux differences $(\Delta f_{k+1/2})^{\pm}$ are now vector quantities and so the ratios $r^{\pm}$ must again be redefined, this time as

$$r_k^+ = \frac{\alpha_{k-1/2}^+\left\langle \left(\Delta f_{k-1/2}\right)^+, v_{k-1/2}\right\rangle}{\alpha_{k+1/2}^+\left\langle \left(\Delta f_{k+1/2}\right)^+, v_{k+1/2}\right\rangle},$$

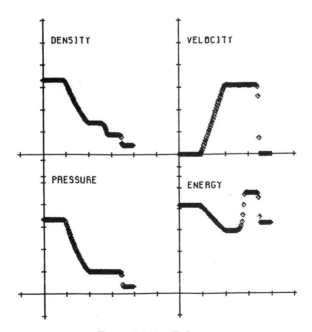

FIGURE 14. ϕ_{CO} limiter.

P. K. SWEBY

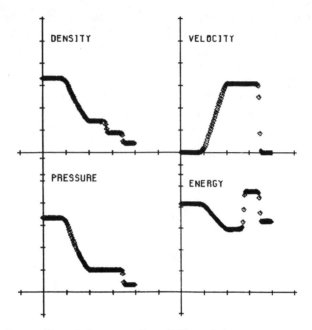

FIGURE 15. ϕ_1 limiter on nonlinear fields, ϕ_2 limiter on linear field.

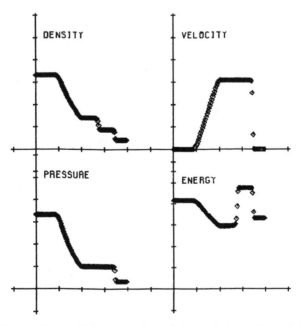

FIGURE 16. ϕ_{VL} limiter on nonlinear fields, ϕ_2 limiter on linear field.

where $\langle \cdot , \cdot \rangle$ denotes an inner product and $v_{k+1/2}$ is some suitable vector. Since there are various different methods for solving systems of equations, the choice of the vector v will vary, e.g. for Roe's decomposition [15] the correct choice for the jth characteristic field is $(R^{-1})_j$ (i.e. the jth row) where the matrix R is the matrix of local right eigenvectors used in the decomposition. (For Osher's scheme, Chakravarthy and Osher [2] propose the use of the grid difference of the gradient of some entropy function.)

The final direct extension mentioned here is to implicit schemes. Owing to the nondifferentiability of the limited antidiffusive flux, a nonconservative in time approach, such as Harten's LNI approach [26] of "freezing" the coefficients $C_{k+1/2}$ and $D_{k+1/2}$ (cf. (2.13)) in time, must be used. This approach is only suitable for steady-state calculations, and this author has found it to not be as reliable as one would hope, sometimes requiring relaxation or averaging of solutions to obtain convergence (even when not using limiters). When using this method with limiters, to remove dependence of the steady-state on the time step Δt, we redefine $\alpha^\pm$ (cf. (3.14)) as $\alpha^\pm \equiv \frac{1}{2}$, making it second order in space but not (artificial) time.

It should be noted that although flux limiter schemes may be used, via dimensional splitting for example, for two-dimensional problems Goodman and LeVeque [4] have shown that 2D TVD schemes are at most first order accurate.

We have not really considered here the important topic of entropy satisfaction [8]. For semidiscrete E-schemes, Osher [13] has shown entropy satisfaction with CFL condition unity and in [14] has shown second order TVD semidiscrete schemes, derived in a similar manner as in this paper, are entropy satisfying *if* artificial compression/rarefaction (ACR) is added. For fully discrete E-schemes, Tadmor [24] has shown entropy satisfaction with CFL condition $\frac{1}{2}$, although many E-schemes are known to satisfy the entropy inequality for the full CFL 1 condition. We have not, as yet, been able to prove the entropy inequality for the fully discrete limiter schemes, but the reduction of the "entropy glitch" noted in the previous section does provide an indication that these schemes may be entropy satisfying.

Finally, we very briefly mention one of the other approaches to obtaining TVD high resolution schemes, that of data preparation, and note the similarity to flux limiters. Harten [5] adopts the approach of applying a first order scheme to a modified flux function which, in fact, involves the use of the ϕ_1 limiter function. LeVeque and Goodman [10] replace the piecewise constant projection of the data onto the mesh with a piecewise linear projection, much as Van Leer did in his MUSCL algorithm [12], except that they use the ϕ_1 limiter function to calculate their gradients. In fact it should be noted that in the case of the linear equation (3.1) all of these methods are identical.

REFERENCES

1. J. P. Boris and D. L. Book, *Flux corrected transport. I. SHASTA, A fluid transport algorithm that works*, J. Comput. Phys. **11** (1973), 38–69.

2. S. Chakravarthy and S. Osher, *High resolution applications of the Osher upwind scheme for the Euler equations*, AIAA paper 83–1943 at 6th CFD Conf. (1983).

3. B. Engquist and S. Osher, *Stable and entropy condition satisfying approximations for transonic flow calculations*, Math. Comp. **34** (1980), 45–75.

4. J. B. Goodman and R. J. LeVeque, *On the accuracy of stable schemes for 2D scalar conservation laws*, Math. Comp., submitted.

5. A Harten, *High resolution schemes for conservation laws*, J. Comput. Phys. **49** (1983), 357–393.

6. A. Harten, J. M. Hyman and P. D. Lax, *On finite difference approximations and entropy conditions for shocks*, Comm. Pure Appl. Math. **29** (1976), 297–322.

7. P. D. Lax and B. Wendroff, *Systems of conservation laws*, Comm. Pure Appl. Math. **13** (1960), 217–237.

8. P. D. Lax, *Hyperbolic systems of conservation laws and the mathematical theory of shock waves*, SIAM Regional Conf. Ser., Lectures in Appl. Math., vol. 11, Soc. Indust. Appl. Math., Philadelphia, Pa., 1972.

9. A. Y. LeRoux, *Convergence of an accurate scheme for first order quasi-linear equations*, RAIRO Anal. Numer. **15** (1981), 151–170.

10. R. LeVeque and J. Goodman, *TVD methods for scalar conservation laws in 1 and 2 space dimensions*, Lecture, AMS-SIAM Summer Seminar (La Jolla, June/July, 1983).

11. B. van Leer, *Towards the ultimate conservative difference scheme. II. Monotonicity and conservation combined in a second order scheme*, J. Comput. Phys. **14** (1974), 361–370.

12. _____ , *Towards the ultimate conservative difference scheme. V: A second order sequel to Godunov's method*, J. Comput. Phys. **32** (1979), 101–136.

13. S. Osher, *Riemann solvers, the entropy condition, and difference approximations*, SIAM J. Numer. Anal. **21** (1984), 217–235.

14. S. Osher and S. Chakravarthy, *High resolution schemes and the entropy condition*, SIAM J. Numer. Anal. (to appear).

15. P. L. Roe, *Approximate Riemann solvers, parameter vectors, and difference schemes*, J. Comput. Phys. **43** (1981), 357–372.

16. _____ , *Numerical algorithms for the linear wave equation*, Royal Aircraft Establishment Technical Report 81047, 1981.

17. P. L. Roe and M. J. Baines, *Algorithms for advection and shock problems*, Proc. 4th GAMM Conf. on Numerical Methods in Fluid Mechanics, 1982, H. Viviand, ed.

18. P. L. Roe, *Some contributions to the modelling of discontinuous flows*, Lecture, AMS-SIAM Summer Seminar on Large Scale Computations in Fluid mechanics (Scripps Institute of Oceanography, University of California, San Diego, 26 June to 8 July, 1983), Amer. Math. Soc., Providence, 1985.

19. R. Sanders, *On convergence of monotone finite difference schemes with variable space differencing*, Math. Comp. **40** (1983), 91–106.

20. G. A. Sod, *A survey of several finite difference methods for systems of non-linear hyperbolic conservation laws*, J. Comput. Phys. **27** (1978), 1–31.

21. P. K. Sweby, *High resolution schemes using flux limiters for hyperbolic conservation laws*, SIAM J. Numer. Anal., 1984.

22. P. K. Sweby and M. J. Baines, *On convergence of Roe's scheme for the general non-linear scalar wave equation*, J. Comput. Phys. **56** (1984), 135–148.

23. P. K. Sweby, *Shock capturing schemes*, Ph. D. thesis, Reading University, 1982.

24. E. Tadmor, *Numerical viscosity and the entropy condition for conservative difference schemes*, NASA Contractor Report 172141 (1983), NASA Langley.

25. R. F. Warming and R. M. Beam, *Upwind second order difference schemes and applications in aerodynamics*, AIAA J. **14** (1976), 1241–1249.

26. H. C. Yee, R. F. Warming and A. Harten, *Implicit total variation diminishing (TVD) schemes for steady-state calculations*, NASA TM 84342, 1983.

27. S. T. Zalesak, *Fully multidimensional flux corrected transport algorithms for fluids*, J. Comput. Phys. **31** (1979), 335–362.

DEPARTMENT OF MATHEMATICS, UNIVERSITY OF CALIFORNIA, LOS ANGELES, CALIFORNIA 90024

Current address: Department of Mathematics, University of Reading, Whiteknights, Reading RG6 2AH, England

Lectures in Applied Mathematics
Volume 22, 1985

Stability of Hyperbolic Finite-Difference Models with One or Two Boundaries

Lloyd N. Trefethen[1]

ABSTRACT. The stability of finite-difference models of hyperbolic partial differential equations depends on how numerical waves propagate and reflect at boundaries. This paper presents an extended numerical example illustrating the key points of this theory.

0. Introduction. In the numerical solution of hyperbolic partial differential equations by finite differences, stability is well known to be a critical issue. As a first step, the difference model must satisfy the von Neumann condition—that is, the basic formula should admit no exponentially growing Fourier modes. For linear problems with smoothly varying coefficients and no boundaries, this is essentially the whole story, and in fact if one rules out algebraically as well as exponentially growing local Fourier modes, then stability is assured. Results of this kind are widely known and are discussed in the superb book by Richtmyer and Morton [8].

When boundaries are introduced, the stability problem becomes more subtle. Even here the literature is copious, and a dozen or more people have made substantial contributions, including Godunov and Ryabenkii, Strang, Osher [7], Kreiss, Gustafsson, Sundström, Tadmor, and Michelson. The best known paper in this area is the one by Gustafsson, Kreiss and Sundström [4] in 1972, which presents what is now often referred to

1980 *Mathematics Subject Classification.* Primary 65M10.

1 Supported by an NSF Mathematical Sciences Postdoctoral Fellowship and by the U.S. Dept. of Energy under Contract DE-AC02-76-ER03077-V while the author was at the Courant Institute of Mathematical Sciences.

as the "GKS stability theory". The great strength of the GKS paper is that it establishes a necessary and sufficient stability condition for difference models of very general form—three-point or multipoint stencil in space, two-level or multilevel in time, explicit or implicit, dissipative or nondissipative. A difficulty with the paper is that it is very hard to read, and this has regrettably limited its influence. Fortunately, some more accessible accounts have appeared recently, including the report of Gustafsson in this volume.

My own work in this field has been concerned with giving the stability question for initial boundary value problem models a physical interpretation based on the ideas of dispersive wave propagation and group velocity. Group velocity effects in finite-difference modeling have been surveyed by me in [9] and by Vichnevetsky and Bowles in [13]; others who have been interested in these matters include Matsuno, Grotjahn and O'Brien in meteorology; Alfold, Bamberger, and Martineau-Nicoletis in geophysics; Kentzer, Giles and Thompkins in aerodynamics; and Hedstrom and Chin in theoretical numerical analysis. I have described a group velocity interpretation of the one-boundary stability problem in [10], showing in particular how the GKS "perturbation test" for unstable "generalized eigensolutions" is equivalent to a test of the sign of a group velocity. In [11] this approach is made rigorous, and various theorems on unstable growth rates are obtained. In [12] I have extended these ideas to problems with two or more boundaries or internal interfaces, where stability depends on what happens when wave packets reflect back and forth. The latter work was motivated in part by ideas of Kreiss (see, e.g., [4, §7]), of Beam, Warming and Yee [1], and of Giles and Thompkins [2].

There is an analogous theory for p.d.e.'s rather than difference approximations. Again, wave radiation from the boundary is a general mechanism of ill-posedness. See Kreiss [6] for the basic theory, and the new survey by Higdon [5] for the wave interpretation. The difference is that for p.d.e.'s, nontrivial cases of ill-posedness do not arise unless the domain contains two or more space dimensions.

The purpose of this paper is to survey these results relating stability and wave propagation by means of an extended example. With the aid of many illustrations we will see exactly how waves can get amplified by reflection at boundaries and how this can lead to instability.

1. No boundaries: dispersive wave propagation. As our basic difference scheme we will take the *Crank-Nicolson (CN)* or *trapezoidal* formula

$$v_j^{n+1} - v_j^n = \frac{\lambda}{4}\left(v_{j+1}^n + v_{j+1}^{n+1} - v_{j-1}^n - v_{j-1}^{n+1}\right) \qquad (1.1)$$

for the p.d.e. $u_t = u_x$. Here j is the space index, n is the time index, and λ is the *mesh ratio* k/h, where k and h are the time and space steps, respectively. Because CN is a nondissipative, two-level formula, it will be easy to analyze how waves propagate. The price we pay for this is that CN is implicit, which means that in principle it cannot be implemented on an unbounded domain, although it still defines a bounded operator on l^2. However, this is no problem for simple test experiments, since we can keep the action away from the boundary when we wish.

A general initial data distribution $v^0 \in l^2$ can be written as a Fourier integral

$$v_j^0 = \frac{1}{2\pi} \int_{-\pi/h}^{\pi/h} e^{i\xi x} \hat{v}^0(\xi)\, d\xi, \qquad x = jh, \tag{1.2}$$

where the dual variable ξ is called the *wave number*. The Fourier transform $\hat{v}^0$ is a $2\pi/h$-periodic function in $L^2[-\pi/h, \pi/h]$ given by the Fourier series

$$\hat{v}^0(\xi) = \lim_{N \to \infty} \sum_{j=-N}^{N} e^{-i\xi x} v_j^0, \qquad x = jh. \tag{1.3}$$

To determine what v^n will look like for $n > 0$, we can first determine what will happen to an initial sine wave $e^{i\xi x}$ and then use Fourier synthesis. To this end, substitute $e^{i(\omega t + \xi x)}$ in (1.1), where ω is a *frequency* to be determined. The result is the equation

$$e^{i\omega k} - 1 = \frac{\lambda}{4} \left(e^{i\xi h} + e^{i(\omega k + \xi h)} - e^{-i\xi h} - e^{i(\omega k - \xi h)} \right),$$

which simplifies to

$$2 \tan \frac{\omega k}{2} = \lambda \sin \xi h. \tag{1.4}$$

This is the *numerical dispersion relation* for CN. This equation defines a unique value $\omega \in [-\pi/k, \pi/k]$ for each $\xi \in [-\pi/h, \pi/h]$, and so it is actually a function

$$\omega(\xi) = \frac{2}{k} \tan^{-1} \left(\frac{\lambda}{2} \sin \xi h \right). \tag{1.4'}$$

Figure 1.1 shows what (1.4′) looks like in the case $\lambda = 1$: For $\xi h \ll \pi$ the dispersion function matches the ideal function $\omega = \xi$ for $u_t = u_x$ closely, but for larger ξ it disagrees markedly.

Now that the dispersion function is known we can synthesize v^n as follows:

$$v_j^n = \frac{1}{2\pi} \int_{-\pi/h}^{\pi/h} e^{i(\omega(\xi)t + \xi x)} \hat{v}^0(\xi)\, d\xi, \qquad x = jh, t = nk. \tag{1.5}$$

Armed with this equation we could duplicate the behavior of CN by computing Fourier integrals. There is little profit in that, but what (1.5) does offer is the prospect of approximate evaluation by a stationary phase argument. For observe that the exponential term introduces an oscillatory behavior that will make the integrand tend to cancel to zero if $\omega(\xi)$ and $\hat{v}^0(\xi)$ are smooth. The exception is that at values of ξ satisfying

$$\frac{d}{d\xi}(\omega(\xi)t + \xi x) = 0, \quad \text{i.e.} \quad \frac{d\omega(\xi)}{d\xi} = -\frac{x}{t},$$

there is no oscillation and no cancellation. In other words, most of the energy associated with wave number ξ travels approximately at the *group velocity*

$$C = -\frac{d\omega}{d\xi}. \tag{1.6}$$

Of course this argument is vague, but it can be made precise in various ways; see, for example, Lemma 5.1 of [11].

Thus wave energy travels at a velocity given by the negative of the slope of the dispersion relation. For CN we can differentiate (1.4) implicitly to obtain

$$C = -\cos \xi h \cos^2 \frac{\omega k}{2}. \tag{1.7}$$

Eliminating ω by means of (1.4′), or differentiating (1.4′) directly, converts this to the functional form

$$C(\xi) = \frac{-\cos \xi h}{1 + (\lambda^2/4)\sin^2 \xi h}. \tag{1.7′}$$

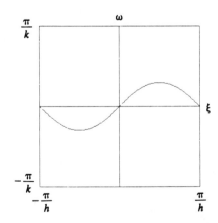

FIGURE 1.1. Dispersion relation for CN with $\lambda = 1$.

This function is plotted in Figure 1.2. One sees that for well-resolved waves—i.e. $\xi h \ll 0$, or many points per wavelength—energy travels at velocity -1, as it should according to the p.d.e. $u_t = u_x$. Less well-resolved waves have lower speeds (less negative velocities), and it is this fact that gives rise to familiar oscillations around discontinuities. At the extreme, the sawtoothed (or *parasitic*) wave $v_j^n = (-1)^j$, i.e. $\xi h = \pm\pi$, has group velocity $+1$, so energy in this mode travels in the physically wrong direction (Figure 1.2).

Let us confirm this last prediction by an experiment. Figure 1.3 shows the evolution under CN with $h = .01$ and $\lambda = 1$ of an initial signal

$$v_j^n = \left[1 + (-1)^j\right] e^{-400(x-.5)^2}$$

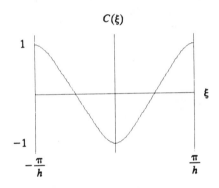

FIGURE 1.2. Group velocity.

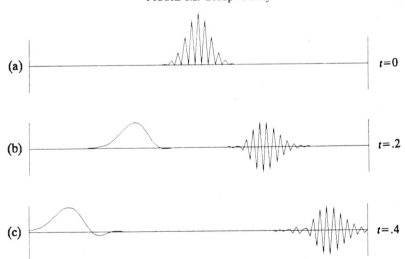

FIGURE 1.3. Propagation of energy at group velocities $C = \pm 1$.

on the interval [0, 1]. This "rectified Gaussian", shown in Figure 1.3(a), contains equal amounts of energy at $\xi h = 0$ and at $\xi h = \pm\pi$. Figures 1.3(b), (c) show the wave forms at times .2 and .4. As predicted, the two wave components have separated and traveled in opposite directions. This backwards motion of the parasitic wave component is of course a purely numerical effect. For further examples see [9 and 13].

In analyzing the behavior of an arbitrary difference model, a useful question to ask is: given a frequency ω_0, what associated wave numbers ξ are admitted by the dispersion relation, and how do the corresponding waves $\exp(i(\omega_0 t + \xi x))$ propagate? To get the answer for CN, imagine drawing a horizontal line at height ω_0 in Figure 1.1. Assuming ω_0 is small enough, it will intersect the dispersion curve at two values ξ_1 and ξ_2. Of the two corresponding sine waves, one has $C \leq 0$ and one has $C \geq 0$. For simplicity, from now on we will write η for the wave number corresponding to $C \leq 0$ and ξ for the other one. By (1.4), ξ and η are related under CN by

$$\xi = \frac{\pi}{h} - \eta \qquad \begin{array}{l} \xi: \text{rightgoing,} \\ \eta: \text{leftgoing.} \end{array} \qquad (1.8)$$

The same relationship holds for any difference model, such as leap frog or backwards Euler, whose spatial discretization consists of the usual second-order centered difference.

As a more complicated example, suppose we had a nondissipative difference formula with the dispersion function plotted in Figure 1.4. (An arbitrary continuous dispersion function defines a bounded operator $v^n \mapsto v^{n+1}$ in l^2, a *Fourier multiplier*, but this operator can be realized by finite differences only when the function is the solution of a trigonometric polynomial equation in ω and ξ.) According to the figure, there are

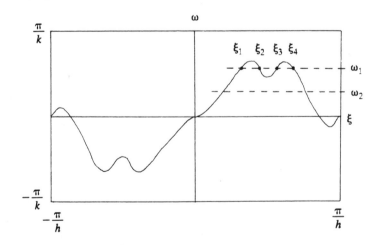

FIGURE 1.4. Dispersion relation for an unknown difference formula.

four wave numbers ξ_ν associated with the frequency ω_1. ξ_1 and ξ_3 correspond to leftgoing waves, and ξ_2 and ξ_4 to rightgoing waves.

On the face of it, the situation looks different at frequency ω_2—there is only one wave propagating in each direction. However, in fact, the missing two wave numbers still exist, but they have become complex. One has negative imaginary part and corresponds to a leftgoing evanescent wave that does not carry energy; the other has positive imaginary part and is evanescent and rightgoing.

Actually, as far as one can tell from Figure 1.4, evanescent modes may exist at frequency ω_1 too. This depends on the size of the stencil of the difference formula. The general rule is this: for a scalar difference formula with stencil extending l points to the left of center and r points to the right, the dispersion relation is a trigonometric polynomial in ξ of degree $l + r$, which for each ω has $l + r$ complex roots ξ that break down into exactly r "leftgoing" and l "rightgoing" linearly independent wave modes. Here we say that a wave $\exp(i(\omega t + \xi x))$ with $\operatorname{Im} \dot\omega = 0$ (or more generally $\operatorname{Im} \omega \leqslant 0$) is *rightgoing* if either

(1) $\operatorname{Im} \xi = 0$ and $C(\xi, \omega) \geqslant 0$, or

(2) $\operatorname{Im} \xi > 0$.

A *leftgoing* wave is defined with the directions of the inequalities in (1) and (2) reversed. These definitions and their consequences are studied in detail in [11].

Thus the general solution of the form $v_j^n = e^{i\omega t}\phi_j$ to an arbitrary scalar finite-difference formula can be written

$$v_j^n = e^{i\omega t} \underset{\substack{\nu=1 \\ (rightgoing)}}{\overset{l}{\sum}} \alpha_\nu e^{i\xi_\nu x} + e^{i\omega t} \underset{\substack{\nu=1 \\ (leftgoing)}}{\overset{r}{\sum}} \beta_\nu e^{i\eta_\nu x}, \qquad x = jh, t = nk.$$

$$(1.9)$$

For convenience I have relabeled $\alpha_{l+1}, \dots, \alpha_{l+r}$ by $\beta_1, \dots, \beta_r$, and $\xi_{l+1}, \dots, \xi_{l+r}$ by $\eta_1, \dots, \eta_r$.

2. One boundary: reflection coefficients. What if we now let more time elapse in Figure 1.3, so that the waves hit the boundaries at $x = \pm 1$? The result will, of course, depend on the boundary conditions there.

The general principle for analyzing such problems is to look for solutions containing only a single frequency ω_0. Different frequencies can be superposed later. If a wave $\exp(i(\omega_0 t + \xi_0 x))$ hits a boundary, then the reflected result after the initial transient has died away will be a linear combination $\sum \alpha_\nu \exp(i\omega_0 t + \xi_\nu x))$ of waves with the same frequency ω_0 but with various new wave numbers ξ_ν. In general, the wave numbers in the reflected waves will be all of those fulfilling the following two conditions:

(1) ω_0, ξ_ν satisfy the dispersion relation.

(2) The wave $\exp(i(\omega_0 t + \xi_v x))$ propagates away from the boundary into the interior (the *radiation condition*). This means that any wave reflected at a left-hand boundary must be rightgoing, and any wave reflected at a right-hand boundary must be leftgoing.

Neither of these statements mentions the boundary conditions. Those do not affect the set of reflected waves, just the coefficients α_v.

To compute numerical reflection coefficients one simply inserts the wave (1.9) into the numerical boundary conditions. For a general formulation consider a left-hand boundary at $x = j = 0$, and let the numerical boundary conditions there consist of l linear homogeneous equations giving $v_0^{n+1}, \ldots, v_{l-1}^{n+1}$ as linear combinations of other values v_j^n. Insertion of (1.9) yields a linear *reflection equation*

$$E(\omega)\alpha = D(\omega)\beta, \tag{2.1}$$

relating leftgoing and rightgoing wave coefficients. Here E and D are matrices of dimensions $l \times l$ and $l \times r$, respectively. If $E(\omega)$ is nonsingular we can solve for the rightgoing wave coefficients to get

$$\alpha = A(\omega)\beta = [E(\omega)]^{-1}D(\omega)\beta. \tag{2.2}$$

$A(\omega)$ is called the *reflection matrix*. Compare [4, (10.2)].

Figure 2.1 shows what happens when the integration of Figure 1.3 is carried to $t > 1/2$ with the following conditions at the boundaries:

$$v_0^{n+1} = v_1^{n+1}, \tag{2.3a}$$

$$v_j^{n+1} = 0. \tag{2.4a}$$

Here $J = 1/h$ is the index of the grid point at the boundary $x = 1$. Figures 2.1(a), (b) show the configuration at times $t = .6$ and $t = .8$. Clearly the leftgoing pulse with $\eta = 0$ has generated a rightgoing reflected wave with $\xi = \pi/h$ at very small amplitude. The rightgoing pulse with $\xi = \pi/h$ has generated a considerably larger reflection with $\eta = 0$. Let us predict these reflected amplitudes after the fact.

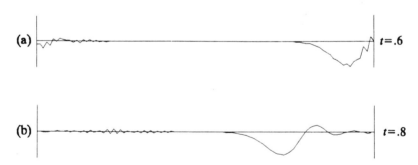

FIGURE 2.1. Continuation of Figure 1.3 showing reflection at boundaries.

First, we compute the reflection coefficient for (2.3a). In the present case, (1.8) implies that (1.9) reduces to

$$v_j^n = \left[\alpha e^{i\xi x} + \beta e^{i\eta x} \right] e^{i\omega t}$$

$$= \left[(-1)^j \alpha e^{-i\eta x} + \beta e^{i\eta x} \right] e^{i\omega t}, \qquad x = jh, \, t = nk. \qquad (2.5)$$

For simplicity in such computations it is convenient to introduce the abbreviations

$$\kappa = e^{i\xi h}, \qquad \mu = e^{i\eta h}, \qquad z = e^{i\omega k}. \qquad (2.6)$$

Then (2.5) becomes

$$v_j^n = \left(\alpha \kappa^j + \beta \mu^j \right) z^n. \qquad (2.7)$$

For CN, (1.8) becomes $\kappa = -1/\mu$. Inserting (2.7) in (2.3a) gives

$$\alpha + \beta = (\alpha \kappa + \beta \mu) = -\alpha/\mu + \beta \mu.$$

In the form of (2.1) this is

$$\left(1 + \frac{1}{\mu} \right) \alpha = (\mu - 1)\beta, \qquad (2.3b)$$

that is,

$$\frac{\alpha}{\beta} = \mu \frac{\mu - 1}{\mu + 1} = i e^{i\eta h} \tan \frac{\eta h}{2}. \qquad (2.3c)$$

For $\eta \approx 0$, as in the present experiment, we get $\alpha/\beta \approx 0$, and this explains the very small amplitude of the pulse reflected from the left-hand boundary in Figure 2.1.

Now the analogous computation for (2.4a): Inserting (2.7) gives

$$\kappa^J \alpha = -\mu^J \beta, \qquad (2.4b)$$

that is,

$$\frac{\alpha}{\beta} = -(-\mu^2)^J = (-1)^{J+1} e^{2i\eta}. \qquad (2.4c)$$

Thus the amplitude of the reflected wave should be equal and opposite to that of the incident wave, regardless of η. Figure 2.1 confirms this nicely.

Let us return to the left-hand boundary and consider some alternative boundary conditions. Suppose we replace (2.3a) by

$$v_0^{n+1} = v_1^n. \qquad (2.8a)$$

Then insertion of (2.7) leads to

$$\left(z + \frac{1}{\mu} \right) \alpha = (-z + \mu)\beta, \qquad (2.8b)$$

or

$$\frac{\alpha}{\beta} = \mu \frac{\mu - z}{1 + \mu z} = \frac{ie^{i\eta h} \sin\left((\eta h - \omega k)/2\right)}{\cos\left((\eta h + \omega k)/2\right)}. \tag{2.8c}$$

Again this predicts a near-zero reflected amplitude for $\eta \approx 0 \approx \omega$. On the other hand, suppose we impose

$$v_0^{n+1} = v_2^{n+1}. \tag{2.9a}$$

The reflection equation is then

$$\left(1 - 1/\mu^2\right)\alpha = \left(\mu^2 - 1\right)\beta. \tag{2.9b}$$

This implies

$$\frac{\alpha}{\beta} = \mu^2 = e^{2i\eta h}, \tag{2.9c}$$

much as in (2.4c), at least for $\eta \neq 0$. At $\eta = 0$, (2.9b) has the form $0\alpha = 0\beta$, so it is not solvable except in a limiting sense.

Figure 2.2 plots results of experiments with boundary conditions (2.8a) and (2.9a). Figure 2.2(a) shows the initial condition, a leftgoing wave with $\eta = 0$ on a mesh with $h = .01$ for $[0, 1]$, and Figures 2.2(b), (c) show the results at $t = 1$ under (2.8a) and (2.9a), respectively. The predicted reflection coefficients 0 and 1 are clearly in evidence. In fact, no reflected

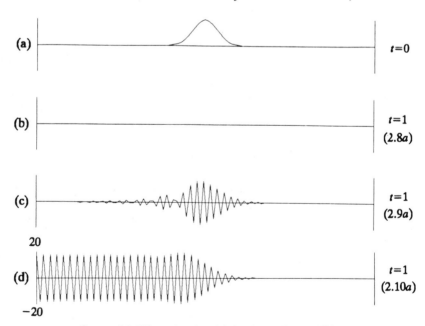

FIGURE 2.2. Effect of various left-hand boundary conditions.

energy at all is visible in Figure 2.2(b); this is because with $\lambda = 1$, i.e. $h = k$, (2.8c) is identically zero.

Figure 2.2(d) shows the response of CN to the same initial data with a more contrived boundary condition:

$$v_0^{n+1} = -v_1^{n+1} + v_2^{n+1} + v_3^{n+1}. \qquad (2.10a)$$

This time a marked instability is evident (note the vertical scale). To see why, compute the reflection equation

$$[1 + \kappa - \kappa^2 - \kappa^3]\alpha = -[1 + \mu - \mu^2 - \mu^3]\beta,$$

which can be simplified to

$$(1 + \mu)(1 - \mu)^2\alpha = -\mu^3(1 + \mu)^2(1 - \mu)\beta, \qquad (2.10b)$$

that is,

$$\frac{\alpha}{\beta} = -\mu^3\frac{1 + \mu}{1 - \mu} = ie^{3i\eta h}\cot\frac{\eta h}{2}. \qquad (2.10c)$$

At $\eta = 0$, this reflection coefficient is *infinite*.

The existence of a frequency ω at which the reflection coefficient at a boundary is infinite implies that a finite-difference model is unstable. In [11] it is shown that under reasonable assumptions, such a model potentially amplifies initial data in the l^2 norm at a rate at least proportional to the time step number n. Obviously this is an unstable situation, since as h and k are decreased, the index n corresponding to a fixed time t increases to ∞.

The presence of an infinite reflection coefficient is a stronger condition than "GKS-instability," a name sometimes given to instability according to Definition 3.3 of [4]. In fact, a difference model is GKS-unstable if and only if the matrix $E(\omega)$ in (2.1) is singular for some ω with $\text{Im } \omega \leqslant 0$, i.e. $|z| \geqslant 1$ (Theorem 1a of [11]). For scalar problems with three-point stencils, this amounts to the condition that the coefficient of α on the left side of a reflection equation, such as (2.3b), (2.4b), (2.8b), (2.9b) or (2.10b), is 0 for some ω. Thus, for example, the boundary condition (2.9a) is GKS-unstable for CN, since the left-hand side of (2.9b) is zero at $\mu = z = 1$, even though no apparent catastrophe occurred in Figure 2.2(c). In such situations the reflection coefficient remains finite if it happens that the right-hand side of the reflection equation has a zero at the same value of ω of at least as high an order. When this occurs, it is shown in [11] that unstable growth of initial data in the l^2 norm need proceed no faster than in proportion to $\sqrt{n}$.

To summarize, there are at least three distinct circumstances that may obtain at the boundary:

(1) nonsingular reflection equation, finite reflection coefficient: stable ($\|v\| = O(1)$);

(2) singular reflection equation, finite reflection coefficient: weakly
unstable ($\|v\| = O(\sqrt{n})$);

(3) singular reflection equation, infinite reflection coefficient: unstable
($\|v\| = O(n)$).

3. Two boundaries: reflection back and forth. What if we let even more
time elapse in Figure 1.3 and 2.1? The answer is simple: new reflections
will occur as the two wave packets bounce back and forth between $x = 0$
and $x = 1$. Each time a rightgoing wave packet with $\xi \approx \pi/h$ hits $x = 1$,
it will reflect as a leftgoing wave packet with $\eta \approx 0$ having essentially the
same amplitude. But each time a leftgoing wave hits $x = 0$, it will reflect
as a rightgoing wave of greatly diminished amplitude. The total energy in
$[0, 1]$ will consequently decay rapidly to near 0. The reason one can only
say "near" is that a certain portion of the energy will have $\xi \approx \pi/2h$,
and since $C = 0$ at this wave number, this portion tends to stay in a fixed
position and never hit the boundaries.

Figures 3.1(a), (b) confirm this prediction by showing the numerical
solutions at $t = 2$ and $t = 4$. At $t = 2$ the signal that remains is very
weak, and at $t = 4$ it is somewhat weaker.

Suppose, more generally, that we set up the CN model on an interval
$[0, L]$ with some pair of stable boundary conditions at $x = 0$ and $x = L$
for which the reflection coefficient functions are $A_1(\omega)$ and $A_2(\omega)$,
respectively. If v^0 is a wave packet with frequency ω, then as t increases
this packet will reflect back and forth, undergoing amplification by $|A_1|$
or $|A_2|$ each time it hits a boundary. Let $C(\omega)$ denote the group speed
$|C(\xi, \omega)| = |C(\eta, \omega)|$. Then the travel time for a complete circuit involving a reflection at each boundary will be

$$T(\omega) = \frac{2L}{C(\omega)}. \tag{3.1}$$

After such a circuit the amplitude will have increased by a factor

$$\alpha(\omega) = |A_1(\omega)A_2(\omega)|. \tag{3.2}$$

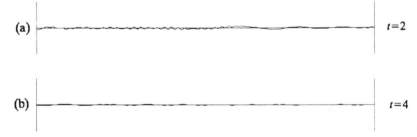

(a) —————————————————————————————— $t=2$

(b) —————————————————————————————— $t=4$

FIGURE 3.1. Further continuation of Figures 1.3 and 2.1.

Combining these formulas, we reach the following conclusion: as $t \to \infty$, energy associated with frequency ω grows approximately at the rate

$$\frac{\|v(t)\|}{\|v(0)\|} \approx \alpha(\omega)^{t/T(\omega)} = \alpha(\omega)^{tC(\omega)/2L}. \tag{3.3}$$

For $\alpha > 1$ this equation predicts exponential growth, and for $\alpha < 1$, exponential decay.

On the basis of this analysis one can now make some very interesting observations about stability and convergence. The following ideas are spelled out in detail in [12].

The first observation is: *stability does not preclude exponential growth in time.* We have seen in the last section that stability for a single boundary implies $|A| < \infty$, but not $|A| \leqslant 1$. Thus it is easily possible for a model composed of two individually stable boundaries to generate growth of truncation errors at a rate exp(const t). One might think that this reveals that the concatenation of two stable boundaries is in general unstable. But in fact, exponential growth does not imply instability in the usual Lax-Richtmyer sense, and it will not prevent convergence. The reason is that the growth factor for any fixed t does not get larger as h, $k \to 0$, whereas the truncation errors that the factor multiplies get smaller (by consistency).

These conclusions are already known. In fact, Theorem 5.4 of [4] asserts that in general, the concatenation of two GKS-stable difference schemes is always GKS-stable, while §7 of [4] investigates the conditions under which exponential growth in t will occur for a particular stable 2×2 example. What is new here is the interpretation by reflection coefficients. This interpretation also helps to clear up a misconception. Some people have thought that although exponential growth can occur for finite h and k, it must vanish as h, $k \to 0$. But the study of reflecting wave packets gives no reason to expect such good fortune, and an example in [12] proves that, indeed, it is not to be expected. To be sure of eliminating growth by refining the mesh, one needs a difference formula that is suitably dissipative [3, 12].

There is a reason why exponential growth in t may be a problem: in practice, time-dependent finite-difference models are often used (especially in aerodynamics) to compute approximate solutions for the steady state $t \to \infty$. Obviously growth of errors at a rate $e^{\text{const } t}$ will make such a procedure fail. With this in mind, Beam, Warming and Yee [1] have defined a difference model to be *P-stable* if it is GKS-stable and, in addition, it admits no exponentially growing modes. In this language the above observation becomes: stability does not imply *P*-stability.

Our second observation now comes as the natural complement to the first: *having reflection coefficients bounded by* 1 *does preclude exponential growth in time.* For in (3.3) we see that if $\alpha(\omega) \leqslant 1$ for every ω, then no frequency exists that can experience repeated amplification by reflection. This argument can be made rigorous, and Proposition 10 of [12] states: if $\alpha(\omega) \leqslant 1$ for every ω, a difference model is *P*-stable.

For a third and final observation, let us turn to the case in which one or both boundaries is individually unstable. Then we find: *a mildly unstable boundary with an infinite reflection coefficient may become catastrophically unstable when a second boundary is introduced.* For consider a boundary condition which has $A(\omega_0) = \infty$ for some ω_0. If a wave packet at frequency ω_0 hits this boundary, it will be amplified by a large factor (typically $O(N)$, where N is the width of the packet in grid points). For a single boundary, this is a one-time-only event, as in Figure 2.2(d). But when there are two boundaries, the reflected wave may reflect at the other boundary, then return to be amplified again, and so on. For boundaries separated by N grid points one gets growth potentially at a very rapid rate

$$\frac{\|v(t)\|}{\|v(0)\|} \approx N^{\text{const } t} = N^{\text{const } nk}.$$

This instability is a far cry from the growth proportional to n mentioned in the last section, and it renders the difference model totally useless.

This phenomenon of catastrophic two-boundary interactions was first noted long ago by Kreiss. The advantage of the present point of view is that it makes it clear that the problem is associated not with all unstable boundaries, but only with those having infinite reflection coefficients. Unstable boundary conditions with finite reflection coefficients typically exhibit only weak instability even when other boundaries are present. Of course, sometimes weak instabilities may be more dangerous than strong ones, since they can more easily go undetected.

Our final numerical experiment, summarized in Figure 3.2, illustrates the difference in two-boundary behavior between unstable boundary conditions with finite and infinite reflection coefficients. Figure 3.2(a) shows the initial data, a uniformly distributed random signal on the usual CN mesh on [0, 1]. At $x = 1$ the boundary condition is (2.4a). First, an integration was performed with the boundary condition (2.9a) at $x = 0$, which is unstable but by (2.9c) has a finite reflection coefficient. In this process nothing dramatic occurs no matter how large t is. Figures 3.2(b) show the result at $t = 10$, and it looks qualitatively much like the initial data. Of course this does not imply that this model will give accurate answers to physical problems, but it does reveal that GKS-instability of a boundary condition does not by itself lead to explosive two-boundary interactions.

In contrast, Figures 3.2(c), (d) show results obtained with boundary condition (2.10a), which has an infinite reflection coefficient. The plots show times $t = 1, 3$. Note the vertical scales.

In summary, for two-boundary problems one has the following possibilities:

(1) stable b.c.'s, $0 \leqslant \alpha \leqslant 1$: stable and P-stable ($\|v\| = O(1)$);
(2) stable b.c.'s, $1 < \alpha < \infty$: stable but P-unstable ($\|v\| = O(e^{\text{const } t})$);
(3) unstable b.c.'s, $0 \leqslant \alpha < \infty$: weakly unstable ($\|v\| = O(\sqrt{n})$);
(4) unstable b.c.'s, $\alpha = \infty$: strongly unstable ($\|v\| = O(N^{\text{const } t})$).

Before closing, a word must be said about dissipation. I have presented a picture in which numerical waves bounce back and forth between boundaries, with changes of amplitude occurring at each reflection but not during the transit in between. Under a dissipative difference formula, however, any wave with $\xi \neq 0$ decays as it travels. The result is that the two-boundary interactions I have spoken of rarely take place, so that in practice, the behavior of a dissipative two-boundary model is not much different from what the individual boundaries would suggest. The exceptions occur when the dissipation is very weak, as in the problems with large mesh ratios considered in [1], or when the number of grid points between boundaries is very small, as in certain cases discussed in §2 of [12].

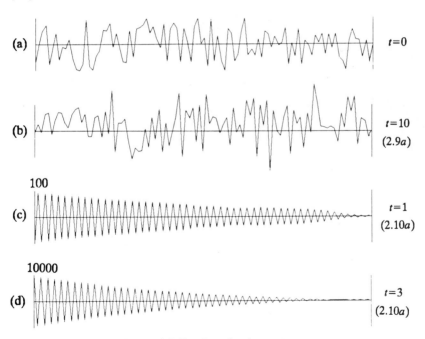

FIGURE 3.2. Two-boundary interactions.

For an overall summary, one may say that difference models for hyperbolic partial differential equations exhibit three important physical processes: propagation, reflection, and dissipation of waves. In general, the waves involved are numerical objects with no physical meaning, but they are still important to stability. An exact mathematical analysis of a finite-difference model is usually too difficult to be practical, but even for complicated models one can often gain insight fairly easily by considering these physical processes.

REFERENCES

1. R. M. Beam, R. F. Warming and H. C. Yee, *Stability analysis for numerical boundary conditions and implicit difference approximations of hyperbolic equations*, Proc. NASA Sympos. on Numerical Boundary Condition Procedures (Moffett Field, CA, October 1981), NASA, 1982, pp. 199–207.

2. M. Giles and W. Thompkins, Jr., *Asymptotic analysis of numerical wave propagation in finite difference equations*, Gas Turbine and Plasma Phys. Lab. Report 171, Mass. Inst. Tech., 1983.

3. B. Gustafsson, *The choice of numerical boundary conditions for hyperbolic systems*, Proc. NASA Sympos on Numerical Condition Boundary Condition Procedures (Moffett Field, CA, October 1981), NASA, 1982, pp. 209–225.

4. B. Gustafsson, H.-O. Kreiss and A. Sundström, *Stability theory of difference approximations for initial boundary value problems. II*, Math. Comp. **26** (1972), 649–686.

5. R. Higdon, *Initial-boundary value problems for linear hyperbolic systems*, Tech. Summer Report 2558, Math. Res. Cent., University of Wisconsin, Madison, 1983.

6. H.-O. Kreiss, *Initial boundary value problems for hyperbolic systems*, Comm. Pure Appl. Math. **23** (1970), 277–298.

7. S. Osher, *Systems of difference equations with general homogeneous boundary conditions*, Trans. Amer. Math. Soc. **137** (1969), 177–201.

8. R. D. Richtmyer and K. W. Morton, *Difference methods for initial-value problems*, Interscience, New York, 1967.

9. L. N. Trefethen, *Group velocity in finite difference schemes*, SIAM Rev. **24** (1982), 113–136.

10. _____, *Group velocity interpretation of the stability theory of Gustafsson, Kreiss, and Sundström*, J. Comput. Phys. **49** (1983), 199–217.

11. _____, *Instability of difference models for hyperbolic initial boundary value problems*, Comm. Pure Appl. Math. **37** (1984), 329–367.

12. _____, *Stability of finite difference models containing two boundaries or interfaces*, Math. Comp., submitted.

13. R. Vichnevetsky and J. B. Bowles, *Fourier analysis of numerical approximations of hyperbolic equations*, SIAM, Philadelphia, Pa., 1982.

DEPARTMENT OF MATHEMATICS, COURANT INSTITUTE OF MATHEMATICAL SCIENCES, NEW YORK UNIVERSITY, NEW YORK, NEW YORK 10012

Current address. Massachusetts Insitute of Technology, Cambridge, Massachusetts 02139

Upwind-Difference Methods for Aerodynamic Problems Governed by the Euler Equations

Bram van Leer

ABSTRACT. The properties of upwind-difference schemes, relevant to the solution of the steady Euler equations, are discussed. The schemes are interpreted as projection-evolution schemes, with a numerical stage (interpolation of discrete initial values) and a physical stage (computation of fluxes). A comparison with a central-difference scheme, stabilized by explicit dissipation, is also given.

1. Introduction. Upwind-difference methods for the Euler equations have mainly been applied to problems of transient high-speed flow, because of their potential in representing moving shock waves [1]. Yet, the representation of steady shocks by these methods is even better [2]. It is not surprising that the growing interest among aerodynamicists in steady solutions of the Euler equations has caused a boom in upwind differencing.

Finding steady solutions actually is simpler than finding unsteady solutions. There are two reasons for this. Firstly, since temporal accuracy is immaterial, the methods reduce to computing the so-called residual, a discrete approximation to the spatial differential operator. A matching relaxation algorithm must be provided, feeding back the value of the residual into the discrete solution in order to reach a steady state. For

1980 *Mathematics Subject Classification.* Primary 76–08, 65N05; Secondary 76H05.

upwind spatial discretizations, up-to-date versions of classic relaxation schemes are pre-eminently suited [3]; these will not be discussed here.

Secondly, since the flow ultimately stabilizes, the recognition of unresolved or unresolvable flow features, crucial in preserving the accuracy elsewhere, simplifies too. In consequence there is room for an extra effort to improve the residual computation, such as detecting the orientation of discontinuities, or even shock fitting.

I shall briefly review the present-day ideas on upwind differencing for conservation laws, in so far as these apply to the solution of problems of steady flow. A comparison of upwind-differencing methods with central-differencing methods stabilized by explicit artificial dissipation [4] seems unavoidable. There is nothing mysterious about the existence of these two distinct techniques. The more elaborate upwind differencing hides a matrix-valued dissipation coefficient [5], whereas central differencing, for the very sake of simplicity, is always combined with a scalar dissipation coefficient.

2. First-order upwind differencing. From my point of view, upwind-differencing schemes consist of a projection stage and an evolution stage [6]. In the projection stage the discrete initial values are interpolated to yield a distribution that is continuous in each computational zone. In the evolution stage the development of the interpolated distribution during a short time interval is computed with aid of the equations for continuous or discontinuous flow.

If only a steady solution is sought, the evolution stage may be restricted to computing the initial time derivative of the interpolated distribution, which is nothing but computing the residual. The accessory relaxation scheme then takes care of using the information contained in the residual to march efficiently toward a steady state.

To fix our thoughts, let us consider the two-dimensional Euler equations in conservation form:

$$\frac{\partial w}{\partial t} = -\frac{\partial}{\partial x}f(w) - \frac{\partial}{\partial y}g(w). \tag{1}$$

Here w is the vector of conserved state quantities; $f(w)$ and $g(w)$ are the vectors of their fluxes in the x- and y-directions. Averaging (1) over a finite volume V with boundary B and unit vector n normal to B, we get

$$\frac{1}{V}\frac{\partial}{\partial t}\iint_V w\,dV = -\frac{1}{V}\oint_B \{f(w)n_x + g(w)n_y\}\,dB. \tag{2}$$

On the left-hand side we see the time derivative of the volume-averaged state vector; it is important to realize that a discrete solution will be nothing but a two-dimensional array of such averages. On the right-hand side appears the residual, which is the boundary integral of the normal flux, per unit volume.

The computation of fluxes across volume boundaries involves the functional dependence of f and g on w and the elementary wave solutions of (1); it is here that physics enters. The fluxes may be computed only if w is known on the boundaries, which explains the need for interpolating the discrete solution. The interpolation is pure "numerics", although some physical knowledge may enter, for instance, in anticipating all sorts of discontinuities while interpolating, and in selecting the most discriminating state quantities for interpolation.

All authors on upwind differencing agree in their interpretation of the first-order upwind method as a projection-evolution scheme, after the example of Godunov [7]. This is only because the projection stage is the simplest possible: the distributions of the state quantities are assumed to be uniform in each zone. At the interface of two zones these uniform states meet in a discontinuity. A unique set of fluxes can be assigned after resolving the discontinuity by a set of elementary waves moving normal to the interface, in other words, after solving a one-dimensional Riemann problem. This may be done exactly or approximately; for a review of Riemann's problem and its approximate solution see [8]. The waves carry information from upwind to the interface, in accordance to the notion of upwind differencing.

I may conclude this section by emphasizing that the first-order upwind schemes encountered in the literature differ only by their "approximate Riemann solver". Osher's [9] scheme includes a simple-wave decomposition as the Riemann solver, Roe's [10] scheme a linear-wave decomposition. Van Leer [11] adopts flux splitting, an approximation based on the collisionless Boltzmann equation (see, again, [8]). Note that Godunov's scheme is associated with the use of the exact Riemann solution. As the projection-evolution approach is due to Godunov, it is fair to speak about "Godunov's scheme" and "Godunov-type schemes", and to associate the names of other authors with the Riemann or Boltzmann solver of their own make.

3. Higher-order upwind differencing. To raise the order of accuracy of upwind differencing, all one needs to do is to raise the order of accuracy of the initial-value interpolation that yields the zone-boundary data, hence, to make an improvement in the projection stage. This is very convenient when it comes to code writing, since it allows a fully modular structure.

Most authors, however, throw away the benefits of the projection-evolution approach, by mixing numerics with physics. Osher [12], for example, first determines interface fluxes by the first-order scheme, then corrects these by an upwind-biased flux-difference term. The correction term is purely numerical in nature, completing the more accurate spatial interpolation required in a second-order scheme. However, since the correction comes after the flux calculation, physics enters, needlessly complicating the interpolation by upwind logic.

Besides complicating the scheme, sandwiching the physics between layers of numerics may cause a loss of accuracy. The discrete distribution of interface fluxes provided by the first-order upwind method is not as smooth as the discrete solution itself, hence less suited for interpolation, as was demonstrated in [13].

Let us therefore adopt the projection-evolution approach and, as an exercise, interpolate the zone averages of w on a uniform Cartesian grid $\{x_i, y_j\}$. As explained in [6], it is useful to express the interpolated distribution zonewise in terms of Legendre polynomials:

$$
\begin{aligned}
w(x, y) = \bar{\tilde{w}}_{ij} &+ (x - x_i)\left(\frac{\partial w}{\partial x}\right)_{ij} + (y - y_j)\left(\frac{\partial w}{\partial y}\right)_{ij} \\
&+ \frac{3\kappa}{2}\left[\left\{(x - x_i)^2 - \frac{(\Delta x)^2}{12}\right\}\left(\frac{\partial^2 w}{\partial x^2}\right)_{ij}\right. \\
&\qquad + 2(x - x_i)(y - y_i)\left(\frac{\partial^2 w}{\partial x \partial y}\right)_{ij} \\
&\qquad \left. + \left\{(y - y_j)^2 - \frac{(\Delta y)^2}{12}\right\}\left(\frac{\partial^2 w}{\partial y^2}\right)_{ij}\right],
\end{aligned}
\tag{3.1}
$$

$$x_{i-1/2} < x < x_{i+1/2}, \quad y_{j-1/2} < y < y_{j+1/2}.$$

Here $\bar{\tilde{w}}_{ij}$ denotes the volume average of w in zone (i, j), with bar and tilde indicating averaging over x and y, respectively; the various partial derivatives of w in (x_i, y_j) are understood to be approximated by centered (!) difference quotients of $\bar{\tilde{w}}$, made from values in the surrounding zones. The parameter κ may be chosen to yield purely linear ($\kappa = 0$) or quadratic ($\kappa = 1/3$) interpolation.

Averaging the above distribution over x and y in zone (i, j) yields the zone average $\bar{\tilde{w}}_{ij}$, as it should. Averaging over y only leads to

$$
\begin{aligned}
\tilde{w}(x, y_j) = \bar{\tilde{w}}_{ij} &+ (x - x_i)\left(\frac{\partial w}{\partial x}\right)_{ij} \\
&+ \frac{3\kappa}{2}\left\{(x - x_i)^2 - \frac{(\Delta x)^2}{12}\right\}\left(\frac{\partial^2 w}{\partial x^2}\right)_{ij},
\end{aligned}
\tag{3.2}
$$

$$x_{i-1/2} < x < x_{i+1/2},$$

a strictly one-dimensional relation if we insert

$$\frac{\tilde{\bar{w}}_{(i+1)j} - \tilde{\bar{w}}_{(i-1)j}}{2\Delta x} := \left(\frac{\partial w}{\partial x}\right)_{ij}, \tag{4.1}$$

$$\frac{\tilde{\bar{w}}_{(i+1)j} - 2\tilde{\bar{w}}_{ij} + \tilde{\bar{w}}_{(i-1)j}}{(\Delta x)^2} := \left(\frac{\partial^2 w}{\partial x^2}\right)_{ij}. \tag{4.2}$$

Equations (3.2) and (4), and similar equations for the y-direction, enable us to compute boundary-averaged values of w on the inside of the four faces of cell (i, j); these will be needed in the evolution stage as the input values for a Riemann solver. Inserting $x = x_{i+1/2}$, for example, gives us a left-side estimate of $\tilde{w}$ at the interface $(x_{i+1/2}, y_j)$:

$$\tilde{w}^L_{(i+1/2)j} = \tilde{\bar{w}}_{ij} + \frac{1}{4}\left\{(1-\kappa)(\tilde{\bar{w}}_{ij} - \tilde{\bar{w}}_{(i-1)j}) + (1+\kappa)(\tilde{\bar{w}}_{(i+1)j} - \tilde{\bar{w}}_{ij})\right\}. \tag{5.1}$$

On the other hand, inserting $x = x_{i+1/2}$ in the formula for zone $(i+1, j)$ yields a right-side estimate:

$$\tilde{w}^R_{(i+1/2)j} = \tilde{\bar{w}}_{(i+1)j} - \frac{1}{4}\left\{(1+\kappa)(\tilde{\bar{w}}_{(i+1)j} - \tilde{\bar{w}}_{ij})\right.$$
$$\left. + (1-\kappa)(\tilde{\bar{w}}_{(i+2)j} - \tilde{\bar{w}}_{(i+1)j})\right\}. \tag{5.2}$$

The difference between these values is $-\frac{1}{4}(1-\kappa)\Delta^3_x\tilde{\bar{w}}_{(i+1/2)j}$, or $O((\Delta x)^3)$. For the moment this is all there is to the projection stage.

In the evolution stage a single numerical flux vector is obtained from each pair of state vectors assigned to one interface. Solving the Riemann problem at $(x_{i+1/2}, y_j)$ with Roe's [10] linearization we get

$$f\left(\tilde{w}^L_{(i+1/2)j}, \tilde{w}^R_{(i+1/2)j}\right) = \frac{1}{2}\left\{f\left(\tilde{w}^L_{(i+1/2)j}\right) + f\left(\tilde{w}^R_{(i+1/2)j}\right)\right\}$$
$$- \frac{1}{2}\left|A_{(i+1/2)j}\right|\left(\tilde{w}^R_{(i+1/2)j} - \tilde{w}^L_{(i+1/2)j}\right)$$
$$= \frac{1}{2}\left\{f\left(\tilde{w}^L_{(i+1/2)j}\right) + f\left(\tilde{w}^R_{(i+1/2)j}\right)\right\}$$
$$+ \frac{1-\kappa}{8}\left|A_{(i+1/2)j}\right|\Delta^3_x\tilde{\bar{w}}_{(i+1/2)j}. \tag{6}$$

Here $A_{(i+1/2)j}$ is a Roe-type average of the Jacobian $A = df(w)/dw$ on the interval $(\tilde{\bar{w}}_{ij}, \tilde{\bar{w}}_{(i+1)j})$ and $|A| = \sqrt{(A^2)}$ is the positive-definite matrix by which downwind differences are annihilated [5].

Equation (6) shows that the upwind-biased numerical flux differs from the average flux, which would lead to central differencing, by a third-order term leading to a fourth-order viscosity. The main purpose of this viscosity is to prevent the checkerboard instability; the value of κ should therefore not be too close to unity. I prefer $\kappa = 1/3$ since it makes the interface averages correct within $O((\Delta x)^3)$. For a one-dimensional problem this choice would lend the scheme third-order accuracy; in two or more dimensions this is spoiled by the nonlinearity of the fluxes (average of flux $\neq$ flux of average). For other choices see Table 1.

TABLE 1. Choices of κ and associated schemes. Note that the one-sided scheme, favored by most authors, incorporates the largest viscosity term.

κ	resulting sheme	used by	viscosity level Re: $\kappa = 1/3$
-1	one-sided, 2nd order	Moretti [21] (nonconservative) Steger and Warming [23] Osher and Chakravarthy [12]	3
0	upwind, 2nd order	Fromm [24] (convection only) Van Leer [25] (MUSCL code) Woodward and Colella [1]	3/2
1/3	upwind, 3rd order	Turkel and Van Leer [26]	1
1	centered, 2nd order	Jameson, Schmidt and Turkel [4]	0

4. Limiting. In practice the projection stage is more complicated than described in §3. Equations (4.1) and (4.2) are actually modified: the difference quotients are limited in value through multiplication by locally evaluated switch functions. Such functions decrease from one to zero with decreasing smoothness of the numerical solution; examples may be found in [14, 6, 15 and 16].

With switch functions, numerical oscillations or "wiggles" caused by differencing across a discontinuity are avoided. For example, if $\tilde{w}_{(i+1)j} - \tilde{w}_{ij}$ is large compared to $\tilde{w}_{ij} - \tilde{w}_{(i-1)j}$, as in a shock profile, limiting yields

$$\tilde{w}^R_{(i+1/2)j} - \tilde{w}^L_{(i+1/2)j} \approx \tilde{w}_{(i+1)j} - \tilde{w}_{ij}. \tag{7}$$

Thus, the viscosity term in (6) locally grows as large as the flux of a second-order viscosity, guaranteeing a monotone shock profile.

A geometrical interpretation of limiting can be enlightening; it is given in Figure 1. The key word is "co-monotone interpolation": the interpolant in zone (i, j) should not go beyond the range spanned by $\tilde{w}_{(i-1)j}$ and $\tilde{w}_{(i+1)}$.

Limiting, when carried out as a part of the projection stage, does not interfere with the evolution stage and may be done smoothly or suddenly, crudely or with great sophistication. Two refinements needed beyond the most elementary switching algorithm are: the recognition of extrema in smooth solutions, where limiting should be avoided, and the restoration of large derivatives inside a discontinuity-like profile after limiting (see e.g. [17]).

Once limiting is introduced (it is standard practice in all higher-order upwind schemes, though often applied in the evolution stage), the conserved quantities are not necessarily best suited for an interpolation like (3.1) or (3.2). For one-dimensional problems [13] the obvious choice is the set of characteristic state quantities (Riemann invariants and entropy); these are also needed in simulating nonreflecting boundaries. A shock or a contact discontinuity will show up as a jump in only one of these quantities, if the latter are properly discretized (as in [16]). In two or more dimensions, where discontinuities generally are oblique to the grid, characteristic quantities are truly superior only when defined in a frame normal to the wave front. A crisp representation of discontinuities therefore requires detection of their orientation; some promising first results have been obtained by Davis [18].

5. Upwind versus centered differences. If, in (5) and (6), we insert $\kappa = 1$, we simply get

$$f\left(\tilde{w}^L_{(i+1/2)j}, \tilde{w}^R_{(i+1/2)j}\right) = f\left(\frac{\tilde{w}_{ij} + \tilde{w}_{(i+1)j}}{2}\right), \qquad (8)$$

leading to the standard centered difference approximation of $\partial f/\partial x$. Explicit artificial dissipation must be added in order to stabilize the solution against checkerboard oscillations. Since the formula (8) is chosen for its simplicity, the extra dissipation should also be simple, that is, have a scalar rather than a matrix coefficient.

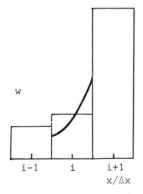
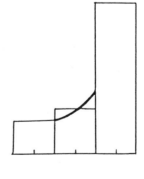

FIGURE 1. Limiting at the foot of a shock profile. Left: quadratic interpolation (heavy line) in zone i according to (3.2). The piecewise uniform distribution shows the zone averages. Right: same after limiting of derivative values.

This is precisely the philosophy underlying Jameson's FLO52 code for integrating the Euler equations [4], which includes explicit fourth-order dissipation. As the flux function (8) does not offer the freedom to include limiting, the effect of limiting must be invoked in his code by another explicit dissipative term, a second-order expression with a scalar coefficient that switches on where the data are not smooth. In the same place the fourth-order dissipation must be switched off.

To achieve a robustness comparable to that of an upwind-biased scheme, the scalar dissipation coefficients in Jameson's scheme should be raised to the level of the spectral radius of $|A|$. This would lead to excessive smearing of flow details, including shocks. In practice the opposite is done: the coefficients are adjusted, by trial and error, such that shock profiles are sharp, yet monotone, while spurious entropy production in smooth flow is low, yet suppresses the checkerboard instability. The optimum dissipation levels clearly are solution-dependent; for flows with little structure, in particular with isolated grid-aligned shocks, they may actually be lower than the level in the standard upwind scheme ($\kappa = 1/3$). This is because almost all the information is in the one characteristic quantity that jumps across the discontinuity; as this discontinuity is steady very little dissipation is required (the relevant eigenvalue of $|A|$ vanishes). The other characteristic quantities are continuous and will not cause wiggles.

TABLE 2. A comparison of the steps taken in an upwind-difference scheme and a central-difference scheme, respectively, in order to compute the flux vector at an interface. Only one dimension is considered.

Upwind-difference scheme [25]	Central-difference scheme [4]
Compute $\bar{\Delta}_x\bar{w}$ (centered), $\Delta_x^2\bar{w}$	Compute $\Delta_x\bar{w}$, $\Delta_x^3\bar{w}$
Compute switch factors for $\bar{\Delta}_x\bar{w}$, $\Delta_x^2\bar{w}$	Compute switch factor for $\Delta_x\bar{w}$, $\Delta_x^3\bar{w}$
(different values for different state quantities)	(same value for all state quantities)
Compute w^L and w^R at same interface	Compute average state at interface
Compute flux vector w^L and w^R	Compute flux vector of interface
with Riemann solver	state, 2nd- and 4th-order
	dissipative fluxes; add together

Table 2 lists the various steps in a higher-order upwind scheme versus the corresponding ones in Jameson's scheme. Because of the matrix-valued multiplication in (6) and the logic entering the computation of the matrix, upwind-biased fluxes are between two and three times as costly to compute as centered fluxes with added dissipation. That is the price one must pay for robustness.

6. Future developments. All schemes discussed before, including Jameson's, have in common that they derive zone-boundary data from zone-center data by interpolation, a typical feature of schemes for calculating transient flows. The interpolation has to introduce some threshold level of dissipation, in order to prevent decoupling of the even and odd grids. While this does not harm in computing moving waves, in calculations of steady flow it is redundant and actually harmful [19].

Box schemes, based on data given in the corner points of zones, do not require the above interpolation, while strongly coupling the even and odd grids without any dissipation. Designers of difference methods for computing steady solutions of the Euler equations should take box schemes very seriously. Some excellent results for one-dimensional transonic flow, including examples of suitable relaxation methods, may be found in [20 and 21]. I hope to be able to show two-dimensional results in the near future.

REFERENCES

1. P. R. Woodward and P. Colella, *The numerical simulation of two-dimensional fluid flow with strong shocks*, J. Comput. Phys. **54** (1984), 115–173.

2. B. Engquist and S. Osher, *One-sided difference approximations for nonlinear conservation laws*, Math. Comp. **36** (1981), 321–352.

3. B. van Leer and W. A. Mulder, *Relaxation methods for hyperbolic equations* (invited lecture), INRIA Workshop on Numerical Methods for the Euler Equations for Compressible Fluids (Le Chesnay, France, Dec. 1983) (to appear).

4. A. Jameson, W. Schmidt and E. Turkel, *Numerical solutions of the Euler equations by finite-volume methods using Runge-Kutta time-stepping*, AIAA Paper 81-1259, 1981.

5. B. van Leer, *Towards the ultimate conservative difference scheme.* III: *Upstream-centered finite-difference schemes for ideal compressible flow*, J. Comput. Phys. **23** (1977), 263–275.

6. _____, *Towards the ultimate conservative difference scheme.* IV: *A new approach to numerical convection*, J. Comput. Phys. **23** (1977), 276–299.

7. S. K. Godunov, *A finite-difference method for the numerical computation of discontinuous solutions of the equations of fluid dynamics*, Mat. Sb. (N.S.) **47** (1959), 271–306. (Russian)

8. A. Harten, P. D. Lax and B. van Leer, *On upstream differencing and Godunov-type schemes for hyperbolic conservation laws*, SIAM Rev. **25** (1983), 35–60.

9. S. Osher and F. Solomon, *Upwind difference schemes for hyperbolic systems of conservation laws*, Math. Comp. **38** (1982), 339–377.

10. P. L. Roe, *The use of the Riemann problem in finite-difference schemes*, Lecture Notes in Phys. **141** (1980), 354–359.

11. B. van Leer, *Flux-vector splitting for the Euler equations*, Lecture Notes in Phys. **170** (1982), 507–512.

12. S. Osher and S. Chakravarthy, *High-resolution applications of the Osher upwind scheme for the Euler equations*, AIAA Paper 83-1943, 1983.

13. W. A. Mulder and B. van Leer, *Implicit upwind methods for the Euler equations* (AIAA Paper 83-1930, 1983); J. Comput. Phys. (to appear).

14. B. van Leer, *Towards the ultimate conservative difference schemes.* II: *Monotonicity and conservation combined in a second-order scheme*, J. Comput. Phys. **14** (1974), 361–370.

15. G. D. van Albada, B. van Leer and W. W. Robers, Jr., *A comparative study of computational methods in cosmic gas dynamics*, Astronom. and Astrophys. **108** (1982), 76–84.

16. P. L. Roe, *Some contributions to the modelling of discontinuous flows*, these PROCEED-INGS.

17. P. Colella and P. R. Woodward, *The piecewise-parabolic method (PPM) for gas-dynamical calculations*, J. Comput. Phys. **54** (1984), 174–201.

18. S. F. Davis, *A rotationally biased upwind difference scheme for the Euler equations*, ICASE Report No. 83-37, 1983; J. Comput. Phys. **56** (1984), 65–92.

19. A. Rizzi, *Experience from the AGARD contest*, Lecture, Minisymposium on Shock Problems—Computational Methods with Applications (Uppsala University, Department of Computer Sciences, Oct. 1983), notes in Internal Report No. 83-07, 1983.

20. F. Casier, H. Deconinck and C. Hirsch, *A class of central bidiagonal schemes with implicit boundary conditions for the solution of Euler's equations*, AIAA Paper 83-0126, 1983.

21. G. Moretti, *The λ-scheme*, Computers and Fluids **7** (1979), 191–205.

22. S. F. Wornom, *Implicit conservative characteristic modeling schemes for the Euler equations*, AIAA Paper 83-1939, 1983.

23. J. L. Steger and R. F. Warming, *Flux-vector splitting of the inviscid gas-dynamic equations with applications to finite-difference methods*, J. Comput. Phys. **40** (1981), 263–293.

24. J. E. Fromm, *A method for reducing dispersion in convective difference schemes*, J. Comput. Phys. **3** (1968), 176–189.

25. B. van Leer, *Towards the ultimate conservative difference schemes. V: A second-order sequel to Godunov's method*, J. Comput. Phys. **32** (1979), 101–136.

26. E. Turkel and B. van Leer, *Flux-vector splitting and Runge-Kutta methods for the Euler equations*, Contributed paper, Ninth Internal. Conf. on Numerical Methods in Fluid Dynamics (Saclay, France, June 1984).

DEPARTMENT OF MATHEMATICS AND COMPUTER SCIENCE, DELFT UNIVERSITY OF TECHNOLOGY, P.O. BOX 356, 2600 AJ DELFT, THE NETHERLANDS

Lectures in Applied Mathematics
Volume **22**, 1985

An MHD Model of the Earth's Magnetosphere

C. C. Wu

I. Introduction. The earth's magnetosphere arises from the interaction of the solar wind with the earth's geomagnetic field. The solar wind is a continuously streaming, hot, collisionless plasma. As it impinges on the earth, it compresses the geomagnetic field until the solar wind dynamic pressure is balanced by the magnetic pressure inside. Thus a magnetosphere is formed. Consequently, the solar wind is deflected by the magnetosphere and flows around the magnetosphere. The flow past the magnetosphere constitutes a problem similar to the familiar aerodynamic problem of supersonic flow past a body. Since the solar wind is both supersonic and super-Alfvénic, a detached bow shock is formed. (In some occasions, the solar wind is observed to be subsonic and sub-Alfvénic.) In the magnetosphere, the magnetic field physically ties the points of the magnetosphere together, guiding charged particles, plasma and electric currents; trapping thermal plasma and energetic particles; and transmitting hydromagnetic stresses between the exterior flow and the earth.

In recent years a global magnetohydrodynamics (MHD) model of the earth's magnetosphere has drawn much attention. It is hoped that this kind of model will help us to achieve a global understanding and self-consistent quantitative description of the cause-and-effect relations among the principal dynamical processes involved. In the MHD model, MHD equations are used to describe the solar wind interaction with the magnetosphere. Since these are highly-nonlinear, time-dependent, three-dimensional equations, current efforts have been towards obtaining numerical solutions, like many other branches of nonlinear physics. The first numerical solutions were carried out by Spreiter and his coworkers

---

1980 *Mathematics Subject Classification*. Primary 65M99, 85A35.

(Spreiter and Alksne [**1969**]). They have formulated the model in terms of MHD equations and have provided some justifications for using the MHD descriptions. However, their numerical calculations modelled only flow outside the magnetosphere by using hydrodynamics.

Recent global MHD models include the magnetosphere as well as the external flow. Both the bow shock and the magnetosphere boundary, called magnetopause, are self-consistently formed as discontinuities in the solutions. These models have been studied by many researchers including Leboeuf et al. [**1978**], Lyon et al. [**1980**], and Wu et al. [**1981**]. However, there are discrepancies among their results. One main reason is that the model is a very large scale numerical work. It will probably require much refinement of the numerical techniques before fully self-consistent, convergent solutions are obtained.

In this paper, some numerical aspects of the model will be presented. In §II, the global MHD model is formulated. In §III, some numerical aspects of the model are discussed. These include shock capturing technique, nonuniform grid system and multiple time scale of the problem. In §IV, some recent results are presented for illustrations.

For a review of the physics of the magnetosphere, please refer to an excellent article by Roederer [**1979**]. The lecture notes entitled "*Solar-terrestrial physics*" provide a good introduction to the magnetospheric physics as well as other aspects of space physics (Carovillano and Forbes [**1983**]).

II. Global MHD model. Our model of the earth's magnetosphere is based on an MHD description of the interaction of the solar wind and the geomagnetic field. The ideal MHD equations are

$$\frac{\partial \rho}{\partial t} = -\nabla \cdot (\rho \mathbf{v}), \tag{1}$$

$$\frac{\partial (\rho \mathbf{v})}{\partial t} = -\nabla \cdot \left[\rho \mathbf{v} \mathbf{v} + \ddot{I}\left(p + \frac{B^2}{2} \right) - \mathbf{B}\mathbf{B} \right], \tag{2}$$

$$\frac{\partial \mathbf{B}}{\partial t} = \nabla \times (\mathbf{v} \times \mathbf{B}), \tag{3}$$

$$\frac{\partial \varepsilon}{\partial t} = -\nabla \cdot \left[\left(\frac{1}{2}\rho v^2 + \frac{p}{\gamma - 1} + p \right) \mathbf{v} - (\mathbf{v} \times \mathbf{B}) \times \mathbf{B} \right]. \tag{4}$$

Here the mass density, pressure, velocity and magnetic field are denoted by ρ, p, $\mathbf{v}$ and $\mathbf{B}$, respectively; γ is the ratio of the specific heats, and the energy density (ε) is given by $\varepsilon = \rho v^2/2 + B^2/2 + p/(\gamma - 1)$. In addition to these equations, we require $\nabla \cdot \mathbf{B} = 0$, which is satisfied if it is satisfied by the initial data in our initial value problems.

In this model the plasma is assumed to have an equation of state that represents the internal energy by the relation

$$I = \frac{p}{(\gamma - 1)\rho}. \qquad (5)$$

For simplicity we use this equation with constant $\gamma = 2$ throughout the whole region. In principle, one could try to use a more complex equation of state to characterize plasma behaviors in various regions.

Since there are discontinuities (bow shock, magnetopause) in the solutions, we should be more precise about the definition of our model. In the smooth region, we require the solutions to satisfy the MHD equations (1)–(4). Across the discontinuities, we require the solutions to satisfy physical jump conditions. Bow shock and magnetopause are treated in our model as "boundaries" only. In the numerical calculations, the jump conditions and the positions of these boundaries are "captured"; this means that they evolve naturally in the solution. In our calculations the geomagnetic field was approximated by a dipole field

$$\mathbf{B}(\mathbf{r}) = \frac{3\hat{r}(\mu \cdot \hat{r}) - \mu}{r^3}, \qquad (6)$$

where μ is the dipole moment and $\mathbf{r}$ is the position relative to the dipole center.

The model is treated as an initial- and boundary-value problem, and thus both initial conditions and boundary conditions must be specified. There are two kinds of boundary conditions: the earth boundary is a physical one where the ionosphere-magnetosphere interaction is treated; the four boundaries marked by either "inflow" or "outflow" in Figure 1 are numerical boundaries which are introduced to limit the computational domain. The physical boundaries are located at infinity. Initially, the plasma is in a static equilibrium with the geomagnetic dipole field. At $t = 0$ the solar wind is introduced at the inflow boundary. Subsequent time evolution is obtained by integrating the MHD equations.

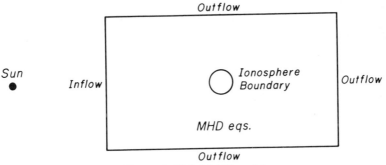

FIGURE 1. Global MHD model.

Since there are no theorems available concerning the existence or the uniqueness of the solutions, the model is defined to be close to (possible) wind tunnel experiments and hopefully, the results will have physical meaning. Exact specifications of the initial and boundary conditions used in the calculations will be given in §IV.

Although the ideal MHD equations are used in the model, there are effective viscosity and resistivity in the calculations due to numerical truncation errors. Because of this resistivity term, magnetic field reconnection can take place. Ideally, one would like to reduce these numerical effects and use viscosity and resistivity determined by physical mechanisms.

Before we go on to the next section, we would like to point out a theorem by Alfvén. The equation (3) is the same as that satisfied by vorticity ω in an inviscid fluid, i.e.,

$$\partial \omega / \partial t = \nabla \times (\mathbf{v} \times \omega). \tag{7}$$

As shown in works on hydrodynamics, (7) implies that the vortex lines move with the fluid. Therefore it follows from (3) that, in a perfectly conducting fluid, the magnetic field lines move with the fluid. This property is often called the "frozen-in" condition and will be used later for determining the shape of the magnetosphere.

III. Numerical methods.

(1) *Shock capturing technique.* In our global MHD model, bow shock and magnetopause boundaries are captured in the calculations, and explicit jump conditions are not required. In the MHD equations, there are five classes of discontinuities. In a frame of reference moving with the discontinuity, those discontinuities that lie along stream lines are called tangential discontinuities or contact discontinuities according to whether or not the normal component of the magnetic field vanishes. Discontinuities across which there is flow are divided into three categories called rotational discontinuities, and fast and slow shocks. Refer to Spreiter et al. [1969] for a discussion of the properties of these discontinuities. In the case that there is no magnetic field component in the solar wind, the bow shock is a fast shock and the magnetopause boundary corresponds to the tangential discontinuity in the MHD description of the magnetosphere.

The capturing technique has been extensively used in gas dynamics, for instance, in studying re-entry problems for space vehicles. It has been one of the main topics in computational fluid dynamics in the last 30 years. Earlier finite difference schemes have been reviewed by Sod [1978]. In our calculations we used one of these schemes, Rusanov's method, for its simplicity and stability. Recent developments of the capturing techniques based on upwind schemes can be found in the review by Harten et al.

[1983] and in many articles in this proceedings, such as the ones by Harten, by Osher, by Roe and by van Leer. The upwind scheme gives a sharp shock jump profile and well-defined contact discontinuities. An upwind scheme for the MHD equations has recently been constructed (Brio and Wu [1983]). Because of the complexity of the wave system in the MHD equations, it is a nontrivial extension of the hydrodynamics codes.

(2) *Computational requirements and techniques for global MHD modelling.* The global MHD modelling requires not only a lot of computer time but also a huge amount of data storage. If it were not for these limitations, the global MHD model might have been attempted long ago. For example, Spreiter and his coworkers (Spreiter and Alksne [1969]) had formulated a similar MHD model in the 1960s, but because of the computational limitations, they had to limit their studies to the flow properties of the magnetosheath by using fluid equations and the Chapman-Ferraro model which will be discussed in §IV. (Spreiter [1982]). The magnetosheath is the region between the bow shock and the magnetosphere. In this section, we discuss the computational requirements of the model and some techniques to alleviate the limitations. The success of the MHD model will strongly depend on our ability in solving these computational problems in addition to having a better and more accurate numerical algorithm. Of course advances in computer technology will play an important role. If the next generation of scientific computers can attain a speed of 10^9 floating point operations per second and have 50 million words of data memory, which represents an order of magnitude faster in CPU speed and larger in memory size than the present Cray-1 computer, some of the limitations imposed in the following may be avoided.

The solutions of the model have two discontinuities: bow shock and magnetopause, as well as two reconnection regions, one on the day side and one on the night side (following Dungey's reconnection model) as sketched in Figure 2. To be self-consistent, the calculation domain should include all of these regions. Therefore, in the solar-magnetospheric coordinate system, in which the x-axis points to the sun, the y-axis to the east and the z-axis to the north with the origin at the earth, the domain should extend from 20 R_E on the day side. (The symbol R_E denotes 1 earth radius.) On the night side, in principle, we can have a well-defined boundary if it is placed some distance beyond the far-tail reconnection region. However, our knowledge about this reconnection process is limited, and one may even dispute the necessity of this process as long as there is one in the tail at about -20 R_E. Thus in calculations, we first place our tail boundary at a sufficiently far distance, for example, at -200 R_E. By carrying out computer experiments with different locations,

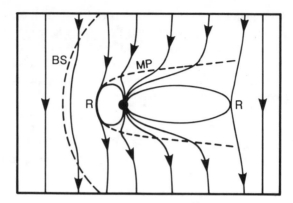

FIGURE 2. Dungey's model of the magnetosphere. *BS* denotes the bow shock, *MP* the magnetopause, *R* the reconnection region. Solid lines are magnetic field lines.

we study the effects of the distant tail on the dynamics near the earth and eventually the model can give us indications concerning the Dungey's far-tail reconnection process. For an estimate, the calculation domain is extended to -200 R_E in the tail. In both the east-west and north-south directions, the calculation domain should be extended far enough that the effects of possible shock reflections from these boundaries on the far-tail is minimized. For a magnetotail of 200 R_E long, boundaries need to be placed at 100 R_E in both north-south and east-west directions.

In the solution of the MHD model, several areas require very fine spatial resolution, such as at the bow shock, the magnetopause, the earth and the reconnection regions. For example, in order to model the reconnection processes in the tail which has a current sheet of $2 \sim 3$ R_E thick, one probably needs about 5 grid points per 1 R_E to model tearing mode instabilities for the magnetic Reynolds number of 1000. For the cases with higher magnetic Reynolds numbers, much finer spatial resolution is required. Another example is that in the study of supersonic flow past a sphere, it is found that 20 grid points are required between the bow shock and the body, therefore one may require the same number of grid points between the bow shock and the magnetopause which is about 3 R_E in physical space in the MHD model. One may also require the same spatial resolution from the magnetopause to the earth.

Obviously, we cannot have fine spatial resolution throughout the whole calculation region. It is more economical to have a nonuniform grid system with more grid points concentrated in the regions where good spatial resolution is required. To get an estimate about the size of the system, let us assume that a nonuniform grid system of 100 by 100 by 100 points is required for the calculations with 50 grid points on the day side and 50 grid points on the night side along the sun-earth line. Since there are 14 data values (ρ, p, $\mathbf{v}$, $\mathbf{B}$, $\mathbf{B}_d$ and $\mathbf{x}$, with $\mathbf{B}_d$ the geomagnetic dipole field, $\mathbf{x}$ the coordinates) for each grid point, the total data storage is 14 million words. In the special case that there is no east-west component of the magnetic field in the solar wind, we can carry out the calculations in one quadrant and we need 100 by 50 by 50 grid points, and 3.5 million words of storage. Actual calculations of this size have been performed.

Grid generation is a difficult numerical problem. Most methods of generating systems of coordinates involve the solution of systems of elliptical partial differential equations and are applied in two dimensions. In our magnetospheric problem, we face a large three-dimensional system; it is important to find an efficient way of generating grids that satisfy our requirements. Figure 3 shows the results of our attempt to

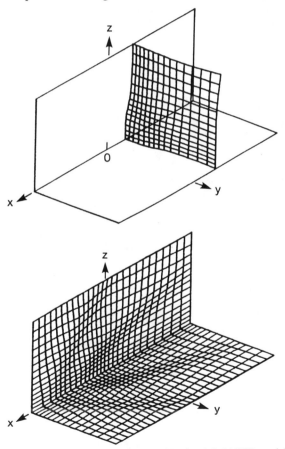

FIGURE 3. A 3D mesh system used in the global MHD model.

generate such a system. As shown in the figure, more grid points are clustered near the bow shock, the magnetopause and the earth. In the future when a more realistic boundary condition at the earth is used, we plan to overlap the coordinate system with a spherical coordinate system near the earth. The grid is generated by using an equispaced grid in the computational domain. First we prescribe two surfaces that correspond to the bow shock surface and the magnetopause boundary, respectively. Then we let points move towards these surfaces. This relocation is achieved by inducing a velocity at each grid point, the magnitude and direction depending on its relative distance to the prescribed surfaces. This method is based on the idea given by Rai and Anderson [1981].

Once the nonuniform coordinate system is generated, the MHD equations are transformed in the computational coordinate system. Since the equations can be written in terms of conservation laws, conservative form for the transformed equations can be obtained in the new coordinate system as follows (Vinokur [1974]). Let us write the conservation laws in the Cartesian system (x, y, z, t),

$$\frac{\partial u}{\partial t} + \frac{\partial E}{\partial x} + \frac{\partial F}{\partial y} + \frac{\partial G}{\partial z} = 0, \tag{8}$$

where E, F and G represent the components of the flux function. Let the new coordinates (computational coordinates) η, ξ and ζ be related to the Cartesian system by the transformation

$$\eta = \eta(x, y, z), \quad \xi = \xi(x, y, z) \quad \text{and} \quad \zeta = \zeta(x, y, z). \tag{9}$$

Equation (8) is then transformed into the conservative form

$$\frac{\partial \bar{u}}{\partial t} + \frac{\partial \bar{E}}{\partial \eta} + \frac{\partial \bar{F}}{\partial \xi} + \frac{\partial \bar{G}}{\partial \zeta} = 0, \tag{10}$$

where

$$\bar{u} = u/\mathcal{T}, \qquad\qquad \bar{E} = (E\eta_x + F\eta_y + G\eta_z)/\mathcal{T},$$

$$\bar{F} = (E\xi_x + F\xi_y + G\xi_z)/\mathcal{T}, \quad \bar{G} = (E\zeta_x + F\zeta_y + G\zeta_z)/\mathcal{T}.$$

The geometric derivatives are given by

$$\eta_x = \mathcal{T}(y_\xi z_\zeta - y_\zeta z_\xi), \qquad \xi_x = \mathcal{T}(y_\zeta z_\eta - y_\eta z_\zeta),$$

$$\zeta_x = \mathcal{T}(y_\eta z_\xi - y_\xi z_\eta), \qquad \eta_y = \mathcal{T}(z_\xi x_\zeta - z_\zeta x_\xi),$$

$$\xi_y = \mathcal{T}(z_\zeta x_\eta - z_\eta x_\zeta), \qquad \zeta_y = \mathcal{T}(z_\eta x_\xi - z_\xi x_\eta), \tag{11}$$

$$\eta_z = \mathcal{T}(x_\xi y_\zeta - x_\zeta y_\xi), \qquad \xi_z = \mathcal{T}(x_\zeta y_\eta - x_\eta y_\zeta),$$

$$\zeta_z = \mathcal{T}(x_\eta y_\xi - x_\xi y_\eta) \quad \text{and} \quad \mathcal{T} = \frac{\partial(\eta, \xi, \zeta)}{\partial(x, y, z)}.$$

For an analytical transformation the geometric derivatives can be evaluated analytically. For a numerical transformation these derivatives will have to be evaluated numerically.

In an "adapted" scheme, the coordinate transformation in (9) is generalized to allow the new coordinates to be functions of time and the transformed equations can also be cast in conservative form. This method will have advantages over the fixed grid system by allowing grid points to move according to the need of the solution; for instance, by having more grid points in the area that has large gradients of some physical quantities such as pressure. However, we have not yet attempted to use the adaptive method in our model for the difficulty of generating a satisfactory grid system.

In our magnetospheric model there are basically three significantly different time scales: the Alfvén transit time, the solar wind flow transit time and the magnetic field diffusion time. By the pressure balance principle at the magnetopause, we can estimate the relative time scales of the first two. At the subsolar point, the stagnation pressure p_{st} is roughly $\rho_{sw} v_{sw}^2$ and the pressure-balance relation is

$$\rho_{sw} v_{sw}^2 = 1/2 \left(2\, B_{dipole} \right)_{mp}^2, \tag{12}$$

where ρ_{sw} and v_{sw} refer to the solar wind density and velocity, respectively. Thus the magnitude of the dipole field at the subsolar point is $(0.5\, \rho_{sw} v_{sw}^2)^{1/2}$. If we set the magnetopause position at 10 R_E, then the magnetic dipole field at 1 R_E is about $700 \times (\rho_{sw} v_{sw}^2)^{1/2}$. Accordingly, the Alfvén velocity at 1 R_E is about $700\, v_{sw}$, if the plasma density at 1 R_E is assumed comparable with the solar wind plasma density. (The observed plasma density near the earth is higher and the Alfvén speed is lower than the estimated value.) Hence the Alfvén transit time is faster than the solar wind flow time by two orders of magnitude.

The magnetic field diffusion time is slower than the solar wind flow transit time. But the estimate depends greatly on our assumptions concerning the mechanism of the reconnection processes. If the reconnection processes at the tail are due to tearing instability, one can expect the time scale to be much slower, perhaps by a factor of 10^4, for the magnetic Reynolds number of 10^6. If, on the other hand, the reconnection processes are "driven" processes, as proposed by Sato et al. [1979], one may expect the time scale to be compatible with the solar wind flow time. Since observationally the magnetospheric dynamics is occurring in the solar wind flow transit time scale, it seems that the reconnection processes should proceed in the same time scale.

In our model we follow the calculations in the solar wind flow time scale. Because of the fast Alfvén wave near the earth, the model is more difficult to solve than the problem of hydrodynamic flow past bodies. In the terminology of differential equations, our model may be considered as a stiff problem.

Realizing the fact that the large Alfvén speed is localized near the earth since the magnetic dipole field drops off like r^{-3}, we can somewhat avoid the stiffness in the problem by devising a numerical scheme which employs two calculation regions:

(i) a small region around the earth with time step size limited by the Alfvén speed, and

(ii) a much larger surrounding region with the time step limited by the solar wind flow velocity.

For a given time step in the large region, many time steps are carried out in the small region near the earth. Since the amount of computer work is proportional to the number of time steps times the number of grid points in the region, the amount of computing in the small region is found to be of the same order as in the larger surrounding region. At the interface of the two regions, numerical stability is maintained. As shown in Figure 4, calculations at both point A in the large region and point B in the small region satisfy the CFL condition.

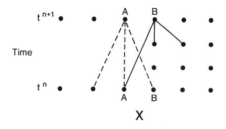

FIGURE 4. Two region scheme.

Using this two-region scheme and using a system of 100 by 50 by 50 grid points in one quadrant as described earlier, we can estimate the total amount of computer time for a run. Suppose that the smallest spatial resolution in the system is $\frac{1}{4}$ R_E and that the total time in the run is for the solar wind to traverse about 250 R_E, then 1000 time steps are required for the time integration. The processing time is about 10 μs per grid for the Rusanov scheme (Rusanov [**1962**]) on the Cray-1 computer. Thus, a total of 1.5 hours of CPU time is required. Due to the requirement of a large data storage, disks are used. During the calculations, data which are arranged plane by plane are transferred between disks and memory at the same time. The I/O charge is an additional important factor.

In this section, we have attempted to state the scale of the problem. Although it is a very large scale numerical work, with some techniques as described above, one can try to carry out the modelling within the power of the present computers. In the next generation of computers, one can certainly carry the task more easily. At that time one will probably go further than the current single fluid MHD model, and use a two fluid description.

IV. Results. In the following, some results from my calculations will be presented to illustrate the global MHD model.

We used Rusanov's scheme (Rusanov [**1962**]) in these calculations. Rusanov's method is a variant of Lax's scheme in which the artificial viscosity is maintained at a minimum value except where a large value is needed. The diffusion term in the Rusanov scheme is

$$\frac{\Delta^2}{\Delta t} \frac{\partial}{\partial x}\left(\alpha \frac{\partial u}{\partial x}\right), \tag{13}$$

where α in the MHD model is proportional to $(v + c_s + c_A)/(v + c_s + c_A)_{max}$, with sound speed c_s and Alfvén speed c_A. The diffusion term is therefore large only in the regions near the bow shock and magnetopause where the change of v is large and in the dipolar region where c_A changes rapidly.

The solar magnetospheric coordinate system is used. The initial data are $p = 0.1$, $\rho = 0.2$, $v = 0$ and $B = B_{dipole}$. At the inflow boundary the values of all variables are fixed to the solar wind parameters $p = 0.5$, $\rho = 1$, $v = -2.5 \hat{x}$ and $B = B_{sw}$. The boundary conditions in both the y and z directions are treated by linear extrapolation along the direction 45° from the x-axis. Linear extrapolation was used at the $-x$ outflow boundary. The boundary conditions at the earth are specified by fixing the values of all variables at their initial values. This does not realistically model the ionospheric effects and is not self-consistent. In our calculations, the numerical diffusion terms are large near the earth and they were used to represent the magnetosphere-ionosphere coupling.

The 3D calculations were performed with $94 \times 53 \times 53$ grid points in a nonuniform mesh covering the quadrant $-38 \leqslant x \leqslant 16$, $0 \leqslant y \leqslant 34$ and $0 \leqslant z \leqslant 34$. All quantities were normalized with respect to the solar wind parameters. Thus, the solar wind sound speed equals 1. The solar wind Mach number was chosen to be 2.5, which is in the range of observed values. The geometric field is given by (6) with $\mu = -350 \hat{z}$. By the pressure-balance principle, we will have the magnetopause at $r = 5.8$ along the sun-earth line, following the relation (12). If this position is identified at 10 R_E then 1 spatial unit is about 1.72 R_E, and 1 unit of magnetic field is about 17.3 gammas.

Figure 5(a), (b) shows pressure contours on both the equatorial and noon-midnight planes. The pressure distribution along the earth-sun line

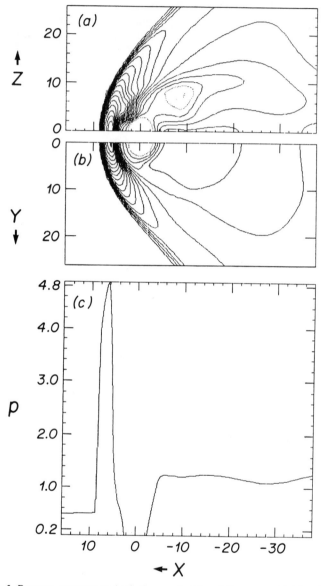

FIGURE 5. Pressure contours on both the equatorial and the noon-midnight planes in (a) and (b), respectively, and the pressure distribution along the earth-sun line in (c) for the case $\mathbf{B}_{sw} = 0$. The solid lines denote $p \geqslant p_{sw}$ and the dotted laines $p < p_{sw}$.

is plotted in Figure 5(c). Similar plots for the density distributions are given in Figure 6. The accuracy of our results can be evaluated by checking against the Rankine-Hugoniot jump conditions. This gives $p = 4$ and $\rho = 2.27$ just inside the shock along the sun-earth line. The

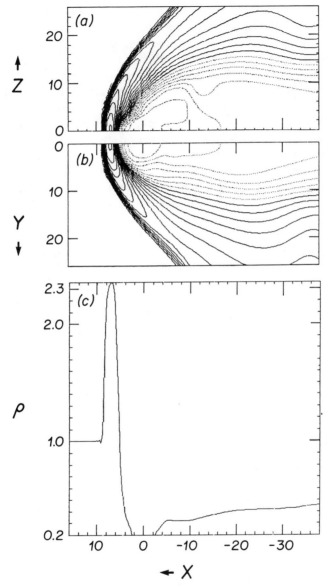

FIGURE 6. Density contours on both the equatorial and the noon-midnight planes in (a) and (b), respectively, and the density distribution along the earth-sun line in (c) for the case $\mathbf{B}_{sw} = 0$. The solid lines denote $\rho \geq \rho_{sw}$ and the dotted lines $\rho < \rho_{sw}$.

calculated values (see Figures 5 and 6) agree well. The calculated stagnation pressure is 4.8 which gives $k = 0.8$ for the Newtonian formula $p_{st} = k\rho_{sw}v_{sw}^2$. The bow shock is located at 8.2 and the magnetopause is at 5.8. The ratio of the standoff distance to the magnetopause is then 0.41, which is larger than the observed average value of 0.33. This difference is due to the small Mach number used in our calculation.

On the night side, the region near the equatorial plane is characterized by a plasma sheet. This is most evident in Figure 5 where the plasma pressure has a local maximum near the equator. The region of low pressure, which contains the dotted contour, corresponds to the tail lobes. At $x \sim -13$ R_E the pressure decreases by about an order of magnitude between $z = 0$ and $z = 10$ R_E. As expected, the region near the equator contains a high beta plasma ($\beta \gg 1$) and that at large z values contains low beta plasma ($\beta \ll 1$). Beta is defined as the ratio of the plasma pressure over the magnetic pressure. The model predicts nearly constant density in both the plasma sheet and lobe regions (Figure 6). Thus, most of the change in plasma pressure across the plasma sheet boundary is caused by a change in plasma temperature.

The magnetospheric current system (from an earlier run, Wu et al. [**1981**]) is presented in Figure 7. These currents are traced from points at the $z = 2$ line on the noon-midnight meridian plane. Their projections on both the equatorial and noon-midnight meridian planes are shown with dotted lines. A typical magnetopause current element at the nose is indicated by AA'. The magnetopause boundary currents are similar to those obtained from Chapman-Ferraro pressure balance models. But as will be discussed later, the MHD model in fact presents a very different physical picture from that of the Chapman-Ferraro model.

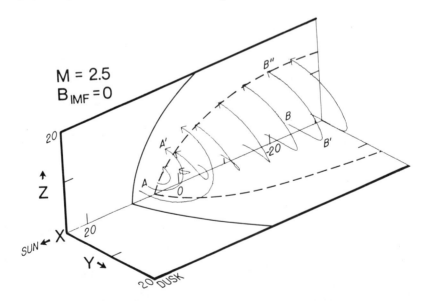

FIGURE 7. Three-dimensional perspective plot of the current system.

The current loop $BB'B''$ is typical of the cross-tail current and its
return current on the magnetopause. The cross-tail current BB' flows
nearly parallel to the equatorial plane and curves towards the earth. The
return current flows almost on a plane perpendicular to the earth-sun line
from B' to B''. Similar currents flow from B' to B'' in the other quadrant
to complete the current loop and a mirror image current loop is formed
in the southern hemisphere to complete the familiar θ shaped current
system. The line BB'' is tilted from the z-axis toward the earth. This tilt is
larger closer to the earth. Many features in our tail-current system are
found in the semiempirical model of Olson and Pfitzer [**1977**]. Olson and
Pfitzer used wire current loops to model the tail field. In their calculation,
they found it necessary to tilt the wire current loops towards the sun and
to bend them around the earth, in order to reproduce the observed field.
The resulting current distribution is very similar to that from our model.
The current density distribution along the earth-sun line obtained from
our model (solid line) and from the Olson-Pfitzer model are presented in
Figure 8. The dotted lines give the current density profile used by Olson

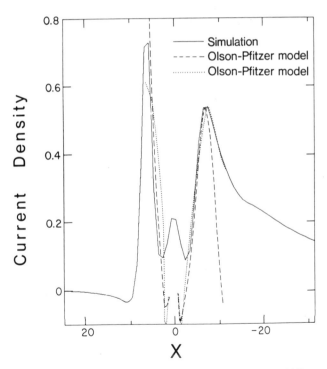

FIGURE 8. Current density distribution along the earth-sun line. The solid lines are from our
model, the dotted lines give the profile used by Olson and Pfitzer [**1977**] to calculate the
field, and the dashed line is from an analytic version of the Olson-Pfitzer model.

and Pfitzer to calculate the field, while the dashed line was generated by calculating $\nabla \times \mathbf{B}$ from an analytic version of the mode for $\mathbf{B}$. These distributions are normalized at the peak at $x \sim -15$ R_E. The agreement is good.

One interesting result from the MHD model is about the shape of the magnetosphere (Wu [**1983**]). Conventional theory for the shape of the magnetosphere boundary was formulated by Chapman and Ferraro over forty years ago. The model uses the pressure-balance principle across the boundary and assumes:

(1) a very simple pressure law on the outside of the boundary,
(2) zero plasma pressure on the inside, and
(3) a zero magnetic field on the outside (e.g. Choe et al. [**1973**]).

Figure 9 shows a sketch of the magnetic field topology in the Chapman-Ferraro model. Note the existence of the neutral points where $\mathbf{B} = 0$ at the boundary, as denoted by C in the figure.

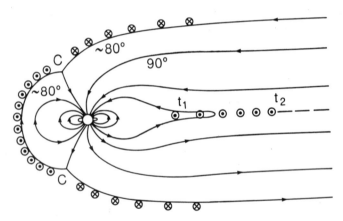

FIGURE 9. A sketch of the magnetic field lines in a conventional Chapman-Ferraro model; C denotes the cusp regions.

In the MHD model, the time development of the solar wind interaction with the dipole magnetic field is sketched in Figure 10. Figure 10(a) represents the initial state where plasma of uniform density and uniform pressure is in static equilibrium with the geomagnetic dipole field. The topology of the magnetic dipole field is represented by the 90° field lines which close at infinity. Figure 10(b) represents a state at a later time when the bow shock and magnetopause have been formed. Because of the "frozen in" condition, the initial field line topology should be maintained. Figure 10(b) represents one of two possible resulting configurations. The other one has the two neutral points on the day side. But in our experiment, the day side magnetosphere was compressed by the solar

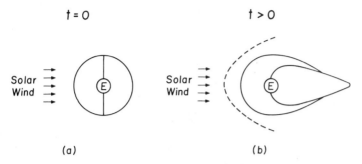

FIGURE 10. A sketch of the time development of the interaction of the solar wind with a line dipole. The dashed line in (b) represents the bow shock.

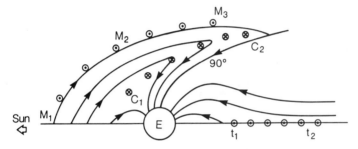

FIGURE 11. Field line pattern and current structure for the configuration Figure 10(b). $M_1 M_2 M_3$ denotes the magnetopause surface, $C_1 C_2$ the cusp current sheet, and $T_1 T_2$ the tail current sheet. C_1 denotes the location of the cusp in our MHD model.

wind and the night side magnetosphere was then compressed by the day side magnetosphere; the configuration of Figure 10(b) results.

The field line pattern and the current structure of the resulting magnetosphere (Figure 10(b)) are sketched in Figure 11. Note the existence of the cusp current sheet, $C_1 C_2$, and that the cusp in this MHD model refers to the region marked by C_1 where field lines converge at the earth. This picture of the magnetopause looks quite different from that of the Chapman-Ferraro model. Especially, the magnetic field topology near the cusps is very different as compared to Figures 9 and 11. But, it can be reconciled in the following way. In Figure 11, curve $(M_1 M_2 C_1 C_2)$ represents the shape based on the Chapman-Ferraro model, curve $(M_1 M_2 M_3)$ represents the magnetopause surface, and curve $(C_1 C_2)$ shows the cusp current sheet from the MHD model. It is clear from this figure that the Chapman-Ferraro model, which is a surface current model, represents the boundary of the magnetopause surface of our model from M_1 to M_2 and the cusp current sheet from C_1 to C_2. The magnetopause surface current from M_2 to M_3 is neglected as it is small in comparison with the cusp

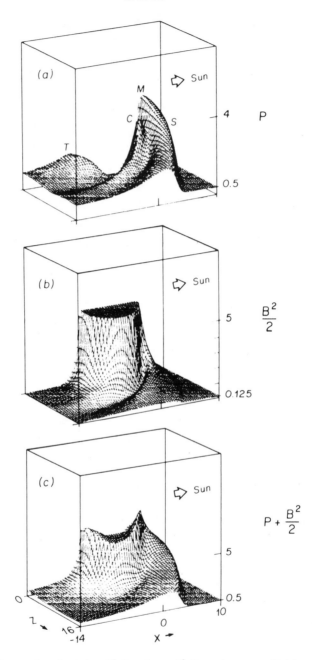

FIGURE 12. Perspective plots of p, $B^2/2$ and $p + B^2/2$ on the noon-midnight meridian plane for the case with northward $\mathbf{B}_{sw}$ in the 3D calculations. S, M, C and T denote bow shock, magnetopause, cusp current sheet and tail-current sheet, respectively.

current sheet. It is not surprising that the current distributions in both the MHD model and the Chapman-Ferraro model should agree. Both models use the pressure balance principle, and the magnetopause currents are used to shield the dipole fields. However, the MHD model presents a very different physical picture from that of the Chapman-Ferraro model (Wu [**1983**]). For instance, in the MHD model, solar wind plasma cannot enter the ionosphere directly through the cusp.

Figure 12(a), (b), (c) shows the perspective plots of p, $B^2/2$ and $p + B^2/2$ on the noon-midnight plane, respectively, from our model. The $p + B^2/2$ plot shows the pressure balance principle (which is true for weak curvature).

V. Conclusion. In this paper an introduction to the global MHD model was given. Despite the efforts of many researchers in the last several years, it is fair to say that this mathematical problem has not been solved with complete satisfaction. However, because of the advances in numerical techniques and the availability of more powerful computers, one can expect rapid progress in the field of quantitative MHD modeling. It is hoped that these models will be invaluable in our understanding of the magnetospheric physics.

This work was supported by National Science Foundation Grant ATM 79-26492, by NASA Solar Terrestrial Theory Grant NAGW-78, and the MHD code was developed with support by the Department of Energy, DE-AM03-76SF00010 PA26, Task VIB.

REFERENCES

M. Brio and C. C. Wu, 1983. Private communication.

R. L. Carovillano and J. M. Forbes, (eds.), 1983. *Solar-terrestrial physics*, Reidel, Dordrecht.

J. Y. Choe, D. B. Beard and E. C. Sullivan, 1973. *Precise calculation of the magnetosphere surface for a tilted dipole*, Planet. Space Sci. **21**, 485.

A. Harten, P. D. Lax and B. van Leer, 1983. *On upstream differencing and Godunov-type schemes for hyperbolic conservation laws*, SIAM Rev. **25**, 35.

L. D. Landau and E. M. Lifshitz, 1959. *Fluid mechanics*, Pergamon Press, New York and Oxford.

P. D. Lax, 1954. *Weak solutions of nonlinear hyperbolic equations and their numerical computation*, Comm. Pure Appl. Math. **7**, 159.

J. N. Leboeuf, T. Tajima, C. F. Kennel and J. M. Dawson, 1978. *Global simulation of the time-dependent magnetosphere*, Geophys. Res. Lett. **5**, 609.

J. S. Lyon, H. Brecht, J. A. Fedder and P. J. Palmadesso, 1980. *The effects on the earth's magnetotail from shocks in the solar wind*, Geophys. Res. Lett. **7**, 721.

W. P. Olson and K. A. Pfitzer, 1977. *Magnetospheric magnetic field modeling*, McDonnell Douglas Astronautics Co., preprint.

S. Osher and F. Solomon, 1982. *Upwind difference schemes for hyperbolic systems of conservation laws*, Math. Comp. **38**, 339.

M. M. Rai and D. A. Anderson, 1981. *Grid evolution in time asymptotic problems*, J. Comput. Phys. **43**, 327.

J. G. Roederer, 1979. *Solar system plasma physics*, (C. F. Kennel, L. J. Lanzerotti and E. N. Parker, eds.), North-Holland, Amsterdam.

V. Rusanov, 1962. *Calculation of interaction of non-steady shock waves with obstacles*, Nat. Res. Council of Canada, Translation No. 1027.

T. Sato and T. Hayashi, 1979. *Externally driven magnetic reconnection and a powerful magnetic energy converter*, Phys. Fluids **22**, 1189.

G. A. Sod, 1978. *A survey of several finite difference methods for systems of nonlinear hyperbolic conservation laws*, J. Comput. Phys. **27**, 1.

J. R. Spreiter, 1982. Private communication.

J. R. Spreiter and A. Y. Alksne, 1968. *Plasma flow around the magnetosphere*, Rev. Geophys. Space Phys. **7**, 11.

M. Vinokur, 1974. *Conservation equations of gasdynamics in curvilinear coordinate systems*, J. Comput. Phys. **14**, 105.

C. Weiland, 1978. *Calculation of three-dimensional stationary supersonic flow fields by applying the "Progonka" process to a conservative formulation of the governing equations*, J. Comput. Phys. **29**, 173.

C. C. Wu, 1983. *Shape of the magnetosphere*, Geophys. Res. Lett. **10**, 545.

C. C. Wu, R. J. Walker and J. M. Dawson, 1981. *A three dimensional MHD model of the earth's magnetosphere*, Geophys. Res. Lett. **8**, 523.

DEPARTMENT OF PHYSICS, UNIVERSITY OF CALIFORNIA, LOS ANGELES, CALIFORNIA 90024

Lectures in Applied Mathematics
Volume **22**, 1985

Application of TVD Schemes for the Euler Equations of Gas Dynamics

H. C. Yee[1], R. F. Warming[1] and Ami Harten[2]

ABSTRACT. Several techniques for the construction of nonlinear, second-order accurate, high-resolution schemes for hyperbolic conservation laws have been developed in recent years. The goal of constructing these highly nonlinear schemes is to simulate complex flow fields more accurately. These schemes were constructed under a common theme; i.e., to achieve second-order accuracy without introducing spurious oscillations near discontinuities by employing some kind of feedback mechanism. Most of these schemes were independently derived, and thus they are very much different in form, methodology and design principle. However, from the standpoint of numerical analysis, these schemes are total variation diminishing (TVD) for nonlinear scalar hyperbolic conservation laws and for constant coefficient hyperbolic systems. The notion of TVD schemes was introduced by Harten in 1981. The purpose of this paper is to review a subset of the class of TVD schemes and to show by numerical experiments the performance of these second-order accurate schemes in solving the Euler equations of gas dynamics.

1. Introduction. In recent years, researchers have been quite active in the area of designing highly accurate and yet stable shock-capturing finite difference schemes for the computation of the Euler equations of

1980 *Mathematics Subject Classification*. Primary 76-08, 65N05.
[1] Research Scientist, Computational Fluid Dynamics Branch.
[2] Associate Professor, School of Mathematical Sciences.

gas dynamics. Of special interest are the methods that generate nonoscillatory but sharp approximations to shocks and contact discontinuities. These schemes are more complicated to use than the classical shock-capturing methods such as variants of the Lax-Wendroff scheme. Although many difference methods were developed with this property in mind [1, 2], there is a continuous need for improvement in efficiency, stability and computational simplicity.

There is a class of shock-capturing second-order accurate almost everywhere algorithms based in part on either an exact or approximate Riemann solver [3–8]. Some of these schemes are constructed under physical and geometric arguments; some are based on more formal mathematical approaches. Most of these schemes are very much different in form, methodology and design principle. However, from the standpoint of numerical analysis, these schemes are total variation diminishing (TVD) for nonlinear scalar hyperbolic conservation laws and for constant coefficient hyperbolic systems. Under certain assumptions, some of these schemes turn out to be identical [9]. For convenience, from here on, second-order will be used to mean second-order accurate almost everywhere.

We are aware of primarily four different (and yet not totally distinct) design principles for the construction of high resolution TVD schemes. They are (a) hybrid schemes such as the flux corrected transport (FCT) of Boris and Book [10], Harten [12], Harten and Zwas [11] and van Leer [13], (b) second-order extension of Godunov's scheme by van Leer [5], Colella and Woodward [6], (c) the modified flux approach of Harten [3, 4] and (d) the numerical fluctuation approach of Roe [7]. Also, Osher [8] have recently extended the first-order scheme of Engquist and Osher to second-order accuracy by using the above ideas. The following is a subjective interpretation of these design principles.

(a) The flux corrected transport scheme is a two-step hybrid scheme consisting of a combined first- and second-order scheme. It computes a provisional update from a first-order scheme, and then filters the second-order corrections to prevent occurrence of new extrema.

The idea of the hybrid scheme of Harten or van Leer is to take a high-order accurate scheme and to switch it explicitly into a monotone first-order accurate scheme when extreme points and discontinuities are encountered.

(b) Van Leer observed that one can obtain second-order accuracy in Godunov's scheme by replacing the piecewise-constant initial data of the Riemann problem with piecewise-linear initial data. The slope of the piecewise-linear initial data is chosen so that no spurious oscillations can occur. Woodward and Colella further refined van Leer's idea by using piecewise-parabolic initial data.

(c) The modified flux TVD scheme is a technique to design a second-order accurate TVD scheme by starting with a first-order TVD scheme and applying it to a modified flux. The modified flux is chosen so that the scheme is second-order at regions of smoothness and first-order at points of extrema. Details of the construction and definition of TVD schemes will be discussed in later sections.

(d) The numerical fluctuation approach of Roe is a variation of the Lax-Wendroff scheme. Roe's variation depends on an average function. The average function is constructed in a way such that spurious oscillations will not occur. This scheme can switch to the second-order upwind scheme of Warming and Beam [14]. As a matter of fact, under certain assumptions, a form of Roe's scheme is equivalent to the modified flux approach [15].

Most of the above methods can also be viewed as three-point central difference schemes with a "smart" numerical dissipation or smoothing mechanism. "Smart" here means automatic feedback mechanism to control the amount of numerical dissipation, unlike the numerical dissipation used in linear theory.

The modified flux approach is relatively simple to understand and easy to implement into a new or existing computer code. One can modify a standard three-point central difference code by simply changing the conventional numerical dissipation term into the one designed for the TVD scheme.

The purpose of this paper is to review the class of second-order TVD schemes via the modified flux approach. Since the construction of these schemes is based on a first-order upwind scheme, a brief review of the fundamentals of first-order TVD schemes is necessary. The rest of the paper is organized as follows: We start with a discussion of first-order TVD schemes, and follow by a brief description of the construction of second-order TVD schemes. The recently developed implicit TVD schemes [4, 16] will also be included. We will conclude with some transient and steady-state calculations to illustrate the applicability of these schemes to the Euler equations.

Before we go on to the next section, we want to emphasize here that all the second-order TVD schemes are constructed so that no spurious oscillations are generated for one-dimensional nonlinear scalar hyperbolic conservation laws and constant coefficient hyperbolic systems. None of the theories have said anything about nonlinear systems or two-dimensional scalar hyperbolic conservation laws. As a matter of fact, we are still very far away from being able to design algorithms for systems of equations in many dimensions. Davis [17] and Roe and Baines [18] have some interesting ideas on the subject. Moreover, Goodman and LeVeque [19] recently have obtained the following result: for a specific norm, a two-dimensional scalar approximation cannot be TVD and still be more than first-order accurate. Maybe what we need is a different norm or a more relaxed definition than the TVD property for one dimension. But, in practice, it is straightforward to formally extend the scheme to one- or two-dimensional nonlinear hyperbolic systems. Surprisingly enough, numerical experiments with these schemes which are based on a one-dimensional concept applied to two-dimensional problems via local one-dimensional splitting show that these types of schemes do perform well for fairly complex shock structures [16, 20].

Therefore, for the rest of the paper, it is understood that the properties of all the schemes under discussion are for one-dimensional nonlinear scalar hyperbolic conservation laws and one-dimensional constant coefficient hyperbolic systems. The schemes are then formally extended to one- or two-dimensional systems of conservation laws and are evaluated by numerical experiments.

2. First-order TVD schemes. Consider the scalar hyperbolic conservation law

$$\frac{\partial u}{\partial t} + \frac{\partial f(u)}{\partial x} = 0, \tag{2.1}$$

where $a(u) = \partial f/\partial u$ is the characteristic speed. Let u_j^n be a numerical solution of (2.1) at $x = j\Delta x$, $t = n\Delta t$, with Δx the spatial mesh size and Δt the time step.

A general three-point explicit difference scheme in conservation form can be written as

$$u_j^{n+1} = u_j^n - \lambda\left(h_{j+1/2}^n - h_{j-1/2}^n\right), \tag{2.2}$$

$h_{j+1/2}^n = h(u_j^n, u_{j+1}^n)$, and $\lambda = \Delta t/\Delta x$. Here, h is commonly called a numerical flux function. The numerical flux function h is required to be consistent with the conservation law in the following sense:

$$h(u_j, u_j) = f(u_j). \tag{2.3}$$

Many popular first-order upwind schemes (with explicit Euler in time) can be cast in the following form:

$$u_j^{n+1} = u_j^n - \lambda\phi^n - \lambda\psi^n. \tag{2.4}$$

Here, ϕ represents some forward difference of the f or u. For example, ϕ can be

$$\phi = \mathscr{A}_1\left(f_j, f_{j+1}, u_j, u_{j+1}\right)\left(f_{j+1} - f_j\right), \tag{2.5a}$$

or

$$\phi = \mathscr{A}_2\left(f_j, f_{j+1}, u_j, u_{j+1}\right)\left(u_{j+1} - u_j\right), \tag{2.5b}$$

and ψ represents some backward difference of the f or u. For example, ψ can be

$$\psi = \mathscr{B}_1(f_{j-1}, f_j, u_{j-1}, u_j)(f_j - f_{j-1}), \qquad (2.6a)$$

or

$$\psi = \mathscr{B}_2(f_{j-1}, f_j, u_{j-1}, u_j)(u_j - u_{j-1}). \qquad (2.6b)$$

Here $\mathscr{A}_1$, $\mathscr{A}_2$, $\mathscr{B}_1$ and $\mathscr{B}_2$ are some known functions of the arguments indicated above. As an example, consider the Engquist and Osher scheme [21], where the ϕ and ψ for any convex flux function f are

$$\phi = f_{j+1}^- - f_j^-, \qquad \psi = f_j^+ - f_{j-1}^+ \qquad (2.7a)$$

with

$$f_j^+ = f(\max(u_j, \bar{u})), \qquad (2.7b)$$

$$f_j^- = f(\min(u_j, \bar{u})), \qquad (2.7c)$$

and $\bar{u}$ is the sonic point of $f(u)$; i.e., $f'(\bar{u}) = 0$.

In [22], Huang introduced a first-order accurate upwind scheme

$$u_j^{n+1} = u_j^n - \frac{\lambda}{2}\left[1 - \operatorname{sgn}(a_{j+1/2}^n)\right](f_{j+1}^n - f_j^n)$$
$$- \frac{\lambda}{2}\left[1 + \operatorname{sgn}(a_{j-1/2}^n)\right](f_j^n - f_{j-1}^n). \qquad (2.8)$$

She was vague in defining $a_{j+1/2}$ for a general flux function f, but for Burgers' equation, she explicitly defined

$$a_{j+1/2} = (a_{j+1} + a_j)/2. \qquad (2.9)$$

Here

$$\phi = \frac{1}{2}\left[1 - \operatorname{sgn}(a_{j+1/2})\right](f_{j+1} - f_j), \qquad (2.10)$$

which has the same form as (2.5a), and

$$\psi = \frac{1}{2}\left[1 + \operatorname{sgn}(a_{j-1/2})\right](f_j - f_{j-1}), \qquad (2.11)$$

which has the same form as (2.6a).

In [23], Roe defined

$$a_{j+1/2} = \begin{cases} (f_{j+1} - f_j)/\Delta_{j+1/2}u, & \Delta_{j+1/2}u \neq 0, \\ a(u_j), & \Delta_{j+1/2}u = 0, \end{cases} \qquad (2.12)$$

with $\Delta_{j+1/2}u = u_{j+1} - u_j$. This is equivalent to Huang's method for Burgers' equation.

With the definition (2.12), scheme (2.8) can be rewritten as a three-point central difference method plus a numerical viscosity term

$$u_j^{n+1} = u_j^n - \frac{\lambda}{2}\left[f_{j+1}^n - f_{j-1}^n - \left|a_{j+1/2}^n\right|\Delta_{j+1/2}u^n + \left|a_{j-1/2}^n\right|\Delta_{j-1/2}u^n\right].$$

(2.13)

Now $\phi = \frac{1}{2}[a_{j+1/2} - |a_{j+1/2}|](u_{j+1} - u_j)$, which has the same form as (2.5b), and $\psi = \frac{1}{2}[a_{j-1/2} + |a_{j-1/2}|](u_j - u_{j-1})$, which has the same form as (2.6b).

Equation (2.8) or (2.13), in Roe's terminology is an approximate Riemann solver to (2.1) and it is sometimes known as the Huang scheme [22], Roe scheme [23] or the Murman scheme [24] for the nonlinear scalar equation (2.1).

The numerical flux function as a function of ϕ can be written as

$$h_{j+1/2} = f_j + \phi.$$

(2.14)

Or, in a more compact form, we can express (2.13) as (2.2) with

$$h_{j+1/2} = \frac{1}{2}\left[f_j + f_{j+1} - Q(a_{j+1/2})\Delta_{j+1/2}u\right]$$

(2.15a)

and

$$Q(a_{j+1/2}) = |a_{j+1/2}|.$$

(2.15b)

Q is sometimes known as the coefficient of the numerical viscosity term. In this paper, we prefer to use equation (2.15a) as the form of the first-order upwind numerical flux function. This form of the numerical flux function (2.15a) is not a common notation. But we can see later that representation (2.15a) is quite useful for the development of a second-order TVD schemes via the modified flux approach.

It is well known that (2.8) or (2.13) is not consistent with an entropy inequality, and the scheme might converge to a nonphysical solution. A slight modification of the coefficient of numerical viscosity term [3]

$$Q(z) = \begin{cases} |z|, & |z| \geqslant \varepsilon, \\ (z^2 + \varepsilon^2)/2\varepsilon, & |z| < \varepsilon, \end{cases}$$

(2.16)

can remedy the entropy violating problem. Other ways of modifying (2.15b) to satisfy an entropy inequality can be found in [7, 8].

If we define

$$C^{\pm}(z) = (1/2)[Q(z) \pm z],$$

(2.17)

then, this upwind scheme can be written as

$$u_j^{n+1} = u_j^n + \lambda C^-(a_{j+1/2}^n)\Delta_{j+1/2}u^n - \lambda C^+(a_{j-1/2}^n)\Delta_{j-1/2}u^n.$$

(2.18)

In other words, this scheme can be viewed as a genearlization of the Courant, Isaacson and Rees upwind scheme [25].

The total variation of a mesh function u_j^n is defined to be

$$TV(u^n) = \sum_{j=-\infty}^{\infty} |u_{j+1}^n - u_j^n| = \sum_{j=-\infty}^{\infty} |\Delta_{j+1/2} u^n|. \tag{2.19}$$

The numerical scheme (2.2) for the initial-value problem of (2.1) is said to be TVD if

$$TV(u^{n+1}) \leqslant TV(u^n), \tag{2.20}$$

It can be shown that a sufficient condition [3] for (2.2), together with (2.15a), to be a TVD scheme is

$$\lambda C^-(a_{j+1/2}) = \frac{\lambda}{2}\left[-a_{j+1/2} + Q(a_{j+1/2})\right] \geqslant 0, \tag{2.21a}$$

$$\lambda C^+(a_{j+1/2}) = \frac{\lambda}{2}\left[a_{j+1/2} + Q(a_{j+1/2})\right] \geqslant 0, \tag{2.21b}$$

$$\lambda\left(C^-(a_{j+1/2}) + C^+(a_{j+1/2})\right) = \lambda Q(a_{j+1/2}) \leqslant 1. \tag{2.21c}$$

Therefore, for the scheme (2.2) together with the general form (2.15a) to be TVD, we have to pick Q such that (2.21a)–(2.21c) are satisfied. Applying the above conditions to (2.13), it can be easily shown that (2.13) is a TVD scheme under the CFL restriction of 1. Here we note that the scheme (2.2) with (2.15a) is conservative, and it is consistent with the original conservation law (2.1).

It should be emphasized that conditions (2.21a)–(2.21c) are only a sufficient condition, i.e., schemes that fail this test might still be TVD.

3. A second-order accurate (almost everywhere) explicit TVD scheme.
The modified flux approach [3, 4] is a technique to design a second-order accurate (almost everywhere) TVD scheme without having to explicitly specify its switching mechanism. To do so we start with the first-order TVD scheme (2.2) together with (2.15a) and (2.16) and apply it to a modified flux $\tilde{f} = (f + g)$. The new numerical flux function $\tilde{h}_{j+1/2}$ depends on $(f + g)$ instead of f alone, the coefficient of the numerical viscosity term Q is a function of a modified characteristic speed instead of the characteristic speed a alone, and $\tilde{h}_{j+1/2}$ can be written as

$$\tilde{h}_{j+1/2} = \frac{1}{2}\left[(f_j + g_j) + (f_{j+1} + g_{j+1}) - Q(a_{j+1/2} + \gamma_{j+1/2})\Delta_{j+1/2}u\right], \tag{3.1a}$$

where g_j is some appropriately chosen function of the $a_{j\pm1/2}$ and $\Delta_{j\pm1/2}u$, to be described shortly, and

$$\gamma_{j+1/2} = \begin{cases} (g_{j+1} - g_j)/\Delta_{j+1/2}u, & \Delta_{j+1/2}u \neq 0, \\ 0, & \Delta_{j+1/2}u = 0. \end{cases} \tag{3.1b}$$

The second-order TVD scheme together with equation (3.1) can be written as

$$u_j^{n+1} = u_j^n + \lambda C^-(\tilde{a}_{j+1/2}^n)\Delta_{j+1/2}u^n - \lambda C^+(\tilde{a}_{j-1/2}^n)\Delta_{j-1/2}u^n, (3.2)$$

with the modified characteristic speed $\tilde{a}_{j+1/2}^n = a_{j+1/2}^n + \gamma_{j+1/2}^n$. This again is an upwind scheme with the modified flux $(f + g)$. Therefore, all we need is to choose γ such that $C^\pm(\tilde{a}_{j+1/2}) \geqslant 0$ and $\lambda Q(\tilde{a}_{j+1/2}) \leqslant 1$; i.e., the same conditions as (2.21a)–(2.21c) except now the independent variable is $(a + \gamma)$ instead of a.

Thus, the requirements on g are: (i) The function g should have a bounded γ in (3.1b), so that the scheme is TVD with respect to the modified flux $(f + g)$. (ii) The modified scheme should be second-order accurate (except at points of extrema). In [3, 4], Harten devised a recipe for g that satisfies the above two requirements. We will use this particular form of g for the discussion here. It can be written as

$$g_j = S \cdot \max\left[0, \min\left(\sigma_{j+1/2}|\Delta_{j+1/2}u|, S \cdot \sigma_{j-1/2}\Delta_{j-1/2}u\right)\right], \quad (3.3a)$$

$$S = \operatorname{sgn}(\Delta_{j+1/2}u),$$

with $\sigma_{j+1/2} = \sigma(a_{j+1/2})$, and we choose

$$\sigma(z) = \frac{1}{2}\left[Q(z) - \lambda z^2\right] \geqslant 0 \qquad (3.3b)$$

for time accurate calculations. With this choice of $\sigma(z)$, the scheme (2.2) together with (3.1) is second-order accurate in both time and space; see [3]. Other more general forms for g can be obtained following a line of argument given by Sweby [15] and Roe [7].

An interesting observation is that equation (3.2) together with (2.15b) can be rewritten as

$$u_j^{n+1} = u_j^n - \frac{\lambda}{2}\left[1 - \operatorname{sgn}(\tilde{a}_{j+1/2}^n)\right]\left(\tilde{f}_{j+1}^n - \tilde{f}_j^n\right)$$

$$- \frac{\lambda}{2}\left[1 + \operatorname{sgn}(\tilde{a}_{j-1/2}^n)\right]\left(\tilde{f}_j^n - \tilde{f}_{j-1}^n\right). \qquad (3.4)$$

This is a straightforward extension of Huang's or Roe's entropy violating first-order upwind scheme to second-order accuracy. The scheme looks identical to their first-order scheme (2.8) except the arguments a's and f's are different.

4. Global order of accuracy of the second-order TVD scheme. The form of g devised by Harten has the property of switching the second-order scheme into first-order at points of extrema. To see this we turn now to examine the behaviour of TVD schemes around points of extrema by considering their application to data where

$$u_{j-1} \leqslant u_j = u_{j+1} \geqslant u_{j+2}. \qquad (4.1)$$

In this case $g_j = g_{j+1} = 0$ in (3.3a) and (3.3b), and thus the numerical flux (3.1) becomes identical to that of the original first-order accurate scheme; consequently, the truncation error of the second-order scheme (2.2) together with (3.1a), (3.1b) deteriorates to $O((\Delta x)^2)$ at j and $j + 1$. This behaviour is common to all TVD schemes, since this is one of the vehicles to prevent spurious oscillations near a shock. Thus, a second-order TVD scheme must have a mechanism that switches itself into a first-order accurate TVD scheme at points of extrema. Because of the above property, second-order accurate TVD schemes are genuinely nonlinear; i.e., they are nonlinear even in the constant coefficient case. Because of the above property of the second-order TVD schemes, we did some numerical experiments on the scheme for Burgers' equation to examine the global order of accuracy. So far, we have used the term "second-order" loosely to mean "second-order accurate almost everywhere".

To examine the global order of accuracy of the "second-order" TVD scheme in the L_1, L_2 and L_∞ norms, we apply the scheme to the Burgers' equation

$$\frac{\partial u}{\partial t} + \frac{\partial}{\partial x} \frac{u^2}{2} = 0. \tag{4.2a}$$

Here the flux function $f(u) = u^2/2$. Since the theory of TVD schemes is only developed for initial value problems at this point, we consider a periodic problem to avoid extra complication. The initial condition is a sine wave

$$u(x, 0) = \sin \pi x, \qquad 0 \le x \le 2. \tag{4.2b}$$

The exact solution at two different times is illustrated in Figure 4.1, at time $t = 0.2$ when the solution is still smooth, and at time $t = 1.0$ when

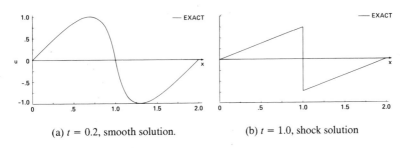

(a) $t = 0.2$, smooth solution.   (b) $t = 1.0$, shock solution

FIGURE 4.1. Exact solution of Burgers' equation at two different times.

the solution has developed into a shock. We define the local error of the computation at each grid point ($j\Delta x, n\Delta t$) as

$$e_j = u_j^n - u(j\Delta x, n\Delta t), \tag{4.3}$$

where $u(j\Delta x, n\Delta t)$ is the exact solution of the differential equation (4.2). Here we assume that there is a fixed relation between Δt and Δx. The global order of accuracy m is determined by

$$\|e\| = O(\Delta x^m) \tag{4.4}$$

as the mesh is refined for some norm.

To obtain the global order of accuracy numerically, we compute the error at a fixed time for a given mesh and repeat the calculation with increasingly finer meshes. Figure 4.2 shows the global order of accuracy of the second-order TVD scheme compared with the Lax-Wendroff method at time $t = 0.2$ when the solution is still smooth. The order of accuracy for the TVD schemes is 2 for the L_1 norm, around $\frac{3}{2}$ for the L_2 norm, and 1 for the L_∞ norm. On the other hand, the order of accuracy for the Lax-Wendroff is 2 for all three norms. The main reason for the

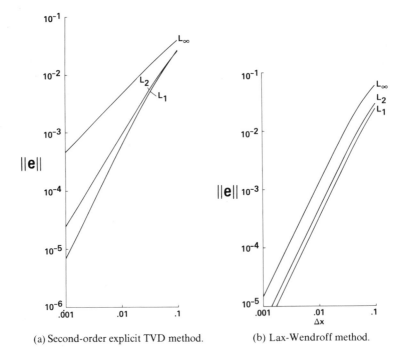

(a) Second-order explicit TVD method.   (b) Lax-Wendroff method.

FIGURE 4.2. Global order of accuracy at $t = 0.2$ when the solution is still smooth.

difference in the order of accuracy on the three norms for the TVD scheme is that the scheme automatically switches itself into first-order whenever extreme points are encountered. In this case there are two extreme points.

Next, we look at the global order of accuracy of the two methods at time $t = 1.0$ when a shock has developed. Figure 4.3 shows the order of accuracy of the TVD scheme at $t = 1.0$, which is identical to the one at

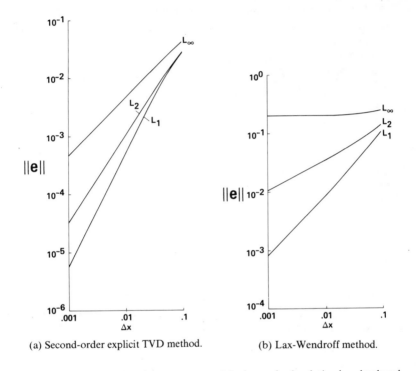

(a) Second-order explicit TVD method.   (b) Lax-Wendroff method.

FIGURE 4.3. Global order of accuracy at $t = 1.0$ when a shock solution has developed.

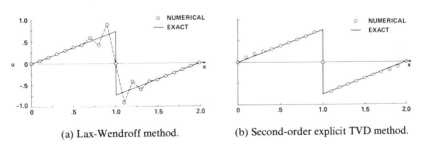

(a) Lax-Wendroff method.   (b) Second-order explicit TVD method.

FIGURE 4.4. Numerical solution of Burgers' equation at $t = 1.0$.

time $t = 0.2$. But the order of accuracy for the Lax-Wendroff is drastically degraded. It is 1 for the L_1 norm, around $\frac{1}{2}$ for the L_2 norm, and 0 for the L_∞ norm. This is due to the inherent characteristic of the Lax-Wendroff method that causes this scheme to generate spurious oscillations near the shock. Figure 4.4 shows the numerical solution of the Lax-Wendroff method compared with the second-order explicit TVD method at $t = 1.0$

5. Extensions to systems and two-dimensional conservation laws.

Systems. At the present development, the concept of TVD schemes, like monotone schemes, is only defined for nonlinear scalar conservation laws or constant coeffcient hyperbolic systems. The main difficulty stems from the fact that, unlike the scalar case, the total variation in x of the solution to a system of nonlinear conservation laws is not necessarily a monotonic decreasing function of time. The total variation of the solution may actually increase at moments of interaction between waves. Not knowing a diminishing functional that bounds the total variation in x in the nonlinear system case makes it impossible to fully extend the theory of the scalar case to the system case. What we can do is to extend the scalar TVD schemes to nonlinear system cases so that the resulting scheme is TVD for the "*locally frozen*" constant coefficient system. To accomplish this, we define at each point a "*local*" system of characteristic fields, then apply the nonlinear scheme to each of the m scalar characteristic equations. Here m is the dimension of the hyperbolic system. This extension technique is a somewhat generalized version of the procedure suggested by Roe in [23].

We can further improve the resolution of the contact discontinuity by an artificial compression method (on the linearly degenerate field). We can do so by increasing the size of g_j by

$$\tilde{g}_j = \left[1 + \omega\theta_j\right] g_j, \qquad \omega > 0, \tag{5.1}$$

with

$$\theta_j = \frac{\left|\Delta_{j+1/2}w - \Delta_{j-1/2}w\right|}{\left|\Delta_{j+1/2}w\right| + \left|\Delta_{j-1/2}w\right|}, \tag{5.2}$$

where $\Delta_{j+1/2}w = w_{j+1} - w_j$, and w_j are the characteristic variables corresponding to the linearly degenerate field [3]. From numerical experiments for the Euler equations, $\omega = 2$ seems to be a good choice., For a detailed implementation of the scheme for one-dimensional Euler equations of gas dynamics, see [16].

Two spatial dimensions. Consider a two-dimensional scalar conservation law

$$\frac{\partial u}{\partial t} + \frac{\partial f(u)}{\partial x} + \frac{\partial g(u)}{\partial y} = 0. \tag{5.3}$$

Let $u_{j,k}^n$ be the numerical solution of (5.3) at $x = j\Delta x$, $y = k\Delta y$ and $t = n\Delta t$, with Δx the mesh size in the x-direction and Δy the mesh size in the y-direction. Also, let $\tilde{h}_{j+1/2,k}$ and $\tilde{q}_{j,k+1/2}$ be the second-order accurate numerical fluxes in the x- and y-directions. The explicit TVD scheme can be implemented in two space dimensions by the method of fractional steps as follows:

$$u_{j,k}^* = u_{j,k}^n - \frac{\Delta t}{\Delta x}\left(\tilde{h}_{j+1/2,k}^n - \tilde{h}_{j-1/2,k}^n\right) = \mathscr{L}_x u_{j,k}^n, \tag{5.4a}$$

$$u_{j,k}^{n+1} = u_{j,k}^* - \frac{\Delta t}{\Delta y}\left(\tilde{q}_{j,k+1/2}^* - \tilde{q}_{j,k-1/2}^*\right) = \mathscr{L}_y u_{j,k}^*, \tag{5.4b}$$

that is,

$$u_{j,k}^{n+1} = \mathscr{L}_y \mathscr{L}_x u_{j,k}^n. \tag{5.5}$$

In order to retain the original time accuracy of the method, we use a Strang type of fractional step operators, namely

$$u_{j,k}^{n+2} = \mathscr{L}_x^* \mathscr{L}_y \mathscr{L}_x \mathscr{L}_y \mathscr{L}_x^* u_{j,k}^n, \tag{5.6}$$

where $\mathscr{L}_x^*$ denotes the operator with the time step equal to $\Delta t/2$.

The above method is a formal extension of the one-dimensional TVD scheme to two dimensions. In order to get a truly two-dimensional TVD scheme, we have to include a two-dimensional modified flux. Davis [17] and Roe and Baines [18] provide some interesting ideas on this subject.

6. Numerical experiments with a second-order explicit TVD scheme. In this section, we show, by numerical experiments, the performance of a second-order accurate explicit TVD scheme (with the numerical flux defined in (3.1a), (3.1b) together with (2.16)) when applied to a two-dimensional gas dynamics problem. The same scheme has been applied to a one-dimensional shock tube problem, a quasi-one-dimensional divergent nozzle problem, and a two-dimensional regular shock reflection problem. All these calculations have been reported separately in references [3, 16, 20, 26]. Here we illustrate the result of a more complicated two-dimensional transient calculation on a shock diffraction problem. This problem was done jointly with P. Kutler of NASA Ames Research Center [20].

A good test problem for assessing the capabilities of any shock-capturing scheme is the shock-diffraction problem; i.e., the computation of the unsteady flow field resulting from a planar moving shock wave striking an obstacle. The shock-diffraction problem contains most of the flow discontinuities of the Euler equations. In the present numerical experiment the diffraction process is determined over a cylinder. The shock patterns at two instances in time t_1 and t_2 after initial impingement are sketched in Figure 6.1.

When the incident shock first collides with the cylinder, regular reflection occurs at the shock impingement point. As the impingement point of the incident shock propagates around the body, the reflection process makes a transition from regular to Mach reflection. It should be pointed out that during the transition process, complex and double Mach reflection shock structures are possible. Their occurrence is dependent on the initial strength of the incident shock wave. For single Mach reflection, a triple point forms and the incident shock no longer touches the body. Emanating from the triple point are three waves: (1) a Mach stem which strikes the body perpendicularly, (2) a slip surface or shear layer which strikes the body and results in a vortical singularity (nodal point of streamlines), and (3) the reflected shock which propagates away from the body. In addition to the above flow field characteristics, a stagnation point (saddle point of streamlines) exists at the plane of symmetry, both forward and aft on the body.

Both MacCormack's explicit method [27] and the explicit TVD scheme were applied to the shock wave-cylinder interaction problem. For a fair

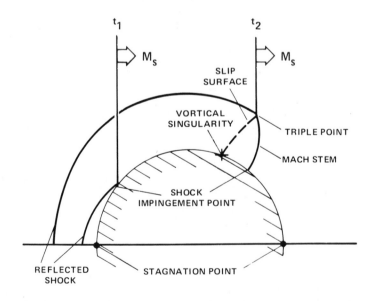

FIGURE 6.1. Shock structure for shock diffraction over cylinder at two different times.

comparison the TVD scheme was implemented in an existing computer code [28] which also contained MacCormack's method so that the same initial conditions, boundary conditions and coordinate transformation were used. A cylindrical grid consisting of 50 point around the half-cylindrical (ξ-direction) body and 51 points between the body and outer boundary (η-direction) was used. The body radius is one and the distance from the body to the outer boundary is 3. Rays from the coordinate system origin are spaced at equal angles with points uniformly placed in the radial direction between the body and the outer boundary (see Figure 6.2).

Figure 6.2 shows a schematic representation of the grid with its boundaries and initial conditions. The nodal points to the right of the planar moving shock are initialized to free stream values while those to the left are set equal to the post moving shock conditions. In the outer boundary, it is necessary to track the moving planar shock as a function of time along this boundary surface.

At the planes of symmetry, the reflection principle is used; i.e., the pressure, density and u-velocity component are treated as even functions across the plane of symmetry while the v-velocity component is treated as an odd function. The boundary condition at the surface of the cylinder must satisfy the tangency condition which requires that the velocity in the radial-direction be equal to zero at the body. Furthermore, for convenience, an image line of nodal points is considered which falls one

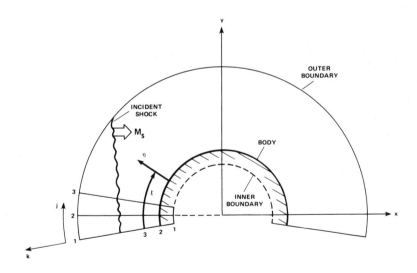

FIGURE 6.2. Schematic of the computational grid.

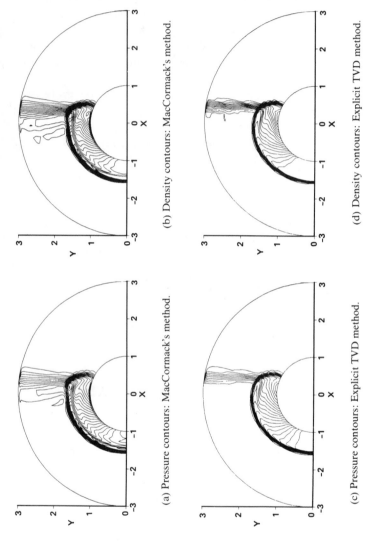

(a) Pressure contours: MacCormack's method.

(b) Density contours: MacCormack's method.

(c) Pressure contours: Explicit TVD method.

(d) Density contours: Explicit TVD method.

FIGURE 6.3. Pressure and density contours for the shock-wave cylinder interaction. ($M_8 = 2$)

mesh interval inside the body, so that the reflection principle can be applied.

MacCormack's method with a fourth-order dissipation term was run at a Courant (CFL) number of 0.6 for stabiliy, while the TVD method (with one-dimensional artificial compression terms added to both the ξ- and η-direction) was run at a Courant numbers of 0.9 for efficiency. The stability and accuracy of the TVD method are insensitive to Courant number between 0.5 and 1. Both methods were run to approximately the same total time (100 steps for the MacCormack's method, 70 steps for the TVD method). The results in the form of pressure and density contour plots are shown in Figure 6.3 at a time for which Mach reflection of the incident shock exists. The incident shock Mach number was 2. The results from MacCormack's method are shown in Figure 6.3(a), (b). Those for the TVD method are shown in Figure 6.3(c), (d). It can be seen that the TVD scheme results in a better defined flow field; i.e., "crisper" shocks and hardly any associated spurious oscillations. The slip surface which emanates from the triple point is not captured by either method. However, if we use more grid points, better control of artificial compression, and more appropriate intermediate boundary conditions procedures for the fractional step method, we think the TVD method will be able to capture the slip surface.

The result shown in Figure 6.3(c), (d) used a simple form of the $Q(z)$ function [3, 16]

$$Q(z) = \lambda z^2 + \tfrac{1}{4}, \tag{6.1a}$$

$$\sigma(z) = \tfrac{1}{8}. \tag{6.1b}$$

We found that with equation (6.1), a slightly better shock resolution was obtained than with $Q(z)$ in (2.15b) or (2.16). Note that one can get (6.1b) by simply substituting (6.1a) into (3.3b).

7. Implicit TVD schemes. Now we consider a one-parameter family of three-point conservative schemes of the form

$$u_j^{n+1} + \lambda\eta\left(h_{j+1/2}^{n+1} - h_{j-1/2}^{n+1}\right) = u_j^n - \lambda(1 - \eta)\left(h_{j+1/2}^n - h_{j-1/2}^n\right),$$
$$\tag{7.1}$$

where η is a parameter, $h_{j+1/2}^n = h(u_j^n, u_{j+1}^n)$, $h_{j+1/2}^{n+1} = h(u_j^{n+1}, u_{j+1}^{n+1})$, and $h(u_j, u_{j+1})$ is the numerical flux (2.15a), (2.15b). This one-parameter family of schemes contains implicit as well as explicit schemes. When $\eta = 0$, (7.1) reduces to (2.2), the explicit method. When $\eta \neq 0$, (7.1) is an implicit scheme. For example, if $\eta = \tfrac{1}{2}$, the time differencing is the trapezoidal formula, and if $\eta = 1$, the time differencing is the backward Euler method. To simplify the notation, we will rewrite equation (7.1) as

$$L \cdot u^{n+1} = R \cdot u^n, \tag{7.2}$$

where L and R are the following finite-difference operators:

$$(L \cdot u)_j = u_j + \lambda \eta \left(h_{j+1/2} - h_{j-1/2} \right), \qquad (7.3a)$$

$$(R \cdot u)_j = u_j - \lambda (1 - \eta) \left(h_{j+1/2} - h_{j-1/2} \right). \qquad (7.3b)$$

Sufficient conditions for (7.1) to be a TVD scheme are that

$$\mathrm{TV}(R \cdot u^n) \leqslant \mathrm{TV}(u^n) \qquad (7.4a)$$

and

$$\mathrm{TV}(L \cdot u^{n+1}) \geqslant \mathrm{TV}(u^{n+1}). \qquad (7.4b)$$

A sufficient condition for (7.4a), (7.4b) is the CFL-like restriction

$$\left| \lambda a_{j+1/2} \right| \leqslant \lambda Q(a_{j+1/2}) \leqslant 1/(1 - \eta), \qquad (7.5)$$

where $a_{j+1/2}$ is defined in (2.12). Therefore, for the scheme to be TVD, we have to pick $Q(a_{j+1/2})$ such that (7.5) is satisfied. For a detailed proof of (7.4a), (7.4b) and (7.5), see [4]. Observe that the backward Euler implicit scheme, $\eta = 1$, in (7.1) is unconditionally TVD, while the trapezoidal formula, $\eta = \frac{1}{2}$, is TVD under the CFL-like restriction of 2. The forward Euler explicit scheme, $\eta = 0$ or (2.2), is TVD under the CFL restriction of 1.

8. Second-order implicit TVD schemes. We can obtain a second-order accurate implicit TVD scheme by replacing the numerical flux function h by $\tilde{h}$ of (3.1a), (3.1b), (3.3a) and (3.3b). However, $\sigma_{j+1/2}$ is different from (3.3b). Instead, we choose

$$\sigma(z) = \begin{cases} \frac{1}{2} Q(z) + \lambda \left(\eta - \frac{1}{2} \right) z^2 & \text{(time dependent calculations)}, \\ \frac{1}{2} Q(z) & \text{(steady-state calculations)}. \end{cases} \qquad (8.1)$$

The second choice in (8.1) makes the scheme second-order accurate in space, but first-order accurate in time. This choice of $\sigma(z)$ ensures that the steady-state solution does not depend on the time step Δt. For example, the unconditionally TVD backward Euler scheme is of the form

$$u_j^{n+1} + \lambda \left(\tilde{h}_{j+1/2}^{n+1} - \tilde{h}_{j-1/2}^{n+1} \right) = u_j^n. \qquad (8.2)$$

This is a highly nonlinear implicit scheme. An efficient procedure to solve this set of nonlinear equations is needed. Here we discuss a linearized form of the implicit scheme that is suitable for steady-state calculations.

For steady-state calculations, we can use the following unconditionally TVD linearized version of (8.2);

$$d_j - \lambda(C^-)^n(d_{j+1} - d_j) + \lambda(C^+)^n(d_j - d_{j-1}) = -\lambda\left[\tilde{h}_{j+1/2} - \tilde{h}_{j-1/2}\right],$$

(8.3a)

with $d_j = u_j^{n+1} - u_j^n$, $\tilde{h}_{j+1/2}$ from (3.1a), (3.1b) (3.3a) and (3.3b), and

$$(C^{\pm})^n = \frac{1}{2}[Q(a + \gamma) \pm (a + \gamma)]_{j+1/2}^n.$$

(8.3b)

One can obtain (8.3) by simply rewriting (8.2) into an upwind form so that the resulting equation is a function $C^{\pm}(a + \gamma)_{j+1/2}^{n+1}$, $C^{\pm}(a + \gamma)_{j+1/2}^n$, $\Delta_{j+1/2}u^{n+1}$ and $\Delta_{j+1/2}u^n$, and then by dropping the time index of the $C^{\pm}$ from $(n + 1)$ to n.

Although (8.3) is formally a five-point scheme, the coefficient matrix associated with it is tridiagonal with a dominant diagonal. See [4, 16] for more details. We can also obtain another TVD linearized form

$$d_j - \lambda(B^-)^n(d_{j+1} - d_j) + \lambda(B^+)^n(d_j - d_{j-1}) = -\lambda\left[\tilde{f}_{j+1/2}^n - \tilde{f}_{j-1/2}^n\right]$$

(8.4a)

with

$$(B^{\pm})^n = \frac{1}{2}[Q(a) \pm a]_{j+1/2}^n.$$

(8.4b)

Scheme (8.4a), (8.4b) is spatially first-order accurate for the implicit operator and spatially second-order accurate for the explicit operator. It can be shown that (8.4a), (8.4b) is still TVD. Detailed implementation for the one- and two-dimensional Euler equations can be found in reference [16].

To show the stability and accuracy of the implicit method for steady-state application, we apply this method to a quasi-one-dimensional

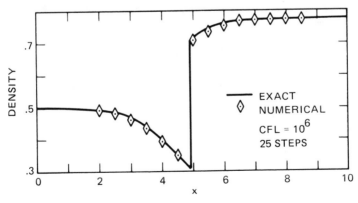

FIGURE 8.1. Second-order implicit TVD method: supersonic inflow, subsonic outflow.

divergent nozzle problem [29]. Figure 8.1 shows the converged density distribution after 25 steps at a CFL number of 10^6 using 20 equal grid spacings. The solid line is the exact solution and the diamonds are the computed solution. Only 14 points are plotted. The 6 points not shown on both ends of the x-axis are equal to the exact solution. A very accurate solution is obtained. See [16] for more details.

Formal extension of the implicit TVD scheme to two dimensions is straightforward. However, the method of solving the resulting nonlinear system of equations efficiently remains an open question. Research is underway to study this problem. Mulder and van Leer [30] suggested some useful ideas in this area.

9. Concluding remarks. Numerical experiments for the Euler equations show that the application of the second-order explicit TVD schemes generate good shock resolution for both transient and steady-state one-dimensional and two-dimensional problems. Numerical experiments for a quasi-one-dimensional nozzle problem show that the second-order implicit TVD scheme produces a fairly rapid convergence rate and remains stable even when running with a CFL of 10^6. Research is underway to investigate an efficient extension of the implicit TVD method to two dimensions.

REFERENCES

1. E. Krause (ed.), Proc. Eighth Internat. Conf. on Numerical Methods in Fluid Dynamics, Lecture Notes in Phys., Vol. 170, Springer-Verlag, Berlin and New York, 1982.

2. Proc. AIAA 6th Computational Fluid Dynamics Conf. (Danvers, Mass., July 1983).

3. A. Harten, *A high resolution scheme for the computation of weak solutions of hyperbolic conservation laws*, New York Univ. Report, October, 1981; J. Comput. Phys. **49** (1983), 357–393.

4. _____, *On a class of high resolution total-variation-stable finite-difference schemes*, New York Univ. Report, October 1982; SIAM J. Numer. Anal. **21** (1984).

5. B. van Leer, *Towards the ultimate conservative difference scheme. V. A second-order sequel to Godunov's method*, J. Comput. Phys. **32** (1979), 101–136.

6. P. Colella and P. R. Woodward, *The piecewise-parabolic method (PPM) for gas-dynamical simulations*, LBL report no. 14661, July 1982.

7. P. L. Roe, *Some contributions to the modelling of discontinuous flows*, Large-scale Computations in Fluid Mechanics (Proc. AMS-SIAM Summer Seminar on Large Scale Computations in Fluid Mechanics, Univ. of Calif., San Diego, June 27–July 8, 1983, B. Engquist, S. Osher and R. Somerville, Eds.), AMS, Providence, 1985.

8. S. Osher, *Shock modeling in transonic and supersonic flow*, Rec. Adv. in Numer. Methods in Fluids, Vol. 4, Advances in Computational Transonics (W. G. Habashi, ed.), Pineridge Press, Swansea, U.K., 1984.

9. A. Harten, *On second-order accurate Godunov-type schemes*, New York Univ. Report, 1982.

10. J. P. Boris and D. L. Book, *Flux-corrected transport. I. SHASTA, a fluid transport algorithm that works*, J. Comput. Phys. **11** (1973), 38–69.

11. A. Harten and G. Zwas, *Self-adjusting hybrid schemes for shock computations*, J. Comput. Phys. **9** (1972), 568–583.

12. A. Harten, *The artificial compression method for computation of shocks and contact discontinuities*. III. *Self-adjusting hybrid schemes*, Math. Comp. **32** (1978), 363–389.

13. B. van Leer, *Towards the ultimate conservative difference scheme*. II. *Monotonicity and conservation combined in a second-order scheme*, J. Comput. Phys. **14** (1974), 361–370.

14. R. F. Warming and R. M. Beam, *Upwind second-order difference schemes and applications in aerodynamic flows*, AIAA J. **14** (1976), 361–370.

15. P. K. Sweby, *High resolution schemes using flux limiters for hyperbolic conservation laws*, U.C.L.A. Report, June 1983.

16. H. C. Yee, R. F. Warming and A. Harten, *Implicit total variation diminishing (TVD) schemes for steady-state calculations*, Proc. AIAA 6th Computational Fluid Dynamics Conf. (Danvers, Mass., July, 1983), AIAA Paper No. 83–1902; J. Comput. Phys. (to appear).

17. S. F. Davis, *A rotationally biased upwind difference scheme for the Euler equations*, ICASE Contractor Report 172179, July 1983, Hampton, Virginia.

18. P. L. Roe and M. J. Baines, *Algorithms for advection and shock problems* Proc. 4th GAMM Conf. on Numerical Methods in Fluid Mechanics (H. Viviand, ed.) (Vieweg, 1982).

19. J. B. Goodman and R. J. LeVeque, *On the accuracy of stable schemes for 2 D scalar conservation laws*, New York Univ. Report, May 1983.

20. H. C. Yee and P. Kutler, *Application of second-order accurate total variation diminishing (TVD) schemes to the Euler equations in general geometries*, NASA TM-85845, August 1983.

21. B. Engquist and S. Osher, *Stable and entropy satisfying approximations for transonic flow calculations*, Math. Comp. **34** (1980), 45–75.

22. L. C. Huang, *Pseudo-unsteady difference schemes for discontinuous solutions of steady-state, one-dimensional fluid dynamics problems*, J. Comput. Phys. **42** (1981), 195–211.

23. P. L. Roe, *Approximate Riemann solvers, parameter vectors, and difference schemes*, J. Comput. Phys. **43** (1981), 357–372.

24. E. M. Murman, *Analysis of embedded shock waves calculated by relaxation methods*, AIAA J. **12** (1974), 626–632.

25. R. Courant, E. Isaacson and M. Rees, *On the solution of nonlinear hyperbolic differential equations by finite differences*, Comm. Pure Appl. Math. **5** (1952), 243–255.

26. H. C. Yee, R. F. Warming and A. Harten, *On the application and extension of Harten's high-resolution scheme*, NASA TM-84256, June 1982.

27. R. W. MacCormack, *The effect of viscosity in hypervelocity impact cratering*, AIAA Paper 69-354, Cincinnati, Ohio, 1969.

28. P. Kutler and A. R. Fernquist, *Computation of blast wave encounter with military targets*, Flow Simulations, Inc. Report No. 80-02, April 1980.

29. G. R. Shubin, A. B. Stephens and H. M. Glaz, *Steady shock tracking and Newton's method applied to one-dimensional duct flow*, J. Comput. Phys. **39** (1981), 364–374.

30. W. A. Mulder and B. van Leer, *Implicit upwind methods for the Euler equations*, Proc. AIAA 6th Computational Fluid Dynamics Conf. (Danvers, Mass., July 1983), AIAA Paper No. 83-1930.

NASA AMES RESEARCH CENTER, MOFFETT FIELD, CALIFORNIA 94035 (Current address of H. Yee and R. Warming)

DEPARTMENT OF MATHEMATICS, TEL - AVIV UNIVERSITY, TEL - AVIV

DEPARTMENT OF MATHEMATICS, NEW YORK UNIVERSITY, NEW YORK, NEW YORK 10012

Current address (A Harten): Department of Mathematics, University of California, Los Angeles, California 90024

Lectures in Applied Mathematics
Volume **22**, 1985

Recent Applications of Spectral Methods in Fluid Dynamics

Thomas A. Zang and M. Yousuff Hussaini[1]

ABSTRACT. Origins of spectral methods, especially their relation to the method of weighted residuals, are surveyed. Basic Fourier and Chebyshev spectral concepts are reviewed and demonstrated through application to simple model problems. Both collocation and tau methods are considered. These techniques are then applied to a number of difficult, nonlinear problems of hyperbolic, parabolic, elliptic and mixed type. Fluid dynamical applications are emphasized.

I. Introduction. Spectral methods may be viewed as an extreme development of the class of discretization schemes known by the generic name of the method of weighted residuals (MWR) [1]. The key elements of the MWR are the trial functions (also called the expansion or approximating functions) and the test functions (also known as weight functions). The trial functions are used as the basis functions for a truncated series expansion of the solution, which, when substituted into the differential equation, produces the residual. The test functions are used to enforce the minimization of the residual.

1980 *Mathematics Subject Classification*. Primary 65P30, 65F10, 65N35; Secondary 76D05, 76N10, 76F10.

Key words and phrases. Spectral methods, multigrid methods, partial differential equations, potential flow, shock-turbulence interaction, shear flow transition.

[1] Research supported by the National Aeronautics and Space Administration under NASA Contract Nos. NAS1-17130 and NASA-17070 while in residence at ICASE, NASA Langley Research Center, Hampton, Virginia 23665.

The choice of trial functions is what distinguishes the spectral methods from the finite element and finite difference methods. The trial functions for spectral methods are infinitely differentiable global functions. (Typically they are tensor products of the eigenfunctions of singular Sturm-Liouville problems.) In the case of finite element methods, the domain is divided into small elements, and a trial function is specified in each element. The trial functions are thus local in character, and well suited for handling complex geometries. The finite difference trial functions are likewise local.

The choice of test function distinguishes between the Galerkin, collocation, and tau approaches. In the Galerkin approach, the test functions are the same as the trial functions, whereas in the collocation approach the test functions are translated Dirac delta functions. In other words, the Galerkin approach is equivalent to a least squares approximation, whereas the collocation approach requires the differential equation to be satisfied exactly at the collocation points. Spectral tau methods are close to Galerkin methods but they differ in the treatment of boundary conditions.

The collocation approach is the simplest of the MWR, and appears to have been first used by Slater [2] in his study of electronic energy bands in metals. A few years later, Barta [3] applied this method to the problem of the torsion of a square prism. Frazer et al. [4] developed it as a general method for solving ordinary differential equations. They used a variety of trial functions and an arbitrary distribution of collocation points. The work of Lanczos [5] established for the first time that a proper choice of trial functions and distribution of collocation points is crucial to the accuracy of the solution. Perhaps he should be credited with laying down the foundation of the orthogonal collocation method. This method was revived by Clenshaw [6], Clenshaw and Norton [7], and Wright [8]. These studies involved application of Chebyshev polynomial expansions to initial value problems. Villadsen and Stewart [9] developed this method for boundary value problems.

The earliest investigations of the spectral collocation method to partial differential equations were those of Kreiss and Oliger [10] (who called it the Fourier method) and Orszag [11] (who termed it pseudospectral). This approach is especially attractive because of the ease with which it can be applied to variable coefficient and even nonlinear problems. The essential details will be furnished below.

The Galerkin approach is perhaps the most esthetically pleasing of the MWR since the trial functions and the test functions are the same. Indeed, the first serious application of spectral methods to PDE's—that of Silberman [12] for meteorological modelling—used the Galerkin approach. However, spectral Galerkin methods only became practical for

high resolution calculations of nonlinear problems after Orszag [13] and
Eliasen et al. [14] developed a transform method for evaluating convolu-
tion sums arising from quadratic nonlinearities. Even in this case spectral
collocation methods retain a factor of 2 in speed. For more complicated
nonlinear terms, high resolution spectral Galerkin methods are still
impractical.

The tau approach is the most difficult to rationalize within the context
of the MWR. Lanczos [5] developed the spectral tau method as a
modification of the Galerkin method for problems with nonperiodic
boundary conditions. Although it, too, is difficult to apply to nonlinear
problems, it has proven quite useful for constant coefficient problems or
subproblems, e.g., for semi-implicit time-stepping algorithms.

The following discussion of spectral methods for PDEs will be
organized around the three basic types of systems—hyperbolic, parabolic
and elliptic—with an additional section for a difficult, nonlinear problem
of mixed type. Simple, one-dimensional, linear examples will be provided
to illustrate the basic principles and details of the algorithms; two-dimensional, nonlinear examples drawn from fluid dynamical applications will
also be furnished to demonstrate the power of the method. The focus will
be on collocation methods, although some discussion of tau methods is
provided.

II. Hyperbolic equations. Linear hyperbolic equations are perhaps the
simplest setting for describing spectral collocation methods. Both Fourier
and Chebyshev schemes have found wide application. This section will
first present the fundamentals of both approaches and then illustrate
them on a nonlinear fluid dynamics problem involving shock waves.

Basic Fourier collocation concepts. The potential accuracy of spectral
methods derives from their use of suitable high-order interpolation
formulae for approximating derivatives. An elementary example is provided by the model problem

$$u_t + u_x = 0, \tag{1}$$

with periodic boundary conditions on $[0, 2\pi]$ and the initial condition

$$u(x, 0) = \sin(\pi \cos x). \tag{2}$$

The exact solution

$$u(x, t) = \sin[\pi \cos(x - t)], \tag{3}$$

has the Fourier expansion

$$u(x, t) = \sum_{k=-\infty}^{\infty} \bar{u}_k(t) e^{ikx}, \tag{4}$$

where the Fourier coefficients

$$\bar{u}_k(t) = \sin(k\pi/2)J_k(\pi)e^{-ikt}, \tag{5}$$

and $J_k(t)$ is the Bessel function of order k. The asymptotic properties of the Bessel functions imply that

$$k^p\bar{u}_k(t) \to 0 \quad \text{as } k \to \infty, \tag{6}$$

for all positive integers p. As a result, the truncated Fourier series

$$u_N(x, t) = \sum_{k=-N/2+1}^{N/2-1} \bar{u}_k(t)e^{ikx}, \tag{7}$$

converges faster than any finite power of $1/N$. This property is often referred to as exponential convergence. A straightforward integration-by-parts argument [15] may be used to show that it applies to any periodic and infinitely differentiable solution.

The standard collocation points are

$$x_j = 2\pi j/N, \quad j = 0, 1, \ldots, N - 1. \tag{8}$$

Let u_j denote the approximation to $u(x_j)$, where the time dependence has been suppressed. Then the spatial discretization of (1) is

$$\frac{\partial u_j}{\partial t} = \frac{\partial \tilde{u}}{\partial x}\bigg|_j, \tag{9}$$

where the right-hand side is determined as follows. First, compute the discrete Fourier coefficients

$$\hat{u}_k = \frac{1}{N}\sum_{j=0}^{N-1} u_j e^{-ikx_j}, \quad k = -\frac{N}{2}, -\frac{N}{2} + 1, \ldots, \frac{N}{2} - 1. \tag{10}$$

Then the interpolating function

$$\tilde{u}(x) = \sum_{k=-N/2}^{N/2-1} \hat{u}_k e^{ikx}, \tag{11}$$

can be differentiated analytically to obtain

$$\frac{\partial \tilde{u}}{\partial x}\bigg|_j = \sum_{k=-N/2+1}^{N/2-1} ik\hat{u}_k e^{ikx_j}. \tag{12}$$

(The term involving $k = -N/2$ makes a purely imaginary contribution to the sum and hence has been dropped.) Note that each derivative approximation uses all available information about the function values. The sums in (10) and (12) can be obtained in $O(N\ln N)$ operations via the fast Fourier transform (FFT).

An illustration of the superior accuracy available from the spectral method for this problem is provided in Table I. Shown there are the maximum errors at $t = 1$ for the truncated series and for the Fourier collocation method as well as for second-order and fourth-order finite difference methods. The time discretization was the classical fourth-order Runge-Kutta method. In all cases the time-step was chosen so small that the temporal discretization error was negligible. Because the solution is infinitely smooth, the convergence of the spectral method on this problem is more rapid than any finite power of $1/N$. (The error for the $N = 64$ spectral result is so small that it is swamped by the round-off error of these single precision CDC Cyber 175 calculations.) In most practical applications the benefit of the spectral method is not the extraordinary accuracy available for large N but rather the small size of N necessary for a moderately accurate solution.

TABLE I. Maximum error for a 1-D periodic problem.

N	Truncated Series	Fourier Collocation	2nd-Order Finite Difference	4th-Order Finite Difference
8	9.87 (-2)	1.62 (-1)	1.11 (0)	9.62 (-1)
16	2.55 (-4)	4.97 (-4)	6.13 (-1)	2.36 (-1)
32	1.05 (-11)	1.03 (-11)	1.99 (-1)	2.67 (-2)
64	6.22 (-13)	9.55 (-12)	5.42 (-2)	1.85 (-3)
128			1.37 (-2)	1.18 (-4)

Basic Chebyshev collocation concepts. Spectral methods for nonperiodic problems can also exhibit exponential convergence. A simple example is again provided by (1) but now on the interval $[-1, 1]$ with initial condition $u(x, 0)$ and boundary condition $u(-1, t)$. Since this is not a periodic problem, a spectral method based upon Fourier series in x would exhibit extremely slow convergence. However, rapid convergence as well as efficient algorithms can be attained for spectral methods based upon Chebyshev polynomials. These are defined on $[-1, 1]$ by

$$\tau_n(x) = \cos(n \cos^{-1} x). \tag{13}$$

The function

$$u(x, t) = \sin \alpha \pi (x - t) \tag{14}$$

is one solution to (1). It has the Chebyshev expansion

$$u(x, t) = \sum_{n=0}^{\infty} \bar{u}_n(t) \tau_n(x), \tag{15}$$

where

$$\bar{u}_n(t) = 2/c_n \sin\left(\frac{n\pi}{2} - \alpha\pi t\right) J_n(\alpha\pi),$$ (16)

with

$$c_n = \begin{cases} 2, & n = 0, \\ 1, & n \geqslant 1. \end{cases}$$ (17)

The truncated series

$$u_N(x, t) = \sum_{n=0}^{N} \bar{u}_n(t)\tau_n(x),$$ (18)

converges at an exponential rate. Note that this result holds whether or not α is an integer. In contrast, the Fourier coefficients of $u(x, t)$ are

$$\bar{u}_k(t) = \frac{i}{2\pi} e^{i\alpha\pi t} \frac{\sin \pi(\alpha + k)}{\alpha + k} - \frac{i}{2\pi} e^{-i\alpha\pi t} \frac{\sin \pi(\alpha - k)}{\alpha - k}.$$ (19)

For noninteger α these decay extremely slowly.

The change of variables

$$x = \cos \theta,$$ (20)

the definition

$$v(\theta, t) = u(\cos \theta, t),$$ (21)

and (13) reduce (15) to

$$v(\theta, t) = \sum_{n=0}^{\infty} \bar{u}_n(t) \cos n\theta.$$ (22)

Thus, the Chebyshev coefficients of $u(x, t)$ are precisely the Fourier coefficients of $v(\theta, t)$. This new function is automatically periodic. If $u(x, t)$ is infinitely differentiable (in x), then $v(\theta, t)$ will be infinitely differentiable (in θ). Hence, straightforward integration-by-parts arguments lead to the conclusion that the Chebyshev coefficients of an infinitely differentiable function will decay exponentially fast. Note that this holds regardless of the boundary conditions.

A Chebyshev spectral method makes use of the interpolating function

$$\tilde{u}(x) = \sum_{n=0}^{N} \hat{u}_n \tau_n(x).$$ (23)

The standard collocation points are

$$x_j = \cos \frac{\pi j}{N}, \qquad j = 0, 1, \ldots, N.$$ (24)

Thus,

$$u_j = \sum_{n=0}^{N} \hat{u}_n \cos \frac{n\pi j}{N},\tag{25}$$

where u_j is the approximation of $u(x_j)$. The inverse relation is

$$\hat{u}_n = \frac{2}{N\bar{c}_n} \sum_{j=0}^{N} \bar{c}_j^{-1} u_j \cos \frac{n\pi j}{N}, \qquad n = 0, 1, \ldots, N,\tag{26}$$

where

$$\bar{c}_j = \begin{cases} 2, & j = 0 \text{ or } N, \\ 1, & 1 \leqslant j \leqslant N - 1. \end{cases}\tag{27}$$

The analytic derivative of this function is

$$\frac{\partial \tilde{u}}{\partial x} = \sum_{n=0}^{N} \hat{u}_n^{(1)} \tau_n(x),\tag{28}$$

where

$$\hat{u}_{N+1}^{(1)} = 0, \qquad \hat{u}_N^{(1)} = 0,\tag{29}$$

$$\bar{c}_n \hat{u}_n^{(1)} = \hat{u}_{n+2}^{(1)} + 2(n + 1)\hat{u}_{n+1}, \qquad n = N - 1, N - 2, \ldots, 0.$$

(See [15] for the derivation of this recursion relation.) The Chebyshev spectral derivatives at the collocation points are

$$\frac{\partial \tilde{u}}{\partial x}\bigg|_j = \sum_{n=0}^{N} \hat{u}_n^{(1)} \cos \frac{\pi j n}{N}.\tag{30}$$

Special verisons of the FFT may be used for evaluating the sums in (26) and (30). The total cost for a Chebyshev spectral derivative is thus $O(N \ln N)$.

The time-stepping scheme for (1) must use the boundary conditions to update u_N (at $x = -1$) and the approximate derivatives from (30) to update u_j for $j = 0, 1, \ldots, N - 1$. Note that no special formula is required for the derivative at $j = 0$ (or $x = +1$).

Results pertaining to $\alpha = 2.5$ at $t = 1$ for a truncated Chebyshev series, a Chebyshev collocation method, a Fourier collocation method, and a second-order finite difference method are given in Table II. For this nonperiodic problem Fourier spectral methods are quite inappropriate, but the Chebyshev spectral method is far superior to the finite difference method.

The Chebyshev collocation points are the extreme points of $\tau_N(x)$. Note that they are not evenly distributed in x, but rather are clustered near the endpoints. The smallest mesh size scales as $1/N^2$. While this distribution contributes to the quality of the Chebyshev approximation and permits the use of the FFT in evaluating the series, it also places a severe time-step limitation on explicit methods for evolution equations.

TABLE II. Maximum error for a 1-D Dirichlet problem.

N	Truncated Series	Chebyshev Collocation	Fourier Collocation	Finite Difference
4	1.24 (0)	1.49 (0)	1.85 (0)	1.64 (0)
8	1.25 (-1)	6.92 (-1)	1.92 (0)	1.73 (0)
16	7.03 (-6)	1.50 (-4)	2.27 (0)	1.23 (0)
32	1.62 (-13)	3.45 (-11)	2.28 (0)	3.34 (-1)
64	1.79 (-13)	9.55 (-11)	2.27 (0)	8.44 (-2)

Application to two-dimensional, supersonic flow. Spectral methods have recently been applied successfuly to the nonlinear hyperbolic system of equations which describes a two-dimensional inviscid gas [16, 17]. The most serious complication over the simple model problems discussed above occurs when shock waves are present. If the shock occurs in the interior of the domain, then the truncated series for the discontinuous flow variables converges very slowly. Elaborate filtering strategies appear necessary to extract useful information from a calculation of such a situation [17, 18]. This difficulty disappears, however, when the shock occurs at the boundary of the domain, as in shock-fitting as opposed to shock-capturing calculations.

A schematic of the type of spectral shock-fitted calculations described below is illustrated in Figure 1. At time $t = 0$ an infinite, normal shock at $x = 0$ separates a rapidly moving, uniform fluid on the left from the fluid on the right which is in a quiescent state except for some specified fluctuation. The initial conditions are chosen so that in the absence of any fluctuation the shock moves uniformly in the positive x-direction with a Mach number (relative to the fluid on the right) denoted by M_s. In the presence of fluctuations the shock front will develop ripples. The shape of the shock is described by the function $x_s(y, t)$. The numerical calculations are used to determine the state of the fluid in the region between the shock front and some suitable left boundary $x_L(t)$, and also to determine the motion and shape of the shock front itself.

Figure 1 is taken from a shock/turbulence calculation [19] in which the downstream fluctuation is a plane vorticity wave that is periodic in y with period y_l. Because of the initial value nature of the calculation, the fluid motion behind the shock is not periodic in x, as Figure 1 makes abundantly clear. The interesting physical domain is given by

$$x_L(t) \leqslant x \leqslant x_s(y, t), \qquad 0 \leqslant y \leqslant y_l, \quad t \geqslant 0. \tag{31}$$

The change of variables

$$X = \frac{x - x_L(t)}{x_s(y, t) - x_L(t)}, \quad Y = \frac{y}{y_l}, \quad T = t, \tag{32}$$

produces the computational domain

$$0 \leqslant X \leqslant 1, \qquad 0 \leqslant Y \leqslant 1, \qquad T \geqslant 0. \tag{33}$$

The fluid motion is modeled by the two-dimensional Euler equations. In terms of the computational coordinates these are

$$Q_T + BQ_X + CQ_Y = 0, \tag{34}$$

where $Q = (P, u, v, S)^T$,

$$B = \begin{bmatrix} U & \gamma X_x & \gamma X_y & 0 \\ \dfrac{a^2}{\gamma} X_x & U & 0 & 0 \\ \dfrac{a^2}{\gamma} X_y & 0 & U & 0 \\ 0 & 0 & 0 & U \end{bmatrix}, \tag{35}$$

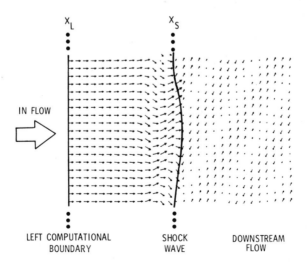

FIGURE 1. Typical shock-fitted time-dependent flow model in the physical plane.

and

$$
C = \begin{bmatrix}
V & \gamma Y_x & \gamma Y_y & 0 \\
\dfrac{a^2}{\gamma} Y_x & V & 0 & 0 \\
\dfrac{a^2}{\gamma} Y_y & 0 & V & 0 \\
0 & 0 & 0 & V
\end{bmatrix}.
\tag{36}
$$

The contravariant velocity components are given by

$$
U = X_t + uX_x + vX_y \quad \text{and} \quad V = Y_t + uY_x + vY_y.
\tag{37}
$$

A subscript denotes partial differentiation with respect to the indicated variable. P, a and S are all normalized by reference conditions at downstream infinity; u and v are velocity components in the x and y directions, both scaled by the characteristic velocity defined by the square root of the pressure-density ratio at downstream infinity. A value $\gamma = 1.4$ has been used.

Let n denote the time level and Δt the time increment. The time discretization of (34) is

$$
\tilde{Q} = [1 - \Delta t L^n]Q^n,
\tag{38}
$$

$$
Q^{n+1} = \frac{1}{2}[Q^n + (1 - \Delta t \tilde{L})\tilde{Q}],
\tag{39}
$$

where L denotes the spatial discretization of $B\,\partial_X + C\,\partial_Y$. The solution Q has the Chebyshev-Fourier series expansion

$$
Q(X, Y, T) = \sum_{p=0}^{M} \sum_{q=-N/2}^{N/2-1} Q_{pq}(T)\tau_p(\xi)e^{2\pi i q Y},
\tag{40}
$$

where $\xi = 2X - 1$. The derivatives Q_X and Q_Y are approximated by

$$
Q_X = 2\sum_{p=0}^{M} \sum_{q=-N/2}^{N/2-1} Q_{pq}^{(1,0)}(T)\tau_p(\xi)e^{2\pi i q Y},
\tag{41}
$$

$$
Q_Y = 2\pi\sum_{p=0}^{M} \sum_{q=-N/2}^{N/2-1} Q_{pq}^{(0,1)}(T)\tau_p(\xi)e^{2\pi i q Y},
\tag{42}
$$

where $Q_{pq}^{(1,0)}$ is computed from Q_{pq} in a manner analogous to (29), and

$$
Q_{qq}^{(0,1)} = iq Q_{pq}.
\tag{43}
$$

As a general rule the correct numerical boundary conditions for a spectral method are the same as the correct analytical boundary conditions. The global nature of the approximation avoids the need for special differentiation formulae at boundaries. At the same time spectral methods are quite unforgiving of incorrect boundary conditions. The inherent dissipation of these methods is so low that boundary errors quickly contaminate the entire solution. In many fluid dynamical applications the computational region must be terminated at some finite, artificial boundary. The difficulty at "artificial" boundaries is that analytically correct, fully nonlinear boundary conditions for systems are seldom known. One example of a workable artificial boundary condition for the Euler equations is given in [20].

The most critical part of the calculation is the treatment of the shock front. The shock-fitting approach used here is desirable because it avoids the severe postshock oscillations that plague shock-capturing methods. The time derivative of the Rankine-Hugoniot relations provides an equation for the shock acceleration. This equation is integrated to update the shock position (see [20] for details). This method is a generalization of the finite difference method developed by Pao and Salas [21] for their study of the shock/vortex interaction.

The nonlinear interaction of plane waves with shocks was examined at length in [19]. The numerical method used there was similar to the one described above but employed second-order finite differences in place of the present Chebyshev-Fourier spectral discretization. Detailed comparisons were made in [19] with the predictions of linear theory [22]. The linear results turned out to be surprisingly robust, remaining valid at very low (but still supersonic) Mach numbers and at very high incident wave amplitudes. The only substantial disagreement occurred for incident waves whose wave fronts were nearly perpendicular to the shock front. This type of shock-turbulence interaction is a useful test of the spectral technique because the method can be calibrated in the regions for which linear theory has been shown to be valid.

The most reliable numerical results can be obtained for the acoustic responses to acoustic waves. Unlike the vorticity responses, these require no differentiation of the flow variables, thus eliminating one extra source of error. Moreover, the acoustic response stretches much further behind the shock that the vorticity response, thus providing greater statistical reliability. Vorticity response results are reported in [23]. The incident pressure wave is taken to be

$$p_1' = A_1' e^{i(\mathbf{k}_1 \cdot \mathbf{x} - \omega_1 t)} \tag{44}$$

where $\mathbf{k}_1 = (k_{1,x}, k_{1,y})$, $\omega = M_s a_1 k_{1,x} + a_1 k_1$ and A_1' is the amplitude. In terms of the incidence angle θ_1, $\mathbf{k}_1 = (k_1 \cos \theta_1, k_1 \sin \theta_1)$. The linearized transmitted acoustic wave can be expressed in the same manner with all subscripts changed from 1 to 2. The amplification coefficient for the transmitted acoustic wave is then the ratio

$$A'_2/A'_1. \tag{45}$$

Figure 2 indicates the transmission coefficient extracted from the computation. At each fixed value of X we perform a Fourier analysis in Y of the pressure. The Fourier coefficient for $q = 1$ provides the amplitude A'_2. In order to reduce the transients that would accompany an abrupt start of the calculation at full wave amplitude, an extra factor of $s(t)$ is inserted into (44), where

$$s(t) = \begin{cases} 3(t/t_s)^2 - 2(t/t_s)^3, & 0 \leqslant t \leqslant t_s, \\ 1, & t \geqslant t_s. \end{cases} \tag{46}$$

The start-up time t_s is some multiple (typically $\frac{1}{2}$) of the time it takes the shock to encounter one full wavelength (in the x-direction) of the incident wave. The ratio A'_2/A'_1 is plotted in Figure 2 as a function of the mean value of the physical coordinate x corresponding to X. The start-up time for this Mach 3 case is $t_s = 0.56$. The average of the x-dependent responses between the start-up interval and the shock produces the computed transmission coefficient. The standard deviation of the individual responses serves as an error estimate.

The dependence upon incidence angle of the acoustic transmission coefficient for $A'_1 = 0.001$ and $M_s = 3$ waves is displayed in Figure 3. As is discussed in [19], linear theory is quite reliable at angles below, say, 45°. Figure 3 contains results from both spectral and finite difference calculations. The finite difference results were obtained with the same

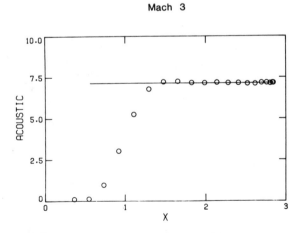

FIGURE 2. Post-shock dependence of the pressure response to a pressure wave incident at 10° to a Mach 3 shock. The solid line is the linear theory prediction. The circles are the spectral solution.

second-order MacCormack's method that was described in [19] except that periodic boundary conditions (rather than stretching) were employed in the y-direction. The finite difference grid was 64×16 and these calculations used a CFL number of 0.70. The spectral grid was 32×8, and the CFL number was 0.50. Figure 3 shows that both methods produce the same results. A head-to-head comparison of both methods for the $\theta_1 = 10°$ case is provided in Table III. The "exact" value is taken from linear theory [22]. Since the amplitude of the incident acoustic wave is so small, it should come as no surprise that four points in the y-direction suffice for the spectral calculation. Note that the standard deviations are substantially smaller for the spectral method. These results suggest that the spectral method requires only half as many grid points in each coordinate direction.

TABLE III. Grid dependence of acoustic transmission coefficient.

Grid	Finite Difference	Chebyshev-Fourier Spectral
16×4	6.403 ± 2.652	7.257 ± 0.587
16×8	6.427 ± 2.626	7.257 ± 0.587
32×4	7.105 ± 0.453	7.158 ± 0.022
32×8	7.134 ± 0.471	7.158 ± 0.022
32×16	7.139 ± 0.497	7.158 ± 0.022
64×16	7.163 ± 0.078	7.157 ± 0.017
128×16	7.152 ± 0.022	
"exact"	7.156	7.156

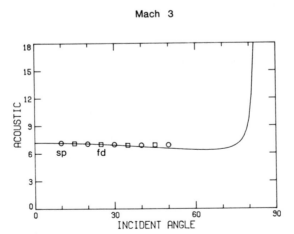

FIGURE 3. Dependence on incident angle of the pressure response to a 0.1% amplitude pressure wave incident on a Mach 3 shock. The solid line is the linear theory result. Circles are spectral solutions; squares are finite difference solutions.

III. Parabolic equations. The nonlinear, parabolic system formed by the incompressible, Navier-Stokes equations was the focus of much of the early development and application of spectral methods to large-scale fluid dynamical problems. Fourier spectral methods have been the obvious choice for the simulation of homogeneous, isotropic turbulence [24]. For shear flows, however, nonperiodic boudnary conditions are required. So far, Chebyshev spectral methods have been favored for these applications [25–27]. This section will present a discussion of the implementation of Chebyshev spectral tau methods and will then illustrate them for the one-dimensional heat equation. This section will close with a description of a promising semi-implicit time-stepping scheme for the Navier-Stokes equations.

Basic Chebyshev tau concepts. The heat equation

$$\frac{\partial u}{\partial t} = \frac{\partial^2 u}{\partial x^2}, \tag{47}$$

is the natural parabolic linear model problem. The spatial domain is $[-1, 1]$, the initial condition is

$$u(x, 0) = \sin \pi x, \tag{48}$$

and the boundary conditions are

$$u(-1, t) = 0, \qquad u(+1, t) = 0. \tag{49}$$

The exact solution is then

$$u(x, 0) = e^{-\pi^2 t} \sin \pi x. \tag{50}$$

The time differencing is again the classical fourth-order Runge-Kutta scheme.

In addition to spectral collocation and series truncation solutions, we will also present spectral tau results. Let $\bar{u}_n(t)$ for $n = 0, 1, \ldots, N$ denote the Chebyshev coefficients of the tau approximation to $u(x, t)$. The semidiscrete tau equations are

$$d\bar{u}_n/dt = \bar{u}_n^{(2)}, \qquad n = 0, 1, \ldots, N - 2, \tag{51}$$

with

$$\sum_{\substack{n=0 \\ n\,\text{even}}}^{N} \bar{u}_n = 0, \qquad \sum_{\substack{n=1 \\ n\,\text{odd}}}^{N} \bar{u}_n = 0. \tag{52}$$

The Chebyshev coefficients of the approximation to the second spatial derivative $\bar{u}_n^{(2)}(t)$ can be obtained from $\bar{u}_n(t)$ by two applications of the recursion relation in (29). In this tau approximation the dynamical equations for the two highest-order coefficients are dropped in favor of the equations for the boundary conditions. Equation (52) follows from the property

$$\tau_n(\pm 1) = (\pm 1)^n. \tag{53}$$

The results at $t = 1$ are given in Table IV. The maximum errors shown there have been boosted up by the factor e^{π^2} so that they represent relative errors. On the whole the collocation results are the best. It goes almost without saying that finite difference results are far inferior to any of these spectral approximations.

TABLE IV. Maximum error for Chebyshev approximations to the heat equation.

N	Truncated Series	Tau	Collection
8	2.44 (-4)	1.61 (-3)	4.58 (-4)
10	5.76 (-6)	2.12 (-5)	8.25 (-6)
12	9.42 (-8)	3.19 (-7)	1.01 (-7)
14	1.14 (-9)	3.35 (-9)	1.10 (-9)
16	1.05 (-11)	8.39 (-11)	2.09 (-11)

The time-step restriction for explicit Chebyshev methods for the heat equation is very severe, scaling as $1/N^4$. This can pose quite a barrier to large-scale calculations for which a relative accuracy of 0.1% or so may be required. Fortunately, many large-scale calculations can be split into one-dimensional, inhomogeneous counterparts of (47) and efficient implicit schemes are available for this linear, constant coefficient equation. They rely on reducing the Chebyshev tau equations to a system which is nearly tridiagonal. The Chebyshev tau equations for a Crank-Nicolson temporal discretization of (47) are

$$\frac{\lambda c_{n-2}}{4n(n-1)}\bar{u}_{n-2} + \left[1 - \frac{\lambda e_{n+2}}{2(n^2-1)}\right]\bar{u}_n + \frac{\lambda e_{n+4}}{4n(n+1)}\bar{u}_{n+2} \tag{54}$$

$$= \frac{c_{n-2}}{4n(n-1)}\bar{f}_{n-2} - \frac{e_{n+2}}{2(n^2-1)}\bar{f}_n + \frac{e_{n+4}}{4n(n+1)}\bar{f}_{n+2},$$

$$n = 2, 3, \ldots, N,$$

where $\lambda = -\Delta t/2$ with Δt the time-step, the coefficients $\bar{u}_n$ on the left-hand side, are at $t + \Delta t$,

$$\bar{f}_n = \bar{u}_n(t) + \tfrac{1}{2}\Delta t\bar{u}_n^{(2)}(t), \tag{55}$$

and

$$e_n = \begin{cases} 1, & 0 \leqslant n \leqslant N, \\ 0 & n > N. \end{cases} \tag{56}$$

Equation (54), for even n, plus the first equation of (52) form a linear system which is tridiagonal except for the boundary condition equation. This is cheap to invert. The odd coefficients display a similar structure.

Application to channel flow. Several three-dimensional Navier-Stokes algorithms have been developed, which incorporate the quasi-tridiagonal structure of the Chebyshev tau equations for the second derivative in semi-implicit schemes which treat the constant coefficient diffusion term implicitly [25–27]. In practice this device has permitted time-steps several orders of magnitude larger than the explicit diffusion limit. Unfortunately, the quasitridiagonal structure is lost even for a linear, variable viscosity coefficient. An effective iterative scheme for this more general case, developed in collaboration with M. Malik, will be described here in its two-dimensional setting.

The rotation form equations for two-dimensional channel flow are

$$u_t - v(v_x - u_y) + P_x = (\mu u_x)_x + (\mu u_y)_y,$$

$$v_t + u(v_x - u_y) + P_y = (\mu v_x)_x + (\mu v_y)y,$$

$$u_x + v_y = 0, \tag{57}$$

with periodic boundary conditions in x and no-slip boundary conditions at $y = \pm 1$. The variable P denotes the total pressure. The viscosity μ is presumed to depend upon y.

A useful discretization employs Fourier series in x and Chebyshev series in y. The pressure gradient term and the incompressibility consraint are best handled implicitly. So, too, are the vertical diffusion terms because of the fine mesh-spacing near the wall. The variable viscosity prevents the standard Poisson equation for the pressure from decoupling from the velocities in the diffusion term. The algorithm described in [26] appears to be a good starting point. A Crank-Nicolson approach is used for the implicit terms and Adams-Bashforth for the remainder. After a Fourier transform in x, the equations for each wave number k have the following implicit structure:

$$\hat{u} - \tfrac{1}{2}\Delta t(\mu \hat{u}_y)_y + \tfrac{1}{2}\Delta t ik\hat{P} = \cdots,$$

$$\hat{v} - \tfrac{1}{2}\Delta t(\mu \hat{v}_y)_y + \tfrac{1}{2}\Delta t\hat{P}_y = \cdots, \tag{58}$$

$$ik\hat{u} + \hat{v}_y = 0.$$

Fourier transformed variables are denoted by hats, the subscript y denotes a Chebyshev spectral derivative, and Δt is the time increment.

The algorithm in [26] was devised for constant viscosity, in which case (58) can be reduced to essentially a block-tridiagonal form. This cannot be done in the present, more general situation. We advocate solving these equations iteratively after applying a finite difference preconditioning.

The interesting physical problems have low viscosity. Thus the first derivative terms in (58) predominate. The effective preconditioning of them is crucial. Four possibilities have been considered. The eigenvalues of preconditioned iterations for the model scalar problem $u_x = f$ with periodic boundary conditions on $[0, 2\pi]$ are given for each possibility in Table V. The term $\alpha \Delta x$ is the product of a wave number α and the grid spacing Δx. It falls in the range $0 \leqslant |\alpha \Delta x| \leqslant \pi$. For the staggered grid case the discrete equation (58) are modified so that the velocities and the

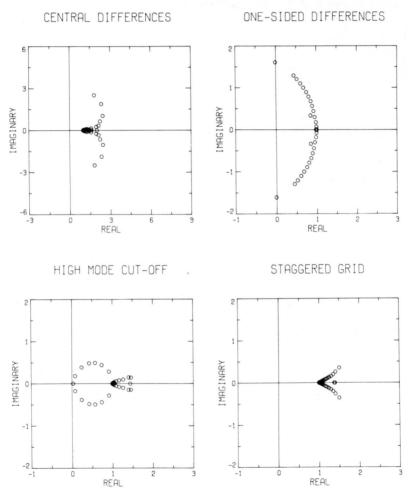

FIGURE 4. Eigenvalues of the preconditioned matrices for semi-implicit channel flow when the streamwise wave number $k = 5$. The grid is 32×16, the viscosity is $(7500)^{-1}$, and the CFL number is 0.10. Note the different scale used for the central differences preconditioning results.

momentum equations are defined at the cell faces $y_j = \cos(\pi j/N)$, $j = 0, 1, \ldots, N$, whereas the pressure and the continuity equation are defined at the cell centers $y_{j-1/2} = \cos \pi (j - \frac{1}{2})/N$, $j = 1, \ldots, N$. Fast cosine transforms enable interpolation between cell faces and cell centers to be implemented efficiently. The staggered grid for the Navier-Stokes equations has the advantage that no artificial boundary condition is required for the pressure at the walls.

The actual eigenvalues for preconditioned iterations of (58) are displayed in Figure 4. The model problem estimates the eigenvalue trends surprisingly well considering that it is just a scalar equation, has only first derivative terms, and uses Fourier series rather than Chebyshev polynomials.

The preceding results indicate that the staggered grid leads to the most effective treatment of the first derivative terms. The condition number of the preconditioned system is reasonably small and no resolution is lost by a high mode cut-off. (Although it is possible to devise a high mode cut-off which avoids the small eigenvalues shown in the figures, some of the spectral resolution is thereby lost.) A simple and effective iterative scheme for this system with its complex eigenvalues is a minimum residual method. At a viscosity of $(7500)^{-1}$ each iteration reduces the residual by almost an order of magnitude.

TABLE V. Preconditioned eigenvalues for a one-dimensional first derivative model problem.

Preconditioning	Eigenvalues
Central Differences	$\dfrac{\alpha\Delta x}{\sin(\alpha\Delta x)}$
One-sided Differences	$e^{-i(\alpha\Delta x/2)}\dfrac{\alpha\Delta x/2}{\sin((\alpha\Delta x)/2)}$
High Mode Cut-off	$\begin{cases} \dfrac{\alpha\Delta x}{\sin(\alpha\Delta x)}, & 0 \leqslant \lvert\alpha\Delta x\rvert \leqslant 2\pi/3, \\ 0, & 2\pi/3 < \lvert\alpha\Delta x\rvert \leqslant \pi, \end{cases}$
Staggered Grid	$\dfrac{(\alpha\Delta x)/2}{\sin((\alpha\Delta x)/2)}$

An equilibrium (time independent) solution to (57) is

$$u(x, y, t) = 1 - y^2, \quad v(x, y, t) = 0,$$

$$P(x, y, t) = -2\mu x + \tfrac{1}{2}(1 - y^2)^2. \tag{59}$$

The evolution of small perturbations from this state has been studied extensively by analytical means. It has been customary to focus on a particular wave number α in the x-direction and to look for solutions of the form

$$u(x, y, t) = (1 - y^2) + \varepsilon \operatorname{Re}\{\phi(y)e^{i\alpha y - i\omega t}\}. \tag{60}$$

The function $\phi(y)$ and the temporal frequency ω are solutions to the Orr-Sommerfeld eigenvalue problem. One would expect the numerical solutions to (57), subject to initial conditions taken from (60) at $t = 0$ for small amplitude ε, to be approximated closely by (60).

This is the sort of problem that will be used here as a more elaborate illustration of the power of spectral methods. The particular problem chosen for study has $\mu = (7500)^{-1}$ and $\alpha = 1$; the eigenfrequency of the only growing mode is $\omega = \omega_r + i\omega_i$, where $\omega_r = .24989154$ and $\omega_i = .00223497$. The amplitude parameter $\varepsilon = 0.0001$. The natural choice of the streamwise periodicity length is $L_x = 2\pi/\alpha = 2\pi$. Three discretizations in x were used: (1) Fourier collocation (FS), (2) second-order finite differences (FD2), and (3) fourth-order finite differences (FD4). In y the options were (1) Chebyshev collocation (CB), and (2) second-order finite differences (FD2). All grids were uniform in x and used the Chebyshev collocation distribution in y. The time-step was chosen so small that the spatial errors predominated. All runs were terminated at $t = 50.2876$. This is twice the time required for the wave to propagate through the horizontal grid. The error in the ratio of the final to the initial perturbation kinetic energy measures the accuracy of the calculation. The linear prediction for this ratio is 1.25205. The results are given in Tables VI and VII. The Fourier-Chebyshev calculation on the 4×64 grid is so accurate that both time-stepping errors and nonlinearities dominate the listed error.

TABLE VI. Accuracy of vertical discretization for $N_x = 4$.

N_y	FS-FD2 Error	FS-CB Error
16	$-.79322$	-1.11895
32	$-.98480$	$.00087$
64	$-.43564$	$.00009$
128	$-.12983$	
256	$-.03384$	

TABLE VII. Accuracy of horizontal discretization for $N_y = 32$.

N_x	FD2-CB Error	FD4-CB Error	FS-CB Error
4	$-.24835$	.04334	.00087
8	.04246	.00804	.00087
16	.02122	.00135	.00087

IV. Elliptic equations. Fruitful nonlinear applications of spectral methods developed the latest for equations of elliptic type. Unlike hyperbolic or parabolic equations, for which explicit schemes can often be tolerated, elliptic equations virtually require implicit iterative schemes in practical situations. It was only a few years ago that Orszag [28] (see also [29]) proposed preconditioning the spectral collocation equations by finite difference operators. More recently still, effective spectral multigrid iterative methods have been developed [30, 31] and applied to the nonlinear potential flow problem of fluid dynamics [32]. These developments will be described in this section.

Poisson's equation. As usual the discussion will begin with a linear model problem, but this time in two spatial dimensions. That problem is the Poisson's equation

$$\frac{\partial^2 u}{\partial x^2} + \frac{\partial^2 u}{\partial y^2} = f, \tag{61}$$

on the square $[-1, 1] \times [-1, 1]$ with homogeneous Dirichlet boundary conditions. The choice

$$f(x, y) = -2\pi^2 \sin \pi x \sin \pi y, \tag{62}$$

corresponds to the analytical solution

$$u(x, y) = \sin \pi x \sin \pi y. \tag{63}$$

Chebyshev spectral methods are appropriate for this problem. Direct solution schemes for the Chebyshev tau method have been described in [33]. They are basically of an alternating direction implicit (ADI) nature and rely on the quasitridiagonal form of the constant coefficient, one-dimensional problem. Haidvogel and Zang [33] report comparisons of the Chebyshev tau method with finite difference methods on numerous problems. They discuss both computational efficiency and accuracy.

These direct solution schemes cannot feasibly be extended to spectral collocation methods because the collocation equations for the one-dimensional components cannot be represented by sparse matrices. However, an ADI iterative scheme based on finite difference preconditioning is an efficient method for obtaining an approximate solution. The description of this scheme in its general nonlinear setting begins by writing the spectral collocation equations as

$$M(U) = 0. \tag{64}$$

Define the Jacobian

$$J(U) = \frac{\partial M}{\partial U}(U). \tag{65}$$

In many cases the Jacobian can be split into the sum of two operators $J_x(U)$ and $J_y(U)$, each involving derivatives in only the one coordinate direction indicated by the subscript. The most straightforward ADI method is

$$\left[\alpha I - J_x(V)\right]\left[\alpha I - J_y(V)\right]\Delta V = \alpha M(V), \tag{66}$$

with the approximate solution V updated by

$$V \leftarrow V + \omega \Delta V. \tag{67}$$

This is just the Douglas-Gunn version of ADI [34]. The term approximate factorization is commonly used for this type of scheme for the nonlinear potential flow problem [35]. This particular scheme is referred to as AF1. For second-order spatial discretizations, the term $[\alpha I - J_x(V)]$ leads to a set of tridiagonal systems, one for each value of y. The second left-hand side factor produces another set of tridiagonal systems. For spectral discretizations, however, these systems are full; hence, (66) is still relatively expansive to invert. A compromise is to replace J_x and J_y with their second-order finite difference analogs, denoted by H_x and H_y, respectively:

$$\left[\alpha I - H_x(V)\right]\left[\alpha I - H_y(V)\right]\Delta V = \alpha M(V). \tag{68}$$

The spectral approximate factorization scheme consists of (67) and (68). The choice of the iteration parameters is discussed in [32].

TABLE VIII. Maximum error for Chebyshev approximations to Poisson's equation.

N	Truncated Series	Tau	Collocation
8	2.88 (-4)	2.79 (-3)	1.17 (-4)
10	6.79 (-6)	5.26 (-5)	2.33 (-6)
12	1.09 (-7)	8.86 (-7)	3.12 (-8)
14	1.34 (-9)	1.09 (-8)	3.27 (-10)
16	1.19 (-11)	9.15 (-11)	2.73 (-12)

The results for the simple model problem are presented in Table VIII. The trend is the same as it was for the heat equation: the collocation method is more accurate than tau. (Since it is not practical to design a spectral method for PDEs using truncated series, those results have been ignored in this comparison.)

Spectral multigrid methods. Iterative schemes for spectral collocation equations, such as AF1, can be accelerated dramatically by applying multigrid concepts. This technique has been extensively developed for finite difference and finite element discretizations [36] and has recently been applied to spectral discretizations [30–32]. Briefly put, multigrid methods take advantage of a property shared by a wide variety of relaxation schemes—potential efficient reduction of the high-frequency error components but unavoidable slow reduction of the low-frequency components.

The fundamentals of spectral multigrid are perhaps easiest to grasp for the simple model problem

$$-\frac{d^2u}{dx^2} = f, \tag{69}$$

on $[0, 2\pi]$ with periodic boundary conditions. The Fourier approximation to the left-hand side of (69) at the collocation points is

$$\sum_{p=-N/2+1}^{N/2-1} p^2 \hat{u}_p e^{ipx_j}. \tag{70}$$

The spectral approximation to (69) may be expressed as

$$LU = F, \tag{71}$$

where

$$U = (u_0, u_1, \ldots, u_{N-1}), \tag{72}$$

$$F = (f_0, f_1, \ldots, f_{N-1}), \tag{73}$$

and L represents the Fourier spectral approximation to $-d^2/dx^2$.

A Richardson's iterative scheme for solving (71) is

$$V \leftarrow V + \omega(F - LV), \tag{74}$$

where ω is a relaxation parameter. On the right side of the replacement symbol ($\leftarrow$), V represents the current approximation to U, and on the left it represents the updated approximation. The eigenfunctions of L are

$$\xi_j(p) = e^{2\pi ijp/N}, \tag{75}$$

with the corresponding eigenvalues

$$\lambda(p) = p^2, \tag{76}$$

where $j = 0, 1, \ldots, N - 1$ and $p = -N/2 + 1, \ldots, N/2 - 1$. The index p has a natural interpretation as the frequency of the eigenfunction.

The error at any stage the iterative process is $V - U$; it can be resolved into an expansion in the eigenvectors of L. Each iteration reduces the pth error component to $\nu(\lambda_p)$ times its previous value, where

$$\nu(\lambda) = 1 - \omega\lambda. \tag{77}$$

The optimal choice of ω results from minimizing $|\nu(\lambda)|$ for $\lambda \in [\lambda_{min}, \lambda_{max}]$, where $\lambda_{min} = 1$ and $\lambda_{max} = N^2/4$. (One need not worry about the $p = 0$ eigenfunction since it corresponds to the mean level of the solution, which is at one's disposal for this problem.) The optimal relaxation parameter for this single-grid procedure is

$$\omega_{SG} = \frac{2}{\lambda_{max} + \lambda_{min}}. \tag{78}$$

It produces the spectral radius

$$\rho_{SG} = \frac{\lambda_{max} - \lambda_{min}}{\lambda_{max} + \lambda_{min}}. \tag{79}$$

Unfortunately, $\rho_{SG} \simeq 1 - 8/N^2$, which implies that $O(N^2)$ iterations are required to achieve convergence.

This slow convergence is the outcome of balancing the damping of the lowest-frequency eigenfunction with that of the highest-frequency one in the minimax problem described after (77). The multigrid approach takes advantage of the fact that the low-frequency modes ($|p| < N/4$) can be represented just as well on coarser grids. It settles for balancing the middle-frequency eigenfunction ($|p| = N/4$) with the highest-frequency one ($|p| = N/2$), and hence damps effectively only those modes which cannot be resolved on coarser grids. In (78) and (79), λ_{min} is replaced with $\lambda_{mid} = \lambda(N/4)$. The optimal relaxation parameter in this context is

$$\omega_{MG} = \frac{2}{\lambda_{max} + \lambda_{mid}}. \tag{80}$$

The multigrid smoothing factor

$$\mu_{MG} = \frac{\lambda_{max} - \lambda_{mid}}{\lambda_{max} + \lambda_{mid}} \tag{81}$$

measures the damping rate of the high-frequency modes. In this example $\mu_{MG} = 0.60$, independent of N. The price of this effective damping of the high-frequency errors is that the low-frequency errors are hardly damped at all. Table IX compares the single-grid and multigrid damping factors for $N = 64$. However, on a grid with $N/2$ collocation points, the modes for $|p| \in [N/8, N/4]$ are now the high-frequency ones. They get damped on this grid. Still coarser grids can be used until relaxations are so cheap that one can afford to damp all the remaining modes, or even to solve the discrete equations exactly. For the case illustrated in Table IX the high-frequency error reduction in the multigrid context is roughly 250 times as fast as the single-grid reduction for $N = 64$.

TABLE IX. Damping factors for $N = 64$.

p	Single Grid	Multigrid
1	.9980	.9984
2	.9922	.9938
4	.9688	.9750
8	.8751	.9000
12	.7190	.7750
16	.5005	.6000
20	.2195	.3750
24	.1239	.1000
28	.5298	.2250
32	.9980	.6000

Let us consider just the interplay between two grids. A general, nonlinear fine-grid problem can be written

$$L^f(U^f) = F^f. \tag{82}$$

The shift to the coarse grid occurs after the fine-grid approximation V^f has been sufficiently smoothed by the relaxation process, i.e., after the high-frequency content of the error $V^f - U^f$ has been sufficiently reduced. The related coarse-grid problem is

$$L^c(U^c) = F^c, \tag{83}$$

where

$$F^c = R[F^f - L^f(V^f)] + L^c(RV^f). \tag{84}$$

The restriction operator R interpolates a function from the fine grid to the coarse grid. The coarse-grid operator and solution are denoted by L^c and U^c, respectively. After an adequate approximation V^c to the coarse-grid problem has been obtained, the fine-grid approximation is corrected via

$$V^f \leftarrow V^f + P(V^c - RV^f). \tag{85}$$

The prolongation operator P interpolates a function from the coarse grid to the fine grid.

A complete multigrid algorithm requires specific choices of the interpolation operators, the coarse-grid operators, and the relaxation schemes. These issues are discussed at length in [30–32] for both Fourier and Chebyshev multigrid methods. Numerous linear, variable coefficient examples are also provided there. The more interesting nonlinear examples from [32] are the subject of the remainder of this paper.

Application to two-dimensional potential flow. Until the recent work of Streett [37], the discretization procedures for the potential equation were invariably based on low-order finite difference or finite element methods. Streett used a spectral discretization of the full potential equation and obtained its solution by a single-grid iterative technique. The application of spectral multigrid techniques by Streett et al. [32] produced a dramatic acceleration of the iterative scheme. Even in its relatively primitive state the spectral multigrid scheme is competitive, and in some cases unequivocally more efficient, than standard finite difference schemes.

After a conformal mapping from the surface of an airfoil to a circle, the potential equation becomes

$$\frac{\partial}{\partial R}\left(R\rho \frac{\partial G}{\partial R} \right) + \frac{\partial}{\partial \Theta}\left(\frac{\rho}{R} \frac{\partial G}{\partial \Theta} \right) = 0. \tag{86}$$

where G is the reduced potential, R and Θ are the computational polar coordinates, and ρ is the fluid density. The reduced potential is periodic in θ, and it satisfies

$$\frac{\partial G}{\partial R} = 0 \quad \text{at } R = 1, \tag{87}$$

$$G \to 0 \quad \text{as } R \to \infty, \tag{88}$$

and the Kutta condition. The density is given by the isentropic relation

$$\rho = \left[1 - \frac{\gamma - 1}{2} M_\infty^2 \left(q_r^2 + q_\theta^2 - 1 \right) \right]^{1/(\gamma - 1)}; \tag{89}$$

the ratio of specific heats is denoted by γ, and M_∞ is the Mach number at infinity. The velocity components in the physical (r, θ) plane are

$$q_r = \frac{1}{\mathcal{H}} \frac{\partial \Phi}{\partial R}, \qquad q_\theta = \frac{1}{R\mathcal{H}} \frac{\partial \Phi}{\partial \Theta}. \tag{90}$$

and the Jacobian between the complex physical plane ($z = re^{i\theta}$) and the complex computational plane ($\sigma = Re^{i\Theta}$) is

$$\mathcal{H} = \left| \frac{dz}{d\sigma} \right|. \tag{91}$$

Further details are provided in [37].

The spectral method employs a Fourier series representation in Θ. Constant grid spacing in Θ corresponds to a convenient dense spacing in the physical plane at the leading and trailing edges. The domain in R (with a large, but finite outer cutoff) is mapped onto the standard Chebyshev domain $[-1, 1]$ by an analytical stretching transformation

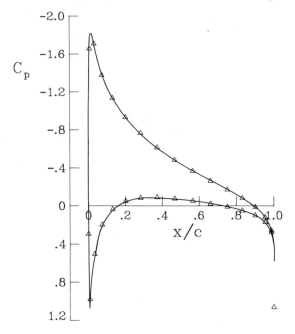

FIGURE 5. Spectral (triangles) and finite difference (solid line) surface pressures for a subcritical flow.

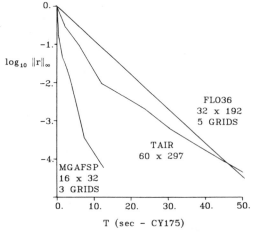

FIGURE 6. Maximum residual versus machine time for a subsonic flow.

that clusters the collocation points near the airfoil surface. The stretching is so severe that the ratio of the largest-to-smallest radial intervals is typically greater than 1000.

The flow past an NACA 0012 airfoil at 4° angle of attack and a freestream Mach number of 0.5 is a challenging subsonic and thus elliptic case. Nevertheless, the spectral solution on a relatively coarse grid captures all the essential details of the flow. The surface pressure coefficient from the spectral code MGAFSP [32] using 16 points in the radial (R) direction, and 32 points in the azimuthal (Θ) direction is displayed in Figure 5. The symbols denote the solution at the collocation points. For comparison, the result from the finite difference, multigrid, approximate factorization code FLO36 [38] is shown as a solid line. The grid used in the benchmark finite difference calculation is so fine (64 × 384 points) that the truncation error is well below plotting accuracy. The FLO36 and MGAFSP results are identical to plotting accuracy. The spectral computation on this mesh yields a lift coefficient with truncation error less than 10^{-4}. Spectral solutions on a 16 × 32 grid are thus of more than adequate resolution and accuracy for subsonic flows.

In Figure 6 are shown convergence histories from FLO36, MGAFSP, and the finite difference, approximate factorization, single-grid code TAIR [39]. Meshes which yield approximately equivalent accuracy were chosen. The surface pressure results are the same to plotting accuracy, the lift coefficient is converged in the third decimal place, and the predicted drag coefficient is less than .001. (Actually, the spectral result is an order magnitude more accurate than these limits, but the TAIR result

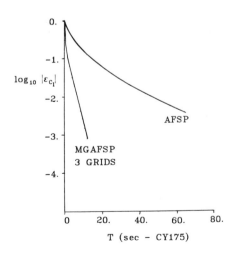

FIGURE 7. Error in lift versus machine time for a subsonic flow from single-grid (AFP) and multigrid (MGAFSP) spectral schemes.

barely meets them.) Figure 7 demonstrates the improvement produced by
the spectral multigrid scheme over the spectral single-grid method (AFSP).
There is well over an order-of-magnitude gain in efficiency.

   V. A mixed equation. The potential flow problem is much more
difficult whenever the flow field contains both supersonic (hyperbolic)
and subsonic (elliptic) regions. Nevertheless, the spectral multigrid algorithm that succeeded for the subsonic flow case requires only a minor
modification in order to succeed for the transonic (mixed) problem as
well.

   The most expedient technique for dealing with the mixed elliptic-hy-
perbolic nature of the transonic problem is to use the artificial density

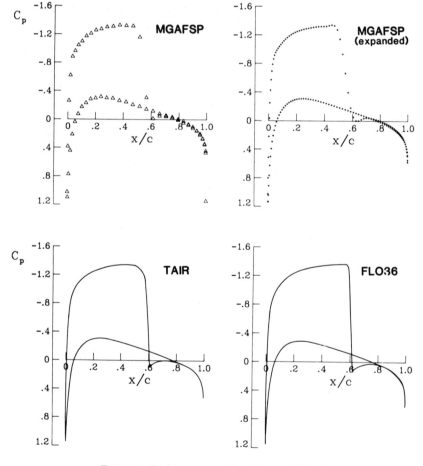

FIGURE 8. Surface pressures for a transonic flow.

approach of Hafez et al. [**40**]. The original artificial density is

$$\tilde{\rho} = \rho - \mu \overleftarrow{\delta}\rho, \tag{92}$$

with

$$\mu = \max\{0, 1 - 1/M^2\}, \tag{93}$$

where M is the local Mach number and $\overleftarrow{\delta}\rho$ is an upwind first-order (undivided) difference. The spectral calculations employed a higher-order artificial density formula. The spectral method also required a weak filtering technique to deal with some high-frequency oscillations generated by the shock. Details are available in [**37**].

A lifting transonic case is provided by the NACA 0012 airfoil at $M_\infty = 0.75$ and 2° angle of attack. A shock appears only on the upper surface for these conditions and is rather strong for a potential calculation; the normal Mach number ahead of the shock is about 1.36. Lifting transonic cases are especially difficult for spectral methods, since the solution will always have significant content in the entire frequency spectrum: the shock populates the highest frequencies of the grid and the lift is predominantly on the scale of the entire domain. An iterative scheme therefore must be able to damp error components across the spectrum.

Surface pressure distributions from MGAFSP, TAIR, and FLO36 are shown in Figure 8. The respective computational grids are 18×64, 30×149 and 32×192. The latter two are the default grids for the

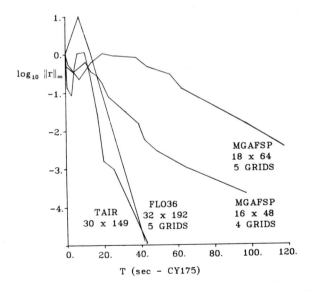

FIGURE 9. Maximum residual versus machine time for a transonic flow.

production finite difference codes. Spectral results obtained by trigono-metrically interpolating the 18×64 grid results onto a much finer grid are included alongside the results at the collocation points. This reveals the wealth of detail that is provided by the rather coarse spectral grid. The shock predicted by TAIR is far more rounded and smeared than that of FLO36, reflecting the coarser mesh and larger artificial viscosity used in the former. The TAIR result shown is also only correct to one decimal place in lift as compared with a finer-grid result. Convergence histories for these three cases are shown in Figure 9 along with the results for MGAFSP on a coarser grid (16×48).

REFERENCES

1. Finlayson, B. A. and L. E. Scriven, *The method of weighted residuals—a review*, Appl. Mech. Rev. **19** (1966), 735–748.

2. Slater, J. C., *Electronic energy bands in metal*, Phys. Rev. **45** (1934), 794–801.

3. Barta, J., *Über die Naherungsweise Lösung einiger Zweidimensionaler Elastizitätsaufgaben*, Z. Angew. Math. Mech. **17** (1937), 184–185.

4. Frazer, R. A., W. P. Jones and S. W. Skan, *Approximation to functions and to the solutions of differential equations*, Report and Memo No. 1799, Great Britain Aero. Res. Council, London, 1937.

5. Lanczos, C. L., *Trigonometric interpolation of empirical and analytic functions*, J. Math. Phys. **17** (1938), 123–199.

6. Clenshaw, C. W., *The numerical solution of linear differential equations in Chebyshev series*, Proc. Cambridge Philos. Soc. **53** (1957), 134–149.

7. Clenshaw, C. W. and H. J. Norton, *The solution of nonlinear ordinary differential equations in Chebyshev series*, Comput. J. **6** (1963), 88–92.

8. Wright, K., *Chebyshev collocation methods for ordinary differential equations*, Comput. J. **6** (1964), 358–365.

9. Villadsen, J. V. and W. E. Stewart, *Solution of boundary value problems by orthogonal collocation*, Chem. Engrg. Sci. **22** (1967), 1483–1501.

10. Kreiss, H.-O. and J. Oliger, *Comparison of accurate methods for the integration of hyperbolic equations*, Report No. 36, Department of Computer Science, Uppsala University, Sweden, 1971.

11. Orszag, S. A., *Comparison of pseudospectral and spectral approximations*, Stud. Appl. Math. **51** (1972), 253–259.

12. Silberman, I., *Planetary waves in the atmosphere*, J. Meteor. **11** (1954), 27–34.

13. Orszag, S. A., *Numerical methods for the simulation of turbulence*, Phys. Fluids, Supplement II, **12** (1969), 250–257.

14. Eliasen, E., B. Machenauer and E. Rasmussen, *On a numerical method for integration of the hydrodynamical equations with a spectral representation of the horizontal fields*, Report No. 2, Department of Meteorology, Copenhagen University, Denmark, 1970.

15. Gottlieb, D. and S. A. Orszag, *Numerical analysis of spectral methods: theory and applications*, CBMS-NSF Regional Conf. Ser. in Appl. Math., SIAM, Philadelphia, Pa., 1977.

16. Salas, M. D., T. A. Zang and M. Y. Hussaini, *Shock-fitted Euler solutions to shock-vortex interactions*, Proc. 8th Internat. Conf. on Numerical Methods in Fluid Dynamics (E. Krause, ed.), Springer-Verlag, Berlin and New York, 1982.

17. Hussaini, M. Y., D. A. Kopriva, M. D. Salas and T. A. Zang, *Spectral methods for the Euler equations*, AIAA J. (to appear).

18. Gottlieb, D., L. Lustman and S. A. Orszag, *Spectral calculations of one-dimensional inviscid compressible flows*, SIAM J. Sci. Statist. Comput. **2** (1981), 296–310.

19. Zang, T. A, M. Y. Hussaini and D. M. Bushnell, *Numerical computations of turbulence amplification in shock wave interactions*, AIAA J. **22** (1984), 13–21.

20. Hussaini, M. Y., M. D. Salas and T. A. Zang, *Spectral methods for inviscid, compressible flows*, Advances in Computational Transonics (W. G. Habashi, ed.), Pineridge Press, Swansea, U. K., 1984.

21. Pao, S. P. and M. D. Salas, *A numerical study of two-dimensional shock vortex interaction*, AIAA Paper 81-1205, 1981.

22. Ribner, H. S., *Shock-turbulence interaction and the generation of noise*, NACA Report 1233, 1955.

23. Zang, T. A., D. A. Kopriva and M. Y. Hussaini, *Pseudospectral calculation of shock turbulence interactions*, Proc. 3rd Internat. Conf. on Numerical Methods in Laminar and Turbulent Flow (C. Taylor, ed.), Pineridge Press, Swansea, U. K., 1983.

24. Orszag, S. A. and G. S. Patterson, *Numerical simulation of three-dimensional homogeneous isotropic turbulence*, Phys. Rev. Lett. **28** (1972), 76–79.

25. Orszag, S. A. and L. C. Kells, *Transition to turbulence in plane Poiseuille and plane Couette flow*, J. Fluid Mech. **96** (1980), 159–205.

26. P. Moin and J. Kim, *On the numerical solution of time-dependent viscous incompressible fluid flows involving solid boundaries*, J. Comput. Phys. **35** (1980), 381–392.

27. Kleiser, L. and U. Schumann, *Spectral simulation of the laminar-turbulent transition process in plane Poiseuille flow*, Proc. ICASE Sympos. on Spectral Methods (R. Voigt, ed.), SIAM-CBMS, Philadelphia, Pa., 1983.

28. Orszag, S. A., *Spectral methods for problems in complex geometries*, J. Comput. Phys. **37** (1980), 70–92.

29. McCrory, R. L. and S. A. Orszag, *Spectral methods for multidimensional diffusion problems*, J. Comput. Phys. **37** (1980), 93–112.

30. Zang, T. A., Y. S. Wong and M. Y. Hussaini, *Spectral multigrid methods for elliptic equations*, J. Comput. Phys. **48** (1982), 485–501.

31. _____ , *Spectral multigrid methods for elliptic equations. II*, J. Comput. Phys. **54** (1984), 489–507.

32. Streett, C. L., T. A. Zang and M. Y. Hussaini, *Spectral multigrid methods with applications to transonic potential flow*, J. Comput. Phys. (to appear).

33. Haidvogel, D. B. and T. A. Zang, *The accurate solution of Poisson's equation by expansion in Chebyshev polynomials*, J. Comput. Phys. **30** (1979), 167–180.

34. Douglas, J. and J. E. Gunn, *A general formulation of alternating direction methods*, Numer. Math. **6** (1964), 428–453.

35. Ballhaus, W. F., A. Jameson and J. Albert, *Implicit approximate factorization schemes for the efficient solution of steady transonic flow problems*, AIAA J. **16** (1978), 573–579.

36. Hackbusch, W. and U. Trottenberg (eds.), *Multigrid methods*, Lecture Notes in Math., Vol. 960, Springer-Verlag, New York, 1982.

37. Streett, C. L., *A spectral method for the solution of transonic potential flow about an arbitrary airfoil*, Proc. Sixth AIAA Computational Fluid Dynamics Conf. (Danvers, Mass., July 1983).

38. Jameson, A., *Acceleration of transonic potential flow calculations on arbitrary meshes by the multiple grid method*, AIAA Paper 79-1458, 1979.

39. Holst, T. L., *A fast, conservative algorithm for solving the transonic full-potential equation*, AIAA Paper 79-1456, 1979.

40. Hafex, M. M., J. C. South and E. M. Murman, *Artificial compressibility methods for numerical solution of transonic full potential equation*, AIAA J. **17** (1979), 838–844.

NASA LANGLEY RESEARCH CENTER, HAMPTON, VIRGINIA 23665

INSTITUTE FOR COMPUTER APPLICATIONS IN SCIENCE AND ENGINEERING, NASA LANGLEY RESEARCH CENTER, HAMPTON, VIRGINIA 23665

ABCDEFGHIJ— AMS/AP— 898765